For my little
Scholar, Riva.
Love,
Mom
Christmas 2004

About the Author

HOWARD ZINN is a historian, playwright, and social activist. He was a shipyard worker and Air Force bombardier before he went to college under the GI Bill and received his Ph.D. from Columbia University. He has taught at Spelman College and Boston University, and has been a visiting professor at the University of Paris and the University of Bologna. He has received the Thomas Merton Award, the Eugene V. Debs Award, the Upton Sinclair Award, and the Lannan Literary Award. He lives in Massachusetts.

Passionate Declarations

Other Books by Howard Zinn

La Guardia in Congress
The Southern Mystique
SNCC: The New Abolitionists
New Deal Thought (editor)
Vietnam: The Logic of Withdrawal
Disobedience and Democracy
The Politics of History
The Pentagon Papers: Critical Essays (editor, with Noam Chomsky)
Postwar America
Justice in Everyday Life (editor)
A People's History of the United States
Failure to Quit: Reflections of an Optimistic Ideology
You Can't Be Neutral on a Moving Train
The Zinn Reader
The Future of History
Marx in Soho: A Play on History
On War
On History
Terrorism and War
Emma: A Play

Howard Zinn

Passionate Declarations

Essays on War and Justice

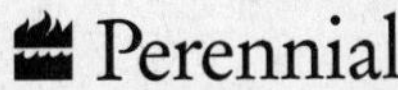

An Imprint of HarperCollins*Publishers*

A hardcover edition of this book was published in 1990 under the title *Declarations of Independence* by HarperCollins Publishers.

First Perennial edition published 2003.

Designed by Barbara DuPree Knowles

Library of Congress Cataloging-in-Publication Data
Zinn, Howard.
[Declarations of independence]
Passionate declarations : essays on war and justice / Howard Zinn.—1st Perennial ed.
p. cm.
Originally published under title: Declarations of independence. New York : HarperCollins, c1990.
Includes bibliographical references and index.
ISBN 0-06-055767-2
1. United States—Politics and government—1945–1989—Philosophy. 2. United States—Politics and government—1989—Philosophy. 3. United States—Foreign relations—Philosophy. 4. War and society—United States. 5. Social justice—United States. 6. Political science—United States. I. Title.
E839.5.Z55 2003
327.73'001—dc21 2003048250

---

03 04 05 06 07 RRD 10 9 8 7 6 5 4 3 2

For my brother Shelly,
who wanted to live in a better world

Contents

Preface

When this book first appeared, over a decade ago, under the title *Declarations of Independence* (claiming that the ideas in the book were fiercely independent of authority), the United States was soon to go to war against Iraq. As I write this now, American troops, planes, ships are massing in the Middle East, the nation poised again for war, once more against Iraq.

As always, in a situation of war or near-war, the air becomes filled with patriotic cries for unity against the enemy. What is supposed to be an opposition party declares its loyalty to the president. The major voices in the media, supposed to be independent of government, join the fray.

Immediately after President Bush declared a "war on terrorism" and told Congress, "Either you are with us or you are with the terrorists," television anchorman Dan Rather (to what, I have wondered, is he anchored?) spoke. He said, "George Bush is the president. He makes the decisions, and, you know, as just one American, if he wants me to line up, just tell me where." Speaking again to a national television audience, Rather said about Bush: "He is our commander in chief. He's the man now. And we need unity. We need steadiness."

Machiavelli would understand this. Writing in sixteenth-century Florence, his concern is to serve the prince. He does not question the aims of the prince. He cares only to give advice to the prince. He does not question the ends, only the means. The end is national power, and the only question is: what means are best to sustain and extend that power?

As I write this, I hear no voices on high questioning that the end of American foreign policy is to maintain the power of the United States. There are such voices, but they are not in positions of authority, either in government or in the media. Wherever they are, I propose to add my voice to theirs. I want to question the premise of Machiavelli, that one must serve the prince, that the all-important thing is national power, and the only issue is how best to augment it.

Thus, I suggest in the following pages that we think about questions other than the goals of states and statesmen. I want to go beyond Machiavellian obedience and discuss dissent and resistance to foreign policies aiming only at national power.

In my first chapter I tell of those Americans who, despite their positions close to power, decided to speak truth to that power. There were the scientists who spoke out against the dropping of the bombs on Hiroshima and Nagasaki. There were the aides of Henry Kissinger who resigned rather than collaborate in the plans he made with President Nixon to invade Cambodia in 1970. There were Daniel Ellsberg and Anthony Russo, who turned on their old employers at the Rand Corporation and the Department of Defense, and turned over to the public 7,000 pages of top-secret documents exposing government lies about the war in Vietnam. At the close of 2002, Ellsberg called on those in high positions to defy protocol and reveal the secrets kept from the public by a government determined to go to war.

In "Violence and Human Nature," I argue against the idea that violence and aggression are inborn, and insist they are determined by culture and indoctrination. I claim that it is possible for people to overcome that indoctrination and act with compassion toward fellow human beings. I now see that claim corroborated in the behavior of some of the families of those who died on September 11, 2001, in the fiery destruction of the Twin Towers in New York City. They reject the idea of retribution, believing we should not react to the terrorism of fanatical groups with the terrorism of war.

In the chapter called "The Use and Abuse of History," I ask that history be more than a cold recitation of facts about the past, that it serve a purpose in shaping the future. Traveling the country this past year, speaking against the drive to war on Iraq, I suggested that history might be useful in showing the futility of war as a solution for fundamental problems in international relations. Studying the past, I believe, would reveal the persistence of governmental deceit in luring the nation into armed conflict.

I argue against the idea of a "just war." This concept was given powerful credence by the struggle against fascism in World War II. But I believe it is no longer morally acceptable given the technology of modern warfare, in which horrific means are used to achieve uncertain ends. Since I wrote this chapter, the United States has used its armed forces against Panama, Iraq, Yugoslavia, Afghanistan. It may be worth considering whether my arguments against "just war" fit those situations.

In "Law and Justice" I question whether obedience to law is morally acceptable when the law protects injustice. I dissect Plato's argument for loyalty to the state and place against him the ideas and actions of Thoreau, Gandhi, Tolstoy, Martin Luther King Jr., and many lesser-known practitioners of civil disobedience. In the eighties and nineties, though the war in Vietnam that had produced so much protest was over, groups of Americans in various parts of the country were willing to risk prison by committing symbolic acts of sabotage of nuclear weapons.

The problem of economic justice, which I discuss in Chapter Seven, remains. Indeed, the nineties saw the gap between the rich and the poor grow dramatically. In the ten years before 1998, the annual income of the poorest 20 percent of the population rose by $110, while the annual income of the richest 20 percent rose by $17,800. The average pay of corporate chief executive officers rose to $12 million dollars, while one of every five children was born in poverty.

I want people to have "second thoughts" on the First Amendment, as I say in one of my chapters, because we grow up naively thinking that the First Amendment guarantees our freedom of speech. I suggest that free speech does not become a reality until people insist on it, struggle for it, practice it, because corporate wealth, governmental power, judicial decisions all limit that right. This is especially true in wartime, and today, the Bush Administration, with the complicity of the Democratic Party, is using its "war on terrorism" as an excuse to pry into the correspondence, the reading habits, the private life of every American.

I use here the experience of black Americans to show the limitations of representative government, of faith in that much over-praised "right to vote" to assure us of the equal rights promised in the Declaration of Independence. The skepticism I express in this book has been enhanced by the experience of the last decade, in which the Democratic Party came to resemble the Republican Party more and more. The "choice" in elections became meaningless as we moved toward the one-party system we have always derided in other countries, and the claim of "free elections" became ludicrous as the major parties depended more and more on the funds provided by wealthy corporations.

My discussion of communism and anti-communism in Chapter Ten may seem outdated after the fall of the Soviet Union and the disappearance of "the Red menace" from our culture. But we would be deceived if we thought so, because "terrorism" has replaced "communism" as the excuse for curtailment of domestic rights, for a bloated military budget, for more armed interventions. Indeed, terrorism is real, not an imaginary thing. So was communism real. But in both cases, the United States has refused to analyze the roots of these phenomena and to deal with the fundamental problems underlying them. Instead the government has chosen to provoke a national hysteria that covers up its violent actions abroad and its failures at home.

Nevertheless, I conclude that we should not be overwhelmed by the power of the Establishment or our own apparent weakness in challenging it. That feeling of powerlessness is always there at the start of a movement for change. We have enough examples—in the history of our own country and that of others—that show it is possible for organized citizens to resist and overcome what seem like hopeless odds. The power of determined people armed with a moral cause is, I believe, "the ultimate power."

Acknowledgments

I want to acknowledge the help I got in producing this book. My patient, encouraging editor at HarperCollins, Hugh Van Dusen. My steadfast literary agent, Rick Balkin. Kitty Benedict for her wise editorial suggestions. *Z Magazine* for letting me work out some ideas in short pieces I wrote there. John Tirman of the Winston Foundation for inviting me to write an essay that became the basis of my final chapter. Nancy Stockford for invaluable help. My students over the years, at Spelman College and Boston University, for making me think about the issues I write about in this book. And finally, Roslyn Zinn, for everything.

Passionate Declarations

CHAPTER

ONE

Introduction: American Ideology

The idea, which entered Western consciousness several centuries ago, that black people are less than human, made possible the Atlantic slave trade, during which perhaps 40 million people died. Beliefs about racial inferiority, whether applied to blacks or Jews or Arabs or Orientals, have led to mass murder.

The idea, presented by political leaders and accepted by the American public in 1964, that communism in Vietnam was a threat to our "national security" led to policies that cost a million lives, including those of 55,000 young Americans.

The belief, fostered in the Soviet Union, that "socialism" required a ruthless policy of farm collectivization, as well as the control of dissent, brought about the deaths of countless peasants and large numbers of political prisoners.

Other ideas—leave the poor on their own ("laissez-faire") and help the rich ("economic growth")—have led the U.S. government for most of its history to subsidize corporations while neglecting the poor, thus permitting terrible living and working conditions and incalculable suffering and death. In the years of the Reagan presidency, "laissez-faire" meant budget cutting for family care, which led to high rates of infant mortality in city ghettos.

We can reasonably conclude that how we *think* is not just mildly interesting, not just a subject for intellectual debate, but a matter of life and death.

If those in charge of our society—politicians, corporate executives, and owners of press and television—can dominate our ideas, they will be secure in their power. They will not need soldiers patrolling the streets. We will control ourselves.

Because force is held in reserve and the control is not complete, we can call ourselves a "democracy." True, the openings and the flexibility make such a society a more desirable place to live. But they also create a more effective form of control. We are less likely to object if we can feel that we have a "pluralist" society, with two parties instead of one, three branches of government instead of one-man rule, and various opinions in the press instead of one official line.[1]

A close look at this pluralism shows that it is very limited. We have the kinds of choices that are given in multiple-choice tests, where you can choose *a, b, c,* or *d.* But *e, f, g,* and *h* are not even listed.

And so we have the Democratic and Republican parties (choose *a* or *b*), but no others are really tolerated or encouraged or financed. Indeed, there is a law limiting the nationally televised presidential debates to the two major parties.

We have a "free press," but big money dominates it; you can choose among *Time, Newsweek,* and *U.S. News & World Report.* On television, you can choose among NBC, CBS, and ABC. There is a dissident press, but it does not have the capital of the great media chains and cannot get the rich corporate advertising, and so it must strain to reach small numbers of people. There is public television, which is occasionally daring, but also impoverished and most often cautious.

We have three branches of government, with "checks and balances," as we were taught in junior high school. But one branch of government (the presidency) gets us into wars and the other two (Congress and the Supreme Court) go sheepishly along.

There is the same limited choice in public policy. During the Vietnam War, the argument for a long time was between those who wanted a total bombing of Indochina and those who wanted a limited bombing. The choice of withdrawing from Vietnam altogether was not offered. Daniel Ellsberg, working for Henry Kissinger in 1969, was given the job of drawing a list of alternative policies on Vietnam. As one possibility on his long list he suggested total withdrawal from the war. Kissinger looked at the possibilities and crossed that one off before giving the list to President Richard Nixon.

In debates on the military budget there are heated arguments about whether to spend $300 billion or $290 billion. A proposal to spend $100 billion (thus making $200 billion available for human needs) is like the *e* or *f* in a multiple-choice test—it is missing. To propose zero billion makes you a candidate for a mental institution.

On the question of prisons there is debate on how many prisons we should have. But the idea of *abolishing* prisons is too outrageous even to be discussed.

We hear argument about *how much* the elderly should have to pay for health care, but the idea that they should not have to pay *anything,* indeed, that no one should have to pay for health care, is not up for debate.

Thus we grow up in a society where our choice of ideas is limited and where certain ideas dominate: We hear them from our parents, in the schools, in the churches, in the newspapers, and on radio and television. They have been in the air ever since we learned to walk and talk. They constitute an American *ideology*—that is, a dominant pattern of ideas. Most people accept them, and if we do, too, we are less likely to get into trouble.

The dominance of these ideas is not the product of a conspiratorial group that has devilishly plotted to implant on society a particular point of view. Nor is it an accident, an innocent result of people thinking freely. There is a process of natural (or, rather *unnatural*) selection, in which certain orthodox ideas are encouraged, financed, and pushed forward by the most powerful mechanisms of our culture. These ideas are preferred because they are safe; they don't threaten established wealth or power.

For instance:

"Be realistic; this is the way things *are;* there's no point thinking about how things *should be.*"

"People who teach or write or report the news should be *objective;* they should not try to advance their own opinions."

"There are unjust wars, but also just wars."

"If you disobey the law, even for a good cause, you should accept your punishment."

"If you work hard enough, you'll make a good living. If you are poor, you have only yourself to blame."

"Freedom of speech is desirable, but not when it threatens national security."

"Racial equality is desirable, but we've gone far enough in that direction."

"Our Constitution is our greatest guarantee of liberty and justice."

"The United States must intervene from time to time in various parts of the world with military power to stop communism and promote democracy."

"If you want to get things changed, the only way is to go through the proper channels."

"We need nuclear weapons to prevent war."

"There is much injustice in the world but there is nothing that ordinary people, without wealth or power, can do about it."

These ideas are not accepted by all Americans. But they are believed widely enough and strongly enough to dominate our thinking. And as long as they do, those who hold wealth and power in our society will remain secure in their control.

In the year 1984 *Forbes* magazine, a leading periodical for high finance and big business, drew up a list of the wealthiest individuals in the United States. The top 400 people had assets totaling $60 billion. At the bottom of the population there were 60 million people who had *no* assets at all.

Around the same time, the economist Lester Thurow estimated that 482 very wealthy individuals controlled (without necessarily owning) over $2,000 billion ($2 trillion).

Consider the influence of such a very rich class—with its inevitable control of press, radio, television, and education—on the *thinking* of the nation.[2]

Dissident ideas can still exist in such a situation, but they will be drowned in criticism and made disreputable, because they are outside the acceptable choices. Or they may be allowed to survive in the corners of the culture—emaciated, but alive—and presented as evidence of our democracy, our tolerance, and our pluralism.

A sophisticated system of control that is confident of its power can permit a measure of dissidence. However, it watches its critics carefully, ready to overwhelm them, intimidate them, and even suppress them should they ever seriously threaten the system, or should the establishment, in a state of paranoia, *think* they do. If readers think I am exaggerating with words such as "*watching . . . overwhelm . . . suppress . . . paranoia,*" they should read the volumes of reports on the FBI and the CIA published in 1975 by the Senate Select Committee on Government Operations.

However, government surveillance and threats are the exception. What normally operates day by day is the quiet dominance of certain ideas, the ideas we are expected to hold by our neighbors, our employers, and our political leaders; the ones we quickly learn are the most acceptable. The result is an obedient, acquiescent, passive citizenry—a situation that is deadly to democracy.

If one day we decide to reexamine these beliefs and realize they do not come naturally out of our innermost feelings or our spontaneous desires, are not the result of independent thought on our part, and, indeed, do not match the real world as we experience it, then we have come to an important turning point in life. Then we find ourselves examining, and confronting, American ideology.

That is what I want to do in this book.

I will be dealing with political ideas. When political ideas are analyzed—issues like violence in human nature, realism and idealism, the best forms of government or whether there should be government at all, a citizen's obligation to the state, and the proper distribution of wealth in society—we are in the area of political theory, or political philosophy. There is a list of famous political thinkers who are traditionally used to initiate discussion on these long-term problems, including Plato, Aristotle, Machiavelli, Hobbes, Locke, Madison, Rousseau, Marx, and Freud.

There are endless arguments that go on in academic circles about what Plato or Machiavelli or Rousseau or Marx *really* meant. Although I taught political theory for twenty years, I don't really care about that. I am interested in these thinkers when it seems to me their ideas are still alive in our time and can be used to illuminate a problem. Readers wanting to know more about some of these writers and the literature will find references in the endnotes of this book. I will assume that our job is not to interpret the great theorists, but to think for ourselves.

I will go back and forth from theory to historical fact (including very recent events), hoping to clarify issues of urgent concern to our time. I will not be too respectful of chronology, but will wander back and forth across the centuries, from Machiavelli to Kissinger, from Socrates in an Athenian prison to a Catholic priest in a Connecticut jail, making whatever connections I find useful.

There is in orthodox thinking a great dependence on experts. Because modern technological society has produced a breed of experts who understand technical matters that bewilder the rest of us, we think that in matters of social conflict, which require *moral* judgments, we must also turn to experts.[3]

There are two false assumptions about experts. One is that they see more clearly and think more intelligently than ordinary citizens. Sometimes they do, sometimes not. The other assumption is that these experts have the same *interests* as ordinary citizens, want the same things, hold the same values, and, therefore, can be trusted to make decisions for all of us.

To depend on great thinkers, authorities, and experts is, it seems to me, a violation of the spirit of democracy. Democracy rests on the idea that, except for technical details for which experts may be useful, the important decisions of society are within the capability of ordinary citizens. Not only *can* ordinary people make decisions about these issues, but they *ought* to, because citizens understand their own interests more clearly than any experts.

In John Le Carré's novel *The Russia House,* a dissident Russian scientist is assured that his secret document has been entrusted "to the authorities. People of discretion. Experts." He becomes angry:

> I do not *like* experts. They are our jailers. I despise experts more than anyone on earth. . . . They solve nothing! They are servants of whatever system hires them. They perpetuate it. When we are tortured, we shall be tortured by experts. When we are hanged, experts will hang us. . . . When the world is destroyed, it will be destroyed not by its madmen but by the sanity of its experts and the superior ignorance of its bureaucrats.[4]

We are expected to believe that great thinkers—experts—are *objective,* that they have no axes to grind and no biases, and that they make pure intellectual judgments. However, the minds of all human beings are powerfully influenced (though not totally bound) by their backgrounds, by whether they are rich or poor, male or female, black or white or Asian, in positions of power, or in lowly circumstances. Even scientists making "scientific" observations know that what they see will be affected by their *position.*[6]

Why should we cherish "objectivity," as if ideas were innocent, as if they don't serve one interest or another? Surely, we want to be objective if that means telling the truth as we see it, not concealing information that may be embarrassing to our point of view. But we don't want to be objective if it means pretending that ideas don't play a part in the social struggles of our time, that we don't take sides in those struggles.

Indeed, it is impossible to be neutral. In a world already moving in certain directions, where wealth and power are already distributed in certain ways, neutrality means accepting the way things are now. It is a world of clashing interests—war against peace, nationalism against internationalism, equality against greed, and democracy against elitism—and it seems to me both impossible and undesirable to be neutral in those conflicts.

Writing this book, I do not claim to be neutral, nor do I want to be. There are things I value, and things I don't. I am not going to present ideas objectively if that means I don't have strong opinions on which ideas are right and which are wrong. I will try to be fair to opposing ideas by accurately representing them. But the reader should know that what appear here are my own views of the world as it is and as it should be.

I do want to influence the reader. But I would like to do this by the strength of argument and fact, by presenting ideas and ways of looking at issues that are outside the orthodox. I am hopeful that given more possibilities people will come to wiser conclusions.

In my years of teaching, I never listened to the advice of people who said that a teacher should be objective, neutral, and professional. All the experiences of my life, growing up on the streets of New York, becoming a shipyard worker at the age of eighteen, enlisting in the Air Force in World War II, participating in the civil rights movement in the Deep South, cried out against that.

It seems to me we should make the most of the fact that we live in a country that, although controlled by wealth and power, has openings and possibilities missing in many other places. The controllers are gambling that those openings will pacify us, that we will not really *use* them to make the bold changes that are needed if we are to create a decent society. We should take that gamble.

We are not starting from scratch. There is a long history in this country of rebellion against the establishment, of resistance to orthodoxy. There has always been a commonsense perception that there are things seriously wrong and that we can't really depend on those in charge to set them right.

This perception has led Americans to protest and rebel. I think of the Boston Bread Rioters and Carolina antitax farmers of the eighteenth century; the black and white abolitionists of slavery days; the working people of the railroads, mines, textile mills, steel mills, and auto plants who went on strike, facing the clubs of policemen and the machine guns of soldiers to get an eight-hour workday and a living wage; the women who refused to stay in the kitchen and marched and went to jail for equal rights; the black protesters and antiwar activists of the 1960s; and the protesters against industrial pollution and war preparations in the 1980s.

In the heat of such movements brains are set stirring with new ideas, which live on through quieter times, waiting for another opportunity to ignite into action and change the world around us.

Dissenters, I am aware, can create their own orthodoxy. So we need a constant reexamination of our thinking, using the evidence of our eyes and ears and the realities of our experience to think freshly. We need declarations of independence from all nations, parties, and programs—all rigid dogmas.

The experience of our century tells us that the old orthodoxies, the traditional ideologies, the neatly tied bundles of ideas—capitalism, socialism, democracy—need to be untied, so that we can play and experiment with all the ingredients, add others, and create new combinations in looser bundles. We know as we come to the twenty-first century that we desperately need to develop new, imaginative approaches to the human problems of our time.

For citizens to do this on their own, to listen with some skepticism to the great thinkers and the experts, and to think for themselves about the great issues of today's world, is to make democracy come alive.

We might begin by confronting one of those great thinkers, Niccolò Machiavelli, and examining the connection between him and the makers of foreign policy in the United States.

CHAPTER

TWO

Machiavellian Realism and U.S. Foreign Policy: Means and Ends

Interests: The Prince and the Citizen

About 500 years ago modern political thinking began. Its enticing surface was the idea of "realism." Its ruthless center was the idea that with a worthwhile end one could justify any means. Its spokesman was Niccolò Machiavelli.

In the year 1498 Machiavelli became adviser on foreign and military affairs to the government of Florence, one of the great Italian cities of that time. After fourteen years of service, a change of government led to his dismissal, and he spent the rest of his life in exile in the countryside outside of Florence. During that time he wrote, among other things, a little book called *The Prince,* which became the world's most famous handbook of political wisdom for governments and their advisers.

Four weeks before Machiavelli took office, something happened in Florence that made a profound impression on him. It was a public hanging. The victim was a monk named Savonarola, who preached that people could be guided by their "natural reason." This threatened to diminish the importance of the Church fathers, who then showed their importance by having Savonarola arrested. His hands were bound behind his back and he was taken through the streets in the night, the crowds swinging lanterns near his face, peering for the signs of his dangerousness.

Savonarola was interrogated and tortured for ten days. They wanted to extract a confession, but he was stubborn. The Pope, who kept in touch with the torturers, complained that they were not getting results quickly enough. Finally the right words came, and Savonarola was sentenced to death. As his body swung in the air, boys from the neighborhood stoned it. The corpse was set afire, and when the fire had done its work, the ashes were strewn in the river Arno.[1]

In *The Prince,* Machiavelli refers to Savonarola and says, "Thus it comes about that all armed prophets have conquered and unarmed ones failed."[2]

Political ideas are centered on the issue of *ends* (What kind of society do we want?) and *means* (How will we get it?). In that one sentence about unarmed prophets Machiavelli settled for modern governments the question of ends: conquest. And the question of means: force.

Machiavelli refused to be deflected by utopian dreams or romantic hopes and by questions of right and wrong or good and bad. He is the father of modern political realism, or what has been called *realpolitik:* "It appears to me more proper to go to the truth of the matter than to its imagination . . . for how we live is so far removed from how we ought to live, that he who abandons what is done for what ought to be done, will rather learn to bring about his own ruin than his preservation."[3]

It is one of the most seductive ideas of our time. We hear on all sides the cry of "be realistic . . . you're living in the real world," from political platforms, in the press, and at home. The insistence on building more nuclear weapons, when we already possess more than enough to destroy the world, is based on "realism."[4] The *Wall Street Journal,* approving a Washington, D.C., ordinance allowing the police to arrest any person on the street refusing to move on when ordered, wrote, "D.C.'s action is born of living in the real world."[5] And consider how often a parent (usually a father) has said to a son or daughter: "It's good to have idealistic visions of a better world, but you're living in the real world, so act accordingly."

How many times have the dreams of young people—the desire to help others; to devote their lives to the sick or the poor; or to poetry, music, or drama—been demeaned as foolish romanticism, impractical in a world where one must "make a living"? Indeed, the economic system reinforces the same idea by rewarding those who spend their lives on "practical" pursuits—while making life difficult for the artists, poets, nurses, teachers, and social workers.

Realism is seductive because once you have accepted the reasonable notion that you should base your actions on *reality,* you are too often led to accept, without much questioning, someone else's version of what that reality is. It is a crucial act of independent thinking to be skeptical of someone else's description of reality.

When Machiavelli claims to "go to the truth of the matter," he is making the frequent claim of important people (writers, political leaders) who press their ideas on others: that their account is "the truth," that they are being "objective."

But his reality may not be our reality; his truth may not be our truth. The real world is infinitely complex. Any description of it must be a partial description, so a choice is made about what part of reality to describe, and behind that choice is often a definite *interest,* in the sense of something useful for a particular individual or group. Behind the claim of someone giving us an objective picture of the real world is the assumption that we all have the same interests, and so we can trust the one who describes the world for us, because that person has our interests at heart.

It is very important to know if our interests are the same, because a description is never simply neutral and innocent; it has consequences. No description is merely that. Every description is in some way a *prescription.* If you describe human nature as Machiavelli does, as basically immoral, it suggests that it is realistic, indeed only human, that you should behave that way too.

The notion that all our interests are the same (the political leaders and the citizens, the millionaire and the homeless person) deceives us. It is a deception useful to those who run modern societies, where the support of the population is necessary for the smooth operation of the machinery of everyday life and the perpetuation of the present arrangements of wealth and power.

When the Founding Fathers of the United States wrote the Preamble to the Constitution, their first words were, "We the People of the United States, in order to form a more perfect union, establish justice . . ." The Constitution thus looked as if it were written by all the people, representing their interests.

In fact, the Constitution was drawn up by fifty-five men, all white and mostly rich, who represented a certain elite group in the new nation.

The document itself accepted slavery as legitimate, and at that time about one of every five persons in the population was a black slave. The conflicts between rich and poor and black and white, the dozens of riots and rebellions in the century before the Revolution, and a major uprising in western Massachusetts just before the convening of the Constitutional Convention (Shays' Rebellion) were all covered over by the phrase "We the people."

Machiavelli did not pretend to a common interest. He talked about what "is necessary for a prince."[6] He dedicated *The Prince* to the rich and powerful Lorenzo di Medici, whose family ruled Florence and included popes and monarchs. (The *Columbia Encyclopedia* has this intriguing description of the Medici: "The genealogy of the family is complicated by the numerous illegitimate offspring and by the tendency of some of the members to dispose of each other by assassination.")

In exile, writing his handbook of advice for the Medici, Machiavelli ached to be called back to the city to take his place in the inner circle. He wanted nothing more than to serve the prince.

In our time we find greater hypocrisy. Our Machiavellis, our presidential advisers, our assistants for national security, and our secretaries of state insist they serve "the national interest," "national security," and "national defense." These phrases put everyone in the country under one enormous blanket, camouflaging the differences between the interest of those who run the government and the interest of the average citizen.

The American Declaration of Independence, however, clearly understood that difference of interest between government and citizen. It says that the purpose of government is to secure certain rights for its citizens—life, liberty, equality, and the pursuit of happiness. But governments may not fulfill these purposes and so "whenever any form of government becomes destructive of these ends, it is the right of the people to alter or abolish it, and to institute new government."

The *end* of Machiavelli's *The Prince* is clearly different. It is not the welfare of the citizenry, but national power, conquest, and control. All is done in order "to maintain the state."[7]

In the United States today, the Declaration of Independence hangs on schoolroom walls, but foreign policy follows Machiavelli. Our language is more deceptive than his; the purpose of foreign policy, our leaders say, is to serve the "national interest," fulfill our "world responsibility." In 1986 General William Westmoreland said that during World War II the United States "inherited the mantle of leadership of the free world" and "became the international champions of liberty."[8] This, from the man who, as chief of military operations in the Vietnam War, conducted a brutal campaign that resulted in the deaths of hundreds of thousands of Vietnamese noncombatants.

Sometimes, the language is more direct, as when President Lyndon Johnson, speaking to the nation during the Vietnam War, talked of the United States as being "number one." Or, when he said, "Make no mistake about it, we will prevail."

Even more blunt was a 1980 article in the influential *Foreign Affairs* by Johns Hopkins political scientist Robert W. Tucker; in regard to Central America, he wrote, "We have regularly played a determining role in making and in unmaking governments, and we have defined what we have considered to be the acceptable behavior of governments." Tucker urged "a policy of a resurgent America to prevent the coming to power of radical regimes in Central America" and asked, "Would a return to a policy of the past work in Central America? . . . There is no persuasive reason for believing it would not. . . . Right-wing governments will have to be given steady outside support, even, if necessary, by sending in American forces."[9]

Tucker's suggestion became the Central America policy of the Reagan administration, as it came into office in early 1981. His "sending in American forces" was too drastic a step for an American public that clearly opposed another Vietnam (unless done on a small scale, like Reagan's invasion of Grenada, and Bush's invasion of Panama). But for the following eight years, the aims of the United States were clear: to overthrow the left-wing government of Nicaragua and to keep in place the right-wing government of El Salvador.

Two Americans who visited El Salvador in 1983 for the New York City Bar Association, described for the *New York Times* a massacre of eighteen peasants by local troops in Sonsonate province:

> Ten military advisers are attached to the Sonsonate armed forces. . . . The episode contains all the unchanging elements of the Salvadoran tragedy—uncontrolled military violence against civilians, the apparent ability of the wealthy to procure official violence . . . and the presence of United States military advisers, working with the Salvadoran military responsible for these monstrous practices . . . after 30,000 unpunished murders by security and military forces and over 10,000 "disappearances" of civilians in custody, the root causes of the killings remain in place and the killing goes on.[10]

The purpose of its policy in Central America, said the U.S. government, was to protect the country from the Soviet threat: a Soviet base in Nicaragua and a possible Soviet base in El Salvador. This was not quite believable. Was the Soviet Union prepared to launch an invasion of the United States from Central America? Was a nation that could not win a war on its borders with Afghanistan going to send an army across the Atlantic Ocean to Nicaragua? And what then? Would that army then march up through Honduras into Guatemala, then through all of Mexico, into Texas, and then . . . ?

It was as absurd as the domino theory of the Vietnam War, in which the falling dominos of Southeast Asia would have had to swim the Pacific to get to San Francisco. Did the Soviet Union, with intercontinental ballistic missiles, with submarines off the coast of Long Island, need Central America as a base for attacking the United States?

Nevertheless, the Kissinger Commission, set up by President Reagan to advise him on Central American policy, warned in its report that our "southern flank" was in danger—a biological reference designed to make all of us nervous.

Even a brief look at history was enough to make one skeptical. How could we explain our frequent interventions in Central America *before* 1917, before the Bolshevik Revolution? How could we explain our taking control of Cuba and Puerto Rico in 1898; our seizure of the Canal Zone in 1903; our dispatch of marines to Honduras, Nicaragua, Panama, and Guatemala in the early 1900s; our bombardment of a Mexican town in 1914; and our long military occupation of Haiti and the Dominican Republic starting in 1915 and 1916?[11] All this before the Soviet Union existed.

There was another official reason given for U.S. intervention in Central America in the 1980s: to "restore democracy." This, too, was hardly believable. Throughout the period after World War II our government had supported undemocratic governments, indeed vicious military dictatorships: in Batista's Cuba, Somoza's Nicaragua, Armas's Guatemala, Pinochet's Chile, and Duvalier's Haiti as well as in El Salvador and other countries of Latin America.

The actual purpose of U.S. policy in Central America was expressed by Tucker in the most clear Machiavellian terms: "The great object of American foreign policy ought to be the restoration of a more normal political world, a world in which those states possessing the elements of great power once again play the role their power entitles them to play."[12]

Undoubtedly, there are Americans who respond favorably to this idea, that the United States should be a "great power" in the world, should dominate other countries, should be number one. Perhaps the assumption is that our domination is benign and that our power is used for kindly purposes. The history of our relations with Latin America does not suggest this. Besides, is it really in keeping with the American ideal of equality of all peoples to insist that we have the right to control the affairs of other countries? Are we the only country entitled to a Declaration of Independence?

Means: The Lion and the Fox

There should be clues to the rightness of the ends we pursue by examining the means we use to achieve those ends. I am assuming there is always some connection between ends and means. All means become ends in the sense that they have immediate consequences apart from the ends they are supposed to achieve. And all ends are themselves means to other ends. Was there not a link, for Machiavelli, between his crass end—power for the prince—and the various *means* he found acceptable?

For a year Machiavelli was ambassador to Cesare Borgia, conqueror of Rome. He describes one event that "is worthy of note and of imitation by others." Rome had been disorderly, and Cesare Borgia decided he needed to make the people "peaceful and obedient to his rule." Therefore, "he appointed Messer Remirro de Orco, a cruel and able man, to whom he gave the fullest authority" and who, in a short time, made Rome "orderly and united." But Cesare Borgia knew his policies had aroused hatred, so,

> in order to purge the minds of the people and to win them over completely, he resolved to show that if any cruelty had taken place it was not by his orders, but through the harsh disposition of his minister. And having found the opportunity he had him cut in half and placed one morning in the public square at Cesena with a piece of wood and blood-stained knife by his side.[13]

In recent American history, we have become familiar with the technique of rulers letting subordinates do the dirty work, which they can later disclaim. As a result of the Watergate scandals in the Nixon administration (a series of crimes committed by underlings in his behalf), a number of his people (former CIA agents, White House aides, and even the attorney-general) were sent to prison. But Nixon himself, although he was forced to resign his office, escaped criminal prosecution, arranging to be pardoned when his vice-president, Gerald Ford, became president. Nixon retired in prosperity and, in a few years, became a kind of elder statesman, a Godfather of politics, looked to for sage advice.

Perhaps as a way of calming the public in that heated time of disillusionment with the government because of Vietnam and Watergate, a Senate committee in 1974–1975 conducted an investigation of the intelligence agencies. It discovered that the CIA and the FBI had violated the law countless times (opening mail, breaking into homes and offices, etc.). In the course of that investigation, it was also revealed that the CIA, going back to the Kennedy administration, had plotted the assassination of a number of foreign rulers, including Cuba's Fidel Castro. But the president himself, who clearly was in favor of such actions, was not to be directly involved, so that he could deny knowledge of it. This was given the term *plausible denial.*

As the committee reported:

> Non-attribution to the United States for covert operations was the original and principal purpose of the so-called doctrine of "plausible denial." Evidence before the Committee clearly demonstrates that this concept, designed to protect the United States and its operatives from the consequences of disclosures, has been expanded to mask decisions of the President and his senior staff members.[14]

In 1988 a story in a Beirut magazine led to information that Ronald Reagan's administration had been secretly selling arms to Iran, the declared enemy of the United States, and using the proceeds to give military aid to counterrevolutionaries (the "contras") in Nicaragua, thus violating an act passed by Congress. Reagan and Vice President Bush denied involvement, although the evidence pointed very strongly to their participation.[15] Instead of impeaching them, however, Congress put their emissaries on the witness stand, and later several of them were indicted. One of them (Robert McFarland) tried to commit suicide. Another, Colonel Oliver North, stood trial for lying to Congress, was found guilty, but was not sentenced to prison. Reagan was not compelled to testify about what he had done. He retired in peace and Bush became the next president of the United States, both beneficiaries of plausible denial. Machiavelli would have admired the operation.

A prince, Machiavelli suggested, should emulate both the lion and the fox.[16] The lion uses force. "The character of peoples varies, and it is easy to persuade them of a thing, but difficult to keep them in that persuasion. And so it is necessary to order things so that when they no longer believe, they can be made to believe by force. . . . Fortune is a woman, and it is necessary, if you wish to master her, to conquer her by force."[17] The fox uses deception.

> If all men were good, this would not be good advice, but since they are dishonest and do not keep faith with you, you, in return, need not keep faith with them; and no prince was ever at a loss for plausible reasons to cloak a breach of faith. . . . The experience of our times shows those princes to have done great things who have had little regard for good faith, and have been able by astuteness to confuse men's brains.[18]

This advice for the prince has been followed in our time by all sorts of dictators and generalissimos. Hitler kept a copy of *The Prince* at his bedside, it is said. (Who says? How do they know?) Mussolini used Machiavelli for his doctoral dissertation. Lenin and Stalin are also supposed to have read Machiavelli.[19] Certainly the Italian Communist Gramsci wrote favorably about Machiavelli, claiming that Machiavelli was not really giving advice to princes, who knew all that already, but to "those who do not know," thus educating "those who must recognize certain necessary means, even if those of tyrants, because they want certain ends."[20]

The prime ministers and presidents of modern democratic states, despite their pretensions, have also admired and followed Machiavelli. Max Lerner, a prominent liberal commentator on the post–World War II period, in his introduction to Machiavelli's writings, says of him: "The common meaning he has for democrats and dictators alike is that, whatever your ends, you must be clear-eyed and unsentimental in pursuit of them." Lerner finds in Machiavelli's *Discourses* that one of its important ideas is "the need in the conduct even of a democratic state for the will to survive and therefore for ruthless instead of half-hearted measures."[21]

Thus the democratic state, behaving like the lion, uses force when persuasion does not work. It uses it against its own citizens when they cannot be persuaded to obey the laws. It uses it against other peoples in the act of war, not always in self-defense, but often when it cannot persuade other nations to do its bidding.

For example, at the start of the twentieth century, although Colombia was willing to sell the rights to the Panama Canal to the United States, it wanted more money than the United States was willing to pay. So the warships were sent on their way, a little revolution was instigated in Panama, and soon the Canal Zone was in the hands of the United States. As one U.S. senator described the operation, "We stole it fair and square."[22]

The modern liberal state, like Machiavelli's fox, often uses deception to gain its ends—not so much deception of the foreign enemy (which, after all, has little faith in its adversaries), but of its own citizens, who have been taught to trust their leaders.

One of the important biographies of President Franklin D. Roosevelt, is titled *Roosevelt: The Lion and the Fox.*[23] Roosevelt deceived the American public at the start of World War II, in September and October 1941, misstating the facts about two instances involving German submarines and American destroyers (claiming the destroyer *Greer,* which was attacked by a German submarine, was on an innocent mission when in fact it was tracking the sub for the British navy). A historian sympathetic to him wrote, "Franklin Roosevelt repeatedly deceived the American people during the period before Pearl Harbor. . . . He was like the physician who must tell the patient lies for the patient's own good."[24]

Then there were the lies of President John Kennedy and Secretary of State Dean Rusk when they told the public the United States was not responsible for the 1961 invasion of Cuba, although in fact the invasion had been organized by the CIA.

The escalation of the war in Vietnam started with a set of lies—in August 1964—about incidents in the Gulf of Tonkin. The United States announced two "unprovoked" attacks on U.S. destroyers by North Vietnamese boats. One of them almost certainly did not take place. The other was undoubtedly provoked by the proximity (ten miles) of the destroyer to the Vietnamese coast and by a series of CIA-organized raids on that coast.[25]

The lies then multiplied. One of them was President Johnson's statement that the U.S. Air Force was only bombing "military targets." Another was a deception by President Richard Nixon; he concealed from the American public the 1969–1970 massive bombing of Cambodia, a country with which we were supposed to be at peace.

The Advisers

Advisers and assistants to presidents, however committed they are in their rhetoric to the values of modern liberalism, have again and again participated in acts of deception that would have brought praise from Machiavelli. His goal was to serve the prince and national power. So was theirs. Because they were advisers to a liberal democratic state, they assumed that advancing the power of such a state was a moral end, which then justified both force and deception. But cannot a liberal state carry out immoral policies? Then the adviser (deceiving himself this time) would consider that his closeness to the highest circles of power put him in a position to affect, even reverse, such policies.

It was a contemporary of Machiavelli, Thomas More, who warned intellectuals about being trapped into service to the state and about the self-deception in which the adviser believes he will be a good influence in the higher councils of the government.[26] In More's book *Utopia*, spokesperson Raphael is offered the advice commonly given today to young people who want to be social critics, prodding the government from outside, like Martin Luther King or Ralph Nader. The advice is to get on the *inside*. Raphael is told, "I still think that if you could overcome the aversion you have to the courts of princes, you might do a great deal of good to mankind by the advice that you would give."

Raphael replies, "If I were at the court of some king and proposed wise laws to him and tried to root out of him the dangerous seeds of evil, do you not think I would either be thrown out of his court or held in scorn?" He goes on,

> Imagine me at the court of the King of France. Suppose I were sitting in his council with the King himself presiding, and that the wisest men were earnestly discussing by what methods and intrigues the King might keep Milan, recover Naples so often lost, then overthrow the Venetians and subdue all Italy, and add Flanders, Brabant, and even all Burgundy to his realm, besides some other nations he had planned to invade. Now in all this great ferment, with so many brilliant men planning together how to carry on war, imagine so modest a man as myself standing up and urging them to change all their plans.[27]

More might have been describing the historian Arthur Schlesinger, Jr., adviser to President Kennedy, who thought it was "a terrible idea" to go ahead with the CIA Bay of Pigs invasion of Cuba in 1961, two years after the revolution there. But he did not raise his voice in protest, because, as he later admitted, he was intimidated by the presence of "such august figures as the Secretaries of State and Defense and the Joint Chiefs of Staff." He wrote, "In the months after the Bay of Pigs I bitterly reproached myself for having kept so silent during those crucial discussions in the Cabinet room."[28]

But the intimidation of Schlesinger-as-adviser went beyond silencing him in the cabinet room—it led him to produce a nine-page memorandum to President Kennedy, written shortly before the invasion of Cuba, in which he is as blunt as Machiavelli himself in urging deception of the public to conceal the U.S. role in the invasion. This would be necessary because "a great many people simply do not at this moment see that Cuba presents so grave and compelling a threat to our national security as to justify a course of action which much of the world will interpret as calculated aggression against a small nation."[29]

The memorandum goes on, "The character and repute of President Kennedy constitute one of our greatest national resources. Nothing should be done to jeopardize this invaluable asset. When lies must be told, they should be told by subordinate officials." It goes on to suggest "that someone other than the President make the final decision and do so in his absence—someone whose head can later be placed on the block if things go terribly wrong." (Cesare Borgia again, only lacking the blood-stained knife.)

Schlesinger included in his memo sample questions and lying answers in case the issue of the invasion came up in a press conference:

> Q. Mr. President, is CIA involved in this affair?
>
> A. I can assure you that the United States has no intention of using force to overthrow the Castro regime.[30]

The scenario was followed. Four days before the invasion, President Kennedy told a press conference, "There will not be, under any conditions, any intervention in Cuba by U.S. armed forces."[31]

Schlesinger was just one of dozens of presidential advisers who behaved like little Machiavellis in the years when revolutions in Vietnam and Latin America brought hysterical responses on the part of the U.S. government. These intellectuals could see no better role for themselves than to serve national power.

Kissinger, secretary of state to Nixon, did not even have the mild qualms of Schlesinger. He surrendered himself with ease to the princes of war and destruction. In private discussions with old colleagues from Harvard who thought the Vietnam War immoral, he presented himself as someone trying to bring it to an end, but in his official capacity he was the willing intellectual tool of a policy that involved the massive killing of civilians in Vietnam.

Kissinger approved the bombing and invasion of Cambodia, an act so disruptive of the delicate Cambodian society that it can be considered an important factor in the rise of the murderous Pol Pot regime in that country. After he and the representatives of North Vietnam had negotiated a peace agreement to end the war in late 1972, he approved the breaking off of the talks and the brutal bombardment of residential districts in Hanoi by the most ferocious bombing plane of the time, the B-52.[32]

Kissinger's biographers describe his role: "If he had disapproved of Nixon's policy, he could have argued against the Cambodian attack. But there is no sign that he ever mustered his considerable influence to persuade the President to hold his fire. Or that he ever considered resigning in protest. Quite the contrary, Kissinger supported the policy."[33]

During the Christmas 1972 bombings *New York Times* columnist James Reston wrote,

> It may be and probably is true, that Mr. Kissinger as well as Secretary of State Rogers and most of the senior officers in the State Department are opposed to the President's bombing offensive in North Vietnam. . . . But Mr. Kissinger is too much a scholar, with too good a sense of humor and history, to put his own thoughts ahead of the president's.[34]

It seems that journalists too, can be Machiavellian.

Serving National Power

Machiavelli never questioned that national power and the position of the prince were proper ends: "And it must be understood that a prince . . . cannot observe all those things which are considered good in men, being often obliged, in order to maintain the state, to act against faith, against charity, against humanity, and against religion."[35]

The end of national power may be beneficial to the prince, and even to the prince's advisers, an ambitious lot. But why should it be assumed as a good end for the average citizen? Why should the citizen tie his or her fate to the nation-state, which is perfectly willing to sacrifice the lives and liberties of its own citizens for the power, the profit, and the glory of politicians or corporate executives or generals?

For a prince, a dictator, or a tyrant national power is an end unquestioned. A democratic state, however, substituting an elected president for a prince, must present national power as benign, serving the interests of liberty, justice, and humanity. If such a state, which is surrounded with the rhetoric of democracy and liberty and, in truth, has some measure of both, engages in a war that is clearly against a vicious and demonstrably evil enemy, then the end seems so clean and clear that any means to defeat that enemy may seem justified.

Such a state was the United States and such an enemy was fascism, represented by Germany, Italy, and Japan. Therefore, when the atomic bomb appeared to be the means for a quicker victory, there was little hesitation to use it.

Very few of us can imagine ourselves as presidential advisers, having to deal with their moral dilemmas (if, indeed, they retain enough integrity to consider them dilemmas). It is much easier, I think, for average citizens to see themselves in the position of the scientists who were secretly assembled in New Mexico during World War II to make the atomic bomb. We may be able to imagine our own trade or profession, our particular skills, called on to serve the policies of the nation. The scientists who served Hitler, like the rocket expert Wernher von Braun, could be as cool as Machiavelli in their subservience; they would serve national power without asking questions. They were professionals, totally consumed with doing "a good job" and they would do that job for whoever happened to be in power. So, when Hitler was defeated and Von Braun was brought by military intelligence agents to the United States, he cheerfully went ahead and worked on rockets for the United States, as he had done for Hitler.

As one satirical songwriter put it:

Once the rockets are up,
Who cares where they come down?
That's not our department,
Says Wernher von Braun.[36]

The scientists who worked on the Manhattan Project were not like that. One cannot imagine them turning to Hitler and working for him if he were victorious. They were conscious, in varying degrees, that this was a war against fascism and that it was invested with a powerful moral cause. Therefore, to build this incredibly powerful weapon was to use a terrible means, but for a noble end.

And yet there was one element these scientists had in common with Wernher von Braun: the sheer pleasure of doing a job well, of professional competence, and of scientific discovery, all of which could make one forget, or at least put in the background, the question of human consequences.

After the war when the making of a thermonuclear bomb was proposed, a bomb a thousand times more destructive than the one dropped on Hiroshima, J. Robert Oppenheimer, personally horrified by the idea, was still moved to pronounce the scheme of Edward Teller and Stanislaw Ulam for producing it as "technically sweet." Teller, defending the project against scientists who saw it as genocidal, said, "The important thing in any science is to do the things that can be done." And, whatever Enrico Fermi's moral scruples were (he was one of the top scientists in the Manhattan Project), he pronounced the plan for making the bombs "superb physics."[37]

Robert Jungk, a German researcher who interviewed many of the scientists involved in the making of the bomb, tried to understand their lack of resistance to dropping the bomb on Hiroshima. "They felt themselves caught in a vast machinery and they certainly were inadequately informed as to the true political and strategic situation." But he does not excuse their inaction. "If at that time they had had the moral strength to protest on purely humane grounds against the dropping of the bomb, their attitude would no doubt have deeply impressed the President, the Cabinet and the generals."[38]

Using the atomic bombs on populated cities was justified in moral terms by American political leaders. Henry Stimson, whose Interim Committee had the job of deciding whether or not to use the atomic bomb, said later it was done "to end the war in victory with the least possible cost in the lives of the men in the armies."[39] This was based on the assumption that without atomic bombs, an invasion of Japan would be necessary, which would cost many American lives.

It was a morality limited by nationalism, perhaps even racism. The saving of American lives was considered far more important than the saving of Japanese lives. Numbers were wildly thrown into the air (for example, Secretary of State James Byrnes talked of "a million casualties" resulting from an invasion) but there was no attempt to seriously estimate American casualties and weigh that against the consequences for Japanese men and women, old people and babies. (The closest to such an attempt was a military estimate that an invasion of the southernmost island of Japan would cause 30,000 American dead and wounded.)

The evidence today is overwhelming that an invasion of Japan was not necessary to bring the war to an end. Japan was defeated, in disarray, and ready to surrender. The U.S. Strategic Bombing Survey, which interviewed 700 Japanese military and political officials after the war, came to this conclusion:

> Based on a detailed investigation of all the facts and supported by the testimony of the surviving Japanese leaders involved, it is the Survey's opinion that certainly prior to 31 December 1945, and in all probability prior to 1 November 1945, Japan would have surrendered even if the atomic bombs had not been dropped, even if Russia had not entered the war, and even if no invasion had been planned or contemplated.[40]

After the war American scholar Robert Butow went through the papers of the Japanese Ministry of Foreign Affairs, the records of the International Military Tribunal of the Far East (which tried Japanese leaders as war criminals), and the interrogation files of the U.S. Army. He also interviewed many of the Japanese principals and came to this conclusion: "Had the Allies given the Prince (Prince Konoye, special emissary to Moscow, who was working on Russian intercession for peace) a week of grace in which to obtain his Government's support for the acceptance of the proposals, the war might have ended toward the latter part of July or the very beginning of the month of August, without the atomic bomb and without Soviet participation in the conflict."[41]

On July 13, 1945, three days before the successful explosion of the first atomic bomb in New Mexico, the United States intercepted Japanese Foreign Minister Togo's secret cable to Ambassador Sato in Moscow, asking that he get the Soviets to intercede and indicating that Japan was ready to end the war, so long as it was not unconditional surrender.

On August 2, the Japanese foreign office sent a message to the Japanese ambassador in Moscow, "There are only a few days left in which to make arrangements to end the war. . . . As for the definite terms . . . it is our intention to make the Potsdam Three-Power Declaration [which called for unconditional surrender] the basis of the study regarding these terms."[42]

Barton Bernstein, a Stanford historian who has studied the official documents closely, wrote,

> This message, like earlier ones, was probably intercepted by American intelligence and decoded. It had no effect on American policy. There is no evidence that the message was sent to Truman and Byrnes [secretary of state], nor any evidence that they followed the intercepted messages during the Potsdam conference. They were unwilling to take risks in order to save Japanese lives.[43]

In his detailed and eloquent history of the making of the bomb, Richard Rhodes says, "The bombs were authorized not because the Japanese refused to surrender but because they refused to surrender unconditionally."[44]

The one condition necessary for Japan to end the war was an agreement to maintain the sanctity of the Japanese emperor, who was a holy figure to the Japanese people. Former ambassador to Japan Joseph Grew, based on his knowledge of Japanese culture, had been trying to persuade the U.S. government of the importance of allowing the emperor to remain in place.

Herbert Feis, who had unique access to State Department files and to records on the Manhattan Project, noted that in the end the United States did give the assurances the Japanese wanted on the emperor. He writes, "The curious mind lingers over the reasons why the American government waited so long before offering the Japanese those various assurances which it did extend later."[45]

Why was the United States in a rush to drop the bomb, if the reason of saving lives turns out to be empty, if the probability was that the Japanese would have surrendered even without an invasion? Historian Gar Alperovitz, after going through the papers of the American officials closest to Truman and most influential in the final decision and especially the diaries of Henry Stimson, concludes that the atomic bombs were dropped to impress the Soviet Union, as a first act in establishing American power in the postwar world. He points out that the Soviet Union had promised to enter the war against Japan on August 8. The bomb was dropped on August 6.[46]

The scientist Leo Szilard had met with Truman's main policy adviser in May 1945 and reported later: "Byrnes did not argue that it was necessary to use the bomb against the cities of Japan in order to win the war. . . . Mr. Byrnes' view was that our possessing and demonstrating the bomb would make Russia more manageable."[47]

The *end* of dropping the bomb seems, from the evidence, to have been not winning the war, which was already assured, not saving lives, for it was highly probable no American invasion would be necessary, but the aggrandizement of American national power at the moment and in the postwar period. For this end, the means were among the most awful yet devised by human beings—burning people alive, maiming them horribly, and leaving them with radiation sickness, which would kill them slowly and with great pain.[48]

I remember my junior-high-school social studies teacher telling the class that the difference between a democracy like the United States and the "totalitarian states" was that "they believe that the end justifies any means, and we do not." But this was before Hiroshima and Nagasaki.

To make a proper moral judgment, we would have to put into the balancing the testimony of the victims. Here are the words of three survivors, which would have to be multiplied by tens of thousands to give a fuller picture.[49]

A thirty-five-year-old man: "A woman with her jaw missing and her tongue hanging out of her mouth was wandering around the area of Shinsho-machi in the heavy, black rain. She was heading toward the north crying for help."

A seventeen-year-old girl: "I walked past Hiroshima Station . . . and saw people with their bowels and brains coming out. . . . I saw an old lady carrying a suckling infant in her arms. . . . I saw many children . . . with dead mothers. . . . I just cannot put into words the horror I felt."

A fifth-grade girl: "Everybody in the shelter was crying out loud. Those voices. . . . They aren't cries, they are moans that penetrate to the marrow of your bones and make your hair stand on end. . . . I do not know how many times I called begging that they would cut off my burned arms and legs."

In the summer of 1966 my wife and I were invited to an international gathering in Hiroshima to commemorate the dropping of the bomb and to dedicate ourselves to a world free of warfare. On the morning of August 6, tens of thousands of people gathered in a park in Hiroshima and stood in total, almost unbearable, silence, awaiting the exact moment—8:16 A.M.—when on August 6, 1945, the bomb had been dropped.

When the moment came, the silence was broken by a sudden roaring sound in the air, eerie and frightening until we realized it was the sound of the beating of wings of thousands of doves, which had been released at that moment to declare the aim of a peaceful world.

A few days later, some of us were invited to a house in Hiroshima that had been established as a center for victims of the bomb to spend time with one another and discuss common problems. We were asked to speak to the group. When my turn came, I stood up and felt I must get something off my conscience. I wanted to say that I had been an air force bombardier in Europe, that I had dropped bombs that killed and maimed people, and that until this moment I had not seen the human results of such bombs, and that I was ashamed of what I had done and wanted to help make sure things like that never happened again.

I never got the words out, because as I started to speak I looked out at the Japanese men and women sitting on the floor in front of me, people with horribly burned faces, people with no eyes in their sockets, without arms, or without legs, but all quietly waiting for me to speak. I choked on my words, could not say anything for a moment, fighting for control, finally managed to thank them for inviting me and sat down.

For the idea that any means—mass murder, the misuse of science, the corruption of professionalism—are acceptable to achieve the end of national power, the ultimate example of our time is Hiroshima. For us, as citizens, the experience of Hiroshima and Nagasaki suggests that we reject Machiavelli, that we do not accept subservience, whether to princes or presidents, and that we examine for ourselves the ends of public policy to determine whose interests they really serve. We must examine the means used to achieve those ends to decide if they are compatible with equal justice for all human beings on earth.

The Anti-Machiavellians

There have always been people who did think for themselves, against the dominant ideology, and when there were enough of them history had its splendid moments: a war was called to a halt, a tyrant was overthrown, an enslaved people won its freedom, the poor won a small victory. Even some people close to the circles of power, in the face of overwhelming pressure to conform, have summoned the moral strength to dissent, ignoring the Machiavellian advice to leave the end unquestioned and the means unexamined.

Not all the atomic scientists rushed into the excitement of building the bomb. When Oppenheimer was recruiting for the project, as he later told the Atomic Energy Commission, most people accepted. "This sense of excitement, of devotion and of patriotism in the end prevailed." However, the physicist I. I. Rabi, asked by Oppenheimer to be his associate director at Los Alamos, refused to join. He was heavily involved in developing radar, which he thought important for the war, but he found it abhorrent, as Oppenheimer reported, that "the culmination of three centuries of physics" should be a weapon of mass destruction.[50]

Just before the bomb was tested and used, Rabi worried about the role of scientists in war:

> If we take the stand that our object is merely to see that the next war is bigger and better, we will ultimately lose the respect of the public. . . . We will become the unpaid servants of the munitions makers and mere technicians rather than the self-sacrificing public-spirited citizens which we feel ourselves to be.[51]

Nobel Prize–winning physical chemist James Franck, working with the University of Chicago Metallurgical Laboratory on problems of building the bomb, headed a committee on social and political implications of the new weapon. In June 1945 the Franck Committee wrote a report advising against a surprise atomic bombing of Japan: "If we consider international agreement on total prevention of nuclear warfare as the paramount objective . . . this kind of introduction of atomic weapons to the world may easily destroy all our chances of success." Dropping the bomb "will mean a flying start toward an unlimited armaments race," the report said.[52]

The committee went to Washington to deliver the report personally to Henry Stimson, but were told, falsely, that he was out of the city. Neither Stimson nor the scientific panel advising him was in a mood to accept the argument of the Franck Report.

Scientist Leo Szilard, who had been responsible for the letter from Albert Einstein to Franklin Roosevelt suggesting a project to develop an atomic bomb, also fought a hard but futile battle against the bomb being dropped on a Japanese city. The same month that the bomb was successfully tested in New Mexico, July 1945, Szilard circulated a petition among the scientists, protesting in advance against the dropping of the bomb, arguing that "a nation which sets the precedent of using these newly liberated forces of nature for purposes of destruction may have to bear the responsibility of opening the door to an era of devastation on an unimaginable scale."[53] Determined to do what he could to stop the momentum toward using the bomb, Szilard asked his friend Einstein to give him a letter of introduction to President Roosevelt. But just as the meeting was being arranged, an announcement came over the radio that Roosevelt was dead.

Would Einstein's great prestige have swayed the decision? It is doubtful. Einstein, known to be sympathetic to socialism and pacifism, was excluded from the Manhattan Project and did not know about the momentous decisions being made to drop the bombs on Hiroshima and Nagasaki.

One adviser to Harry Truman took a strong position against the atomic bombing of Japan: Undersecretary of the Navy Ralph Bard. As a member of Stimson's Interim Committee, at first he agreed with the decision to use the bomb on a Japanese city, but then changed his mind. He wrote a memorandum to the committee talking about the reputation of the United States "as a great humanitarian nation" and suggesting the Japanese be warned and that some assurance about the treatment of the emperor might induce the Japanese to surrender. It had no effect.[54]

A few military men of high rank also opposed the decision. General Dwight Eisenhower, fresh from leading the Allied armies to victory in Europe, met with Stimson just after the successful test of the bomb in Los Alamos. He told Stimson he opposed use of the bomb because the Japanese were ready to surrender. Eisenhower later recalled, "I hated to see our country be the first to use such a weapon."[55] General Hap Arnold, head of the army air force, believed Japan could be brought to surrender without the bomb. The fact that important military leaders saw no need for the bomb lends weight to the idea that the reasons for bombing Hiroshima and Nagasaki were political.

In the operations of U.S. foreign policy after World War II, there were a few bold people who rejected Machiavellian subservience and refused to accept the going orthodoxies. Senator William Fulbright of Arkansas was at the crucial meeting of advisers when President Kennedy was deciding whether to proceed with plans to invade Cuba. Arthur Schlesinger, who was there, wrote later that "Fulbright, speaking in an emphatic and incredulous way, denounced the whole idea."[56]

During the Vietnam War, advisers from MIT and Harvard were among the fiercest advocates of ruthless bombing, but a few rebelled. One of the earliest was James Thomson, a Far East expert in the State Department who resigned his post and wrote an eloquent article in the

Atlantic Monthly criticizing the U.S. presence in Vietnam.[57]

While Henry Kissinger was playing Machiavelli to Nixon's prince, at least three of his aides objected to his support for an invasion of Cambodia in 1970. William Watts, asked to coordinate the White House announcement on the invasion of Cambodia, declined and wrote a letter of resignation. He was confronted by Kissinger aide General Al Haig, who told him, "You have an order from your Commander in Chief." He, therefore, could not resign, Haig said. Watts replied, "Oh yes, I can—and I have!" Roger Morris and Anthony Lake, asked to write the speech for President Nixon justifying the invasion, refused and instead wrote a joint letter of resignation.[58]

The most dramatic action of dissent during the war in Vietnam came from Daniel Ellsberg, a Ph.D. in economics from Harvard who had served in the marines and held important posts in the Department of Defense, the Department of State, and the embassy in Saigon. He had been a special assistant to Henry Kissinger and then worked for the Rand Corporation, a private "think tank" of brainy people who contracted to do top-secret research for the U.S. government. When the Rand Corporation was asked to assemble a history of the Vietnam War, based on secret documents, Ellsberg was appointed as one of the leaders of the project. But he had already begun to feel pangs of conscience about the brutality of the war being waged by his government. He had been out in the field with the military and what he saw persuaded him that the United States did not belong in Vietnam. Then, reading the documents and helping to put together the history, he saw how many lies had been told to the public and was reinforced in his feelings.

With the help of a former Rand employee he had met in Vietnam, Anthony Russo, Ellsberg secretly photocopied the entire 7,000-page history—the "Pentagon Papers" as they came to be called—and distributed them to certain members of Congress as well as to the *New York Times.* When the *Times,* in a journalistic sensation, began printing this "top-secret" document, Ellsberg was arrested and put on trial. The counts against him could have brought a prison sentence of 130 years. But while the jury deliberated the judge learned, through the Watergate scandal, that Nixon's "plumbers" had tried to break into Ellsberg's psychiatrist's office to find damaging material and he declared the case tainted and called off the trial.

Ellsberg's was only one of a series of resignations from government that took place during and after the Vietnam War. A number of operatives of the CIA quit their jobs in the late sixties and early seventies and began to write and speak about the secret activities of the agency—for example, Victor Marchetti, Philip Agee, John Stockwell, Frank Snepp, and Ralph McGehee.

For the United States, as for other countries, Machiavellianism dominates foreign policy, but the courage of a small number of dissenters suggests the possibility that some day the larger public will no longer accept that kind of "realism." Machiavelli himself might have smiled imperiously at this suggestion, and said, "You're wasting your time. Nothing will change. It's human nature."

That claim is worth exploring.

CHAPTER THREE

Violence and Human Nature

I remember three different incidents of violence in three different parts of my life. In two of them I was an observer, in one a perpetrator.

In the fall of 1963 I was in Selma, Alabama, and saw two young black civil rights workers clubbed to the ground by state troopers and then attacked with electric prods, because they tried to bring food and water to black people standing in line waiting to register to vote.

As a twenty-two-year-old Air Force bombardier, I flew a bombing mission in the last weeks of World War II, which can only be considered an atrocity. It was the napalm bombing of a small French village, for purposes that had nothing to do with winning the war, leaving a wasteland of death and destruction five miles below our planes.

Years before that, while a teenager on the streets of Brooklyn, I watched a black man in an argument with an old Jewish man, a pushcart peddler who seemed to be his employer. It was an argument over money the black man claimed he was owed, and he seemed desperate, by turns pleading and threatening, but the older man remained adamant. Suddenly the black man picked up a board and hit the other over the head. The older man, blood trickling down his face, just kept pushing his cart down the street.

I have never been persuaded that such violence, whether of an angry black man or a hate-filled trooper or of a dutiful Air Force officer, was the result of some natural instinct. All of those incidents, as I thought about them later, were explainable by social circumstances. I am in total agreement with the statement of the nineteenth-century English philosopher John Stuart Mill: "Of all the vulgar modes of escaping from the consideration of the effect of social and moral influences upon the human mind, the most vulgar is that of attributing the diversities of conduct and character to inherent natural differences."[1]

Yet, at an early point in any discussion of human violence, especially a discussion of the causes of war, someone will say, "It's human nature."[2] There is ancient, weighty intellectual support for that common argument. Machiavelli, in *The Prince,* expresses confidently his own view of human nature, that human beings tend to be *bad.* This gives him a good reason, being "realistic," to urge laying aside moral scruples in dealing with people: "A man who wishes to make a profession of goodness in everything must necessarily come to grief among so many who are not good. Therefore it is necessary for a prince, who wishes to maintain himself, to learn how not to be good."[3]

The seventeenth-century philosopher Thomas Hobbes said, "I put forth a general inclination of all mankind, a perpetual and restless desire for power after power, that ceaseth only in death." This view of human nature led Hobbes to favor any kind of government, however authoritarian, that would keep the peace by blocking what he thought was the natural inclination of people to do violence to one another. He talked about "the dissolute condition of masterless men" that required "a coercive power to tie their hands from rapine and revenge."[4]

Beliefs about human nature thus become self-fulfilling prophecies. If you believe human beings are naturally violent and bad, you may be persuaded to think (although not *required* to think) that it is "realistic" to be that way yourself. But is it indeed realistic (meaning, "I regret this, but it's a *fact* . . .") to blame *war* on human nature?

In 1932, Albert Einstein, already world famous for his work in physics and mathematics, wrote a letter to another distinguished thinker, Sigmund Freud. Einstein was deeply troubled by the memory of World War I, which had ended only fourteen years before. Ten million men had died on the battlefields of Europe, for reasons that no one could logically explain. Like many others who had lived through that war, Einstein was horrified by the thought that human life could be destroyed on such a massive scale and worried that there might be another war. He considered that Freud, the world's leading psychologist, might throw light on the question Why do men make war?

"Dear Professor Freud," he wrote. "Is there any way of delivering mankind from the menace of war?" Einstein spoke of "that small but determined group, active in every nation, composed of individuals who . . . regard warfare, the manufacture and sale of arms, simply as an occasion to advance their personal interests and enlarge their personal authority." And then he asked, "How is it possible for this small clique to bend the will of the majority, who stand to lose and suffer by a state of war, to the service of their ambitions?"

Einstein volunteered an answer, "Because man has within him a lust for hatred and destruction." And then he put his final question to Freud, "Is it possible to control man's mental evolution so as to make him proof against the psychoses of hate and destructiveness?"

Freud responded, "You surmise that man has in him an active instinct for hatred and destruction, amenable to such stimulations. I entirely agree with you. . . . The most casual glance at world-history will show an unending series of conflicts between one community and another." Freud pointed to two fundamental instincts in human beings: the erotic, or love, instinct and its opposite, the destructive instinct. But the only hope he could hold for the erotic triumphing over the destructive was in the cultural development of the human race, including "a strengthening of the intellect, which tends to master our instinctive life."[5]

Einstein had a different view of the value of intellect in mastering the instincts. After pointing to "the psychoses of hate and destructiveness," Einstein concluded, "Experience proves that it is rather the so-called 'Intelligentsia' that is most apt to yield to these disastrous collective suggestions."

Here are two of the greatest minds of the century, helpless and frustrated before the persistence of war. Einstein, venturing that aggressive instincts are at the root of war, asks Freud, the expert on instincts, for help in coming to a solution. Note, however, that Einstein has jumped from "man has within him a lust" to "disastrous collective suggestions." Freud ignores this leap from instinct to culture and affirms that the "destructive instinct" is the crucial cause of war.

But what is Freud's evidence for the existence of such an instinct? There is something curious in his argument. He offers no proof from the field of his expertise, psychology. His evidence is in "the most casual glance at world-history."

Let's move the discussion forward, fifty years later, to a school of thought that did not exist in Freud's time, sociobiology. The leading spokesperson in this group is E. O. Wilson, a Harvard University professor and distinguished scientist. His book *Sociobiology* is an impressive treatise on the behavior of various species in the biological world that have social inclinations, like ants and bees.[6]

In the last chapter of *Sociobiology,* Wilson turned to human beings, and this drew so much attention that he decided to write a whole book dealing with this subject: *On Human Nature.* In it there is a chapter on aggression. It starts off with the question: "Are human beings innately aggressive?" Two sentences later: "The answer to it is yes." (No hesitation here.) And the next sentence explains why: "Throughout history, warfare, representing only the most organized technique of aggression, has been endemic to every form of society, from hunter-gatherer bands to industrial states."[7]

Here is a peculiar situation. The psychologist (Freud) finds his evidence for the aggressive instinct not in psychology but in history. The biologist (Wilson) finds his evidence not in biology but in history.

This suggests that the evidence from neither psychology nor biology is sufficient to back up the claim for an aggressive instinct, and so these important thinkers turn to history. In this respect, they are no different from the ordinary person, whose thinking follows the same logic: history is full of warfare; one cannot find an era free of it; this must mean that it comes out of something deep in human nature, something biological, a drive, an instinct for violent aggression.[8]

This logic is widespread in modern thought, in all classes of people, whether highly educated or uneducated. And yet, it is almost certainly wrong. And furthermore, it's dangerous.

Wrong, because there is no real evidence for it. Not in genetics, not in zoology, not in psychology, not in anthropology, not in history, not even in the ordinary experience of soldiers in war. Dangerous because it deflects attention from the nonbiological causes of violence and war.

It turns out, however, that Wilson's firm assent to the idea that human beings are "innately aggressive" depends on his redefinitions of *innately* and *aggressive.* In *On Human Nature* he says, "Innateness refers to the measurable probability that a trait will develop in a specified set of environments. . . . By this criterion human beings have a marked hereditary predisposition to aggressive behavior." And the word *aggression* takes in a variety of human actions, only some of which are violent.

In other words, when Wilson speaks of people being "innately aggressive" he does not mean that we are all born with an irresistible drive to become violent—it depends on our environment. And even if we become aggressive, that need not take the form of violence. Indeed,

Wilson says that "the more violent forms of human aggression are not the manifestations of inborn drives." We now have, he says, "a more subtle explanation based on the interaction of genetic potential and learning."

The phrase *genetic potential* gets us closer to a common ground between Wilson and his radical critics, who have attributed to him sometimes more extreme views about innate aggression than he really holds. That is, human beings certainly have, from the start (genetically) a *potential* for violence, but also a potential for peacefulness. That leaves us open to all sorts of possibilities, depending on the circumstances we find ourselves in and the circumstances we create for ourselves.

There is no known gene for aggression. Indeed, there is no known gene for any of the common forms of human behavior (I am allowing the possibility that a genetic defect of the brain might make a person violent, but the very fact that it is a defect means it is not a normal trait). The science of genetics, the study of that hereditary material carried in the forty-odd chromosomes in every human cell and transmitted from one generation to the next, knows a good deal about genes for physical characteristics, very little about genes for mental ability, and virtually nothing about genes for personality traits (violence, competitiveness, kindness, surliness, a sense of humor, etc.).

Wilson's colleague at Harvard, scientist Stephen Jay Gould, a specialist in evolution, says very flatly (in *Natural History Magazine*, 1976): "What is the direct evidence for genetic control of specific human social behavior? At the moment, the answer, is none whatever."

The distinguished biologist P. W. Medawar puts it this way, "By far the most important characteristic of human beings is that we have and exercise moral judgement and are not at the mercy of our hormones and genes."[9]

In the spring of 1986 an international conference of scientists in Seville, Spain, issued a statement on the question of human nature and violent aggression, concluding, "It is scientifically incorrect to say that war is caused by 'instinct' or any single motivation. . . . Modern war involves institutional use of personal characteristics such as obedience, suggestibility, and idealism. . . . We conclude that biology does not condemn humanity to war."

What about the evidence of psychology? This is not as "hard" a science as genetics. Geneticists can examine genes, even "splice" them into new forms. What psychologists do is look at the way people behave and think, test them, psychoanalyze them, conduct experiments to see how people react to different experiences, and try to come to reasonable conclusions about why people behave the way they do. There is nothing in the findings of psychologists to make any convincing argument for an instinct for the violent aggressivenes of war. That's why Freud, the founder of modern psychology, had to look for evidence of the destructive instinct in history.[10]

There was a famous "Milgram experiment" at Yale in the 1960s, named after the psychologist who supervised it.[11] A group of paid volunteers were told they were helping with an experiment dealing with the effects of punishment on learning. Each volunteer was seated in a position to observe someone taking a test, wearing electrodes connected to a control panel operated by the volunteer. The volunteer was told to monitor the test and, whenever a wrong answer was given, to pull a switch that would give a painful electrical jolt to the person taking the test, each wrong answer leading to a greater and greater electrical charge. There were thirty switches, with labels ranging from "Slight Shock" to "Danger—Severe Shock."

The volunteer was *not* told, however, that the person taking the test was an actor and that no real jolt was given. The actor would pretend to be in pain when the volunteer pulled the switch. When a volunteer became reluctant to continue causing pain, the experimenter in charge would say something like "The experiment requires that you continue." Under these conditions, two-thirds of the volunteers continued to pull the electrical switches on wrong answers, even when the subjects showed agonizing pain. One-third refused.

The experiment was tried with the volunteers at different distances from the subjects. When they were not physically close to the subject, about 35 percent of the volunteers defied authority even when they could not see or talk with the subject. But when they were right next to the subject, 70 percent refused the order.

The behavior of the people who were willing to inflict maximum pain can certainly be explained without recourse to "human nature." Their behavior was learned, not inborn. What they learned is what most people learn in modern culture, to follow orders, to do the job you are hired to do, to obey the experts in charge. In the experiment the supervisors, who had a certain standing and a certain legitimacy as directors of a "scientific" experiment, kept assuring the volunteers that they should go ahead, even if the subjects showed pain. When they were distant from the subjects, it was easier to obey the experimenters. But seeing or hearing the pain close up brought out some strong *natural* feeling of empathy, enough to disobey even the legitimate, confident, scientific supervisors of the experiment.

Some people interpreted the results of the experiment as showing an innate cruelty in human beings, but this was not the conclusion of Stanley Milgram, who directed the study. Milgram sums up his own views, "It is the extreme willingness of adults to go to almost any lengths on the command of an authority that constitutes the chief finding of the study. . . . This is, perhaps, the most fundamental lesson of our study: ordinary people, simply doing their jobs, and without any particular hostility on their part, can become agents in a terrible destructive process."

So it is a learned response—"always obey," "do your job"—and not a natural drive, that caused so many of the people to keep pulling the pain switches. What is remarkable in the Milgram experiment, given the power of "duty . . . obedience" taught to us from childhood, is not that so many obeyed, but that so many refused.

C. P. Snow, a British novelist and scientist, wrote in 1961,

> When you think of the long and gloomy history of man, you will find more hideous crimes have been committed in the name of obedience than have ever been committed in the name of rebellion. The German Officer Corps were brought up in the most rigorous code of obedience . . . in the name of obedience they were party to, and assisted in, the most wicked large scale actions in the history of the world.[12]

What about the evidence from anthropology—that is, from the behavior of "primitive" people, who are supposed to be closest to the "natural" state and, therefore, give strong clues about "human nature"? There have been many studies of the personality traits of such people: African Bushmen, North American Indians, Malay tribes, the Stone Age Tasaday from the Philippines, etc.

The findings can be summed up easily: There is no single pattern of warlike or peaceable behavior; the variations are very great. In North America, the Plains Indians were warlike, the Cherokee of Georgia were peaceful.

Anthropologist Colin Turnbull conducted two different studies in which he lived for a while with native tribes. In *The Forest People,* he describes the Pygmies of the Ituri rain forest in central Africa, wonderfully gentle and peaceful people whose idea of punishing a wrongdoer was to send him out into the forest to sulk. When he observed the Mbuti tribe of Zaire, he found them cooperative and pacific. However, when Turnbull spent time with the Ik people of East Africa, whom he describes in *The Mountain People*, he found them ferocious and selfish.[13]

The differences in behavior Turnbull found were explainable, not by genetics, not by the "nature" of these people, but by their environment, or their living conditions. The relatively easy life of the forest people fostered goodwill and generosity. The Ik, on the other hand, had been driven from their natural hunting grounds by the creation of a national game reserve into an isolated life of starvation in barren mountains. Their desperate attempt to survive brought out the aggressive destructiveness that Turnbull saw.

There have been many attempts to use the evidence of ethology (the study of the behavior of animals) to "prove" innate aggressiveness in human beings. We find Robert Ardrey using animal protection of their territory to argue for a "territorial imperative," which drives human beings to war against one another, or Desmond Morris, who uses the evidence of primates (*The Naked Ape*) to see human beings as deeply influenced by their evolutionary origins as tribal hunters.

But the study of animal behavior turns up all kinds of contradictory evidence. Baboons observed in a London zoo were found to be violent, but when studied on the plains of South Africa their behavior was peaceful. The difference was easily explainable by the fact that in the zoo baboons were strangers to one another, brought together by man. Even when baboons were aggressive, this consisted mostly of yelling and squabbling, not doing serious damage to one another.

We might note the work of Konrad Lorenz, an important zoologist and a specialist in the study of birds who could not resist the temptation to turn to human behavior in his book *On Aggression.* Lorenz is often cited to support the idea that aggressive instincts in human beings derive from evolutionary origins in animal behavior. But Lorenz was not that certain. Indeed, he said at one point that none of our so-called instincts are as dangerous as our "emotional allegiance to cultural values."[14]

It is a big jump, in any case, from bees or ducks or even baboons to human beings. Such a jump does not take account of the critically different factor of the human brain, which enables learning and culture and which creates a whole range of possibilities—good and bad. Those wide possibilities are not available to creatures with limited intelligence whose behavior is held close to their genetic instincts (although even with them different environments bring different characteristics).

The psychologist Erik Erikson, moving away from Freud's emphasis on biological instinct and on impressions gained in infancy, has pointed to the fact that, unlike most animals, human beings have a long childhood, a period for learning and cultural influence. This creates the possibility for a much wider range of behaviors.[15] Erikson says that our cultures have created "pseudospecies," that is, false categories of race and nation that obliterate our sense of ourselves as one species and thus encourage the hostility that turns violent.

Animals other than human beings do not make war. They do not engage in organized violence on behalf of some abstraction. That is a special gift of creatures with advanced brains and cultures. The animal commits violence for a specific, visible reason, the needs for food and for self-defense.

Genetics, psychology, anthropology, and zoology—in none of these fields is there evidence of a human instinct for the kind of aggressive violence that characterizes war. But what about history, which Freud pointed to?

Who can deny the frequency of war in human history? But its persistence does not prove that its origin is in human nature. Are there not persistent facts about human society that can explain the constant eruption of war without recourse to those mysterious instincts that science, however hard it tries, cannot find in our genes? Is not one of those facts the existence of ruling elites in every culture, who become enamored of their own power and seek to extend it? Is not another of those facts the greed, not of the general population, but of powerful minorities in society who seek more raw materials or more markets or more land or more favorable places for investment? Is there not a persistent ideology of nationalism, especially in the modern world, a set of beliefs taught to each generation in which the Motherland or the Fatherland is an object of veneration and becomes a burning cause for which one becomes willing to kill the children of other Motherlands or Fatherlands?

Surely we do not need human nature to explain war; there are other explanations. But human nature is simple and easy. It requires very little thought. To analyze the social, economic, and cultural factors that throughout human history have led to so many wars—that is hard work. One can hardly blame people for avoiding it.

But we should take another look at the proposition that the persistence of war in history *proves* that war comes from human nature. The claim requires that wars be not only frequent, but perpetual, that they not be limited to some nations but be true of all. Because if wars are only intermittent—if there are periods of war and periods of peace and if there are nations that go to war and other nations that don't—then it is unreasonable to attribute war to something as universal as human nature.

Whenever someone says, "history proves . . ." and then cites a list of historical facts, we should beware. We can always select facts from history (there are lots to choose from) to prove almost anything about human behavior. Just as one can select from a person's life just those instances of mean and aggressive behavior to prove the person *naturally* mean and aggressive, one can also select from that same person's life only those instances of kind and affectionate behavior to prove him or her naturally nice.

Perhaps we should turn from these scholarly studies of history, genetics, anthropology, psychology, and zoology to the plain reality of war itself. We surely have a lot of experience with that in our time.

I remember reading John Hersey's novel, *The War Lover.* It interested me greatly, partly because I am an admirer of Hersey's writing, but even more because his subject was the crew of a Flying Fortress, the B-17 heavy bomber in World War II. I had been a bombardier on such a crew in just that war. The novel's main character is a pilot who loves war. He also loves women. He is a braggart and a bully in regard to both. It turns out that his boasted sex exploits are a fraud and, in fact, he is impotent; it appears that his urge to bomb and kill is connected to that impotence.

When I finished reading the novel, I thought, Well, that may explain this piss-poor (a phrase left over from that war) fellow Hersey has picked as his subject and *his* lust for violence and death. But it doesn't explain war.

The men I knew in the air force—the pilots, navigators, bombardiers, and gunners on the crews flying over Europe, dropping bombs, and killing lots of people—were not lusting to kill, were not enthusiasts for violence, and were not war lovers. They—we—were engaged in large-scale killing, mostly of noncombatants, the women, children, and elderly people who happened to inhabit the neighborhoods of the cities that we bombed (officially, these were all "military targets"). But this did not come out of our *natures,* which were no different than when we were peacefully playing, studying, and living the lives of American boys back in Brooklyn, New York, or Aurora, Missouri.

The bloody deeds we did came out of a set of experiences not hard to figure out: We had been brought up to believe that our political leaders had good motives and could be trusted to do right in the world;

we had learned that the world had good guys and bad guys, good countries and bad countries, and ours was good. We had been trained to fly planes, fire guns, operate bombsights, and to take pride in doing the job well. And we had been trained to follow orders, which there was no reason to question, because everyone on our side was good, and on the other side, bad. Besides, we didn't have to watch a little girl's legs get blown off by our bombs; we were 30,000 feet high and no human being on the ground was visible, no scream could be heard. Surely that is enough to explain how men can participate in war. We don't have to grope in the darkness of human nature.

Indeed, when you look at modern war, do you find men rushing into it with a ferocious desire to kill? Hardly. You find men (and some women) joining the armed forces in search of training, careers, companionship, glamour, and psychological and economic security. You find others being conscripted by law, under penalty of prison if they refuse. And all of them suddenly transported into a war, where the habit of following orders and the dinning into their ears of the rightness of their cause can overcome the fear of death or the moral scruples of murdering another human being.

Many observers of war, and former soldiers too, have spoken of the lures of war for men, its attractions and enticements, as if something in men's *nature* makes war desirable for them. J. Glenn Gray, who was in army intelligence and close to combat situations in the European theater during World War II, has a chapter in his book *The Warriors* called "The Enduring Appeals of Battle." He writes of the "powerful fascination" of war. He says, "The emotional environment of warfare has always been compelling. . . . Many men both hate and love combat."[16] What are these "appeals" of war according to Gray? "The delight in seeing, the delight in comradeship, the delight in destruction."

He recalls the biblical phrase "the lust of the eye" to describe the sheer overpowering spectacle of war, the astounding scenes, the images, the vignettes—things never before experienced by young men who lived ordinary lives on ordinary farms or ordinary streets. That is certainly true. I had never seen the innards of a fifty-caliber machine gun; had never flown in an airplane miles high, in the night and close to the stars, overwhelmed by the beauty of that, and operated my bombsight and watched specks of fire flare like tiny torches on the ground below; and had never seen at close range the black puffs that were the explosions of antiaircraft shells, threatening my life. But that is not a love of war; it is an aesthetic need for visual and emotional excitement that comes, unrequested, with war and that can also be produced by other experiences.

Gray is also certainly right about the extraordinary comradeship of men in combat. But they don't *seek* combat because of that, any more than men in prison seek imprisonment because in prison they often forge human ties with fellow prisoners far stronger than any they have on the outside.

As for the "delight in destruction," I am skeptical about that. Granted, there is something visually exciting about explosions and something satisfying about hitting your target efficiently, as you were trained to do. But the delight that comes in a job well done would accompany any kind of job, not just destroying things.

All of the elements Gray and others have talked about as "the enduring appeals" of war are appeals not of violence or murder but of the concomitants of the war situation. It is sad that life is so drab, so unsatisfying for so many that combat gives them their first ecstatic pleasures, whether in "seeing" or companionship or work done well. It challenges us to find what the philosopher William James called "the moral equivalent of war," ways to make life outside of war vivid, affectionate, even thrilling.

Gray himself, although he tries to understand and explain those "enduring appeals," is offended by war. *The Warriors* recalls an entry in his own wartime journal, made December 8, 1944, which reflects not only his own feelings, but that of so many other veterans of war, that war *is* an affront to our nature as human beings. He wrote,

> Last night I lay awake and thought of all the inhumanity of it, the beastliness of the war. . . . I remembered all the brutal things I had seen since I came overseas, all the people rotting in jail, some of whom I had helped to put there. . . . I thought of Plato's phrase about the wise man caught in an evil time who refuses to participate in the crimes of his fellow citizens, but hides behind a wall until the storm is past. And this morning, when I rose, tired and distraught from bed, I knew that in order to survive this time I must love more. There is no other way.

When the U.S. government decided to enter World War I, it did not find an eager army of males, just waiting for an opportunity to vent their "natural" anger against the enemy, to indulge their "natural" inclination to kill. Indeed, there was a large protest movement against entrance into the war, leading Congress to pass punitive legislation for antiwar statements (2,000 people were prosecuted for criticizing the war). The government, besides conscripting men for service on threat of prison and jailing antiwar protesters, had to organize a propaganda campaign, sending 75,000 speakers to give 750,000 speeches in hundreds of towns and cities to persuade people of the rightness of the war.

Even with all that, there was resistance by young men to the draft. In New York City, ninety of the first hundred draftees claimed exemption. In Minnesota, the *Minneapolis Journal* reported, "Draft Opposition Fast Spreading in State." In Florida, two black farm workers went into the woods with a shotgun and mutilated themselves to avoid the draft; one blew off four fingers of his hand, the other shot off his arm below the elbow. A senator from Georgia reported "general and widespread opposition . . . to the enactment of the draft. . . . Mass meetings held in every part of the State protested against it." Ultimately, over 330,000 men were classified as draft evaders.[17]

We have an enormous literature of war. Much of it was written by men who experienced combat: Erich Remarque and Ernest Hemingway on World War I; Norman Mailer, James Jones, Kurt Vonnegut, Joseph Heller, and Paul Fussell on World War II; Philip Caputo, Tim O'Brien, John DelVecchio, Bill Ehrhart, and Ron Kovic on Vietnam. The men they write about are not (with occasional exceptions) bloodthirsty killers, consumed by some ferocious instinct to maim and destroy other human beings. They connect across a whole century with the young scared kid in *Red Badge of Courage;* they experience fear more than hate, fatigue more than rage, and boredom more than vengefulness. If any of them turn into crazed killers for some moment or some hour, it is not hard to find the cause in the crazed circumstances of war, coming on top of the ordinary upbringing of a young man in a civilized country.

A GI named John Ketwig wrote a letter to his wife:

> After all those years of preparation in the schools, you walked out the door, and they told you it was your duty to kill the commies in South Vietnam. If you wouldn't volunteer, they would draft you, force you to do things against your will. Put you in jail. Cut your hair, take away your mod clothes, train you to kill. How could they do that? It was directly opposite to everything your parents had been saying, the teachers had been saying, the clergymen had been saying. You questioned it, and your parents said they didn't want you to go, but better that than jail. The teacher said it was your duty. The clergy said you wouldn't want your mother to live in a commu-

> nist country, so you'd best go fight them in Asia before they landed in California. You asked about 'Thou shalt not kill', and they mumbled.[18]

It was no instinct to kill that led John Ketwig into military duty, but the pressure of people around him, the indoctrination of his growing up. So it is not remarkable that he joined the military. What is remarkable is that at a certain point he rebelled against it.

While 2 million men served in Vietnam at one time or another, another 0.5 million evaded the draft in some way. And of those who served, there were perhaps 100,000 deserters. About 34,000 GIs were court-martialed and imprisoned. If an instinct really was at work, it was not for war, but against it.

Once in the war, the tensions of combat on top of the training in obedience produced atrocities. In the My Lai Massacre we have an extreme example of the power of a culture in teaching obedience. In My Lai, a hamlet in South Vietnam, a company of U.S. soldiers landed by helicopter early one morning in March 1968, with orders to kill everybody there. In about one hour, although not a single shot was fired at them, they slaughtered about 400 Vietnamese, most of them old people, women, and children. Many of them were herded into ditches and then mowed down with automatic rifles.

One of the American soldiers, Charles Hutto, said later, "The impression I got was that we was to shoot everyone in the village. . . . An order came down to destroy all of the food, kill all the animals and kill all the people . . . then the village was burned. . . . I didn't agree with the killings but we were ordered to do it."[19]

It is not at all surprising that men go to war, when they have been cajoled, bribed, propagandized, conscripted, threatened, and also not surprising that after rigorous training they obey orders, even to kill unarmed women and children. What is surprising is that some refuse.

At My Lai a number of soldiers would not kill when ordered to: Michael Bernhardt, Roy Wood, Robert Maples, a GI named Grzesik. Warrant Officer Hugh Thompson commanded a helicopter that flew over the scene and, when he saw what was happening, he landed the helicopter and rescued some of the women and children, ordering his crewmen to fire on GIs if they fired on the Vietnamese. Charles Hutto, who participated in the My Lai Massacre, said afterward,

> I was 19 years old, and I'd always been told to do what the grown-ups told me to do. . . . But now I'll tell my sons, if the government calls, to go, to

> serve their country, but to use their own judgment at times . . . to forget about authority . . . to use their own conscience. I wish somebody had told me that before I went to Vietnam. I didn't know. Now I don't think there should be even a thing called war . . . 'cause it messes up a person's mind.[20]

In British novelist George Orwell's essay, "Shooting an Elephant," he recalls his experience in Burma, when he was a minor official of the British Empire. An elephant ran loose, and he finally shot it to death, but notes he did this not out of any internal drive, not of malice, but because people around him expected him to do that, as part of his job. It was not in his "nature."

The American feminist and anarchist Emma Goldman, writing at the beginning of the twentieth century before so much of the scientific discussion of the relationship between violence and human nature, said,

> Poor human nature, what horrible crimes have been committed in thy name! Every fool, from king to policeman, from the flathead parson to the visionless dabbler in science, presume to speak authoritatively of human nature. The greater the mental charlatan, the more definite his insistence on the wickedness and weaknesses of human nature. Yet how can any one speak of it today, with every soul a prison, with every heart fettered, wounded, and maimed?[21]

Her point about "the visionless dabbler in science" was affirmed half a century later by Nobel Prize–winning biologist Salvadore E. Luria, who points to the misuse of science in attributing violent behavior to our genes. Moving away from genetic determinism and its mood of inevitability (as too often interpreted, the inevitability of war and death), Luria says that biologists have a nobler role for the future: to explore "the most intriguing feature—the creativity of the human spirit."[22]

That creativity is revealed in human history, but it is a history that Machiavelli and a succession of scholarly pessimists ignore as they concentrate on the worst aspects of human behavior. There is another history, of the rejection of violence, the refusal to kill, and the yearning for community. It has shown itself throughout the past in acts of courage and sacrifice that defied all the immediate pressures of the environment.

This was true even in the unspeakable conditions of the German death camps in World War II, as Terence des Pres pointed out in his book *The Survivor.* He wrote, "The depth and durability of man's social nature may be gauged by the fact that conditions in the concentration camps were designed to turn prisoners against each other, but that in a multitude of ways, men and women persisted in social acts."[23]

It is true that there is an infinite human capacity for violence. There is also an infinite potential for kindness. The unique ability of humans to *imagine* gives enormous power to idealism, an imagining of a better state of things not yet in existence. That power has been misused to send young men to war. But the power of idealism can also be used to attain justice, to end the massive violence of war.

Anyone who has participated in a social movement has seen the power of idealism to move people toward self-sacrifice and cooperation. I think of Sam Block, a young black Mississippian, very thin and with very bad eyes, taking black people to register to vote in the murderous atmosphere of Greenwood, Mississippi, in the early 1960s. Block was accosted by a sheriff (another civil rights worker, listening, recorded their conversation):

SHERIFF: Nigger, where you from?
BLOCK: I'm a native of Mississippi.
SHERIFF: I know all the niggers here.
BLOCK: Do you know any colored people?
(The sheriff spat at him.)
SHERIFF: I'll give you till tomorrow to get out of here.
BLOCK: If you don't want to see me here, you better pack up and leave, because I'll be here.[24]

History, so diligent at recording disasters, is largely silent on the enormous number of courageous acts by individuals challenging authority and defying death.

The question of history, its use and abuse, deserves a discussion of its own.

CHAPTER

FOUR

The Use and Abuse of History

Before I became a professional historian, I had grown up in the dirt and dankness of New York tenements, had been knocked unconscious by a policeman while holding a banner in a demonstration, had worked for three years in a shipyard, and had participated in the violence of war. Those experiences, among others, made me lose all desire for "objectivity," whether in living my life, or writing history.

This statement is troubling to some people. It needs explanation.

I mean by it that by the time I began to study history formally I knew I was not doing it because it was simply "interesting" or because it meant a solid, respectable career. I had been touched in some way by the struggle of ordinary working people to survive, by the glamour and ugliness of war, and by the reading I had done on my own trying to understand fascism, communism, capitalism, and socialism. I could not possibly study history as a neutral. For me, history could only be a way of understanding and helping to change (yes, an extravagant ambition!) what was wrong in the world.

That did not mean looking only for historical facts to reinforce the beliefs I already held. It did not mean ignoring data that would change or complicate my understanding of society. It meant asking questions that were important for social change, questions relating to equality, liberty, peace, and justice, but being open to whatever answers were suggested by looking at history.

I decided early that I would be biased in the sense of holding fast to certain fundamental values: the equal right of all human beings—whatever race, nationality, sex, religion—to life, liberty, and the pursuit of happiness, Jefferson's ideals. It seemed to me that devoting a life to the study of history was worthwhile only if it aimed at those ideals.

But I wanted to be flexible in arriving at the *means* to achieve those ends. Scrupulous honesty in reporting on the past would be needed, because any decision on means (tactics, avenues, and instruments) had to be tentative and had to be open to change based on what one could learn from history. The values, ends, and ideals I held need not be discarded, whatever history disclosed. So there would be no incentive to distort the past, fearing that an honest recounting would hurt the desired ends.

Does this mean that our values, our most cherished ideals, have no solid basis in *fact,* that desires for freedom and justice have the lightness of personal whims and subjective desires? On the contrary, our powerful impulses for freedom and community come from deep, dependable internal drives (these too are *facts*), often deflected or overcome by terrible pressures in our culture, but never extinguished. Does this not account for the way peoples long oppressed and apparently beaten into silence, suddenly rebel, demanding their freedom?[1]

Professional philosophers refer to the "fact-value" problem. That is Do your basic values depend on certain facts, so that if you discover your facts are wrong, you are compelled to change your values? I am arguing here for holding on to certain basic values—and insisting that whatever facts you discover in history may change your means without dislodging your ends.[2]

I can illustrate that with my own experience. At seventeen or eighteen, I was reading lots of novels. Some were pure entertainment. Others were novels of social criticism, like Upton Sinclair's *The Jungle,* and John Steinbeck's *The Grapes of Wrath.*

I don't know exactly when I decided that I believed in the socialism described by Sinclair in the last pages of *The Jungle.* Or that I wouldn't be afraid of the epithet "Communist," because, as someone said (I recall it approximately) in *The Grapes of Wrath,* "A Communist is anyone who asks for twenty cents an hour when the boss is paying fifteen."

When I encountered young Communists in my working-class neighborhood and they bombarded me with literature on the Soviet Union, I was persuaded (like many Americans in the Depression years) that here was a model for a future society of equality and justice, the rational planning of production and distribution, the creation of a "workers' state." But while flying bombing missions in World War II, I became friends with a young gunner on another crew who, like me, was a constant reader. He gave me a book I had never heard of, by a writer I had never heard of. It was *The Yogi and the Commissar* by Arthur Koestler.

That book, written by a former Communist who had fought against fascism in Spain, was a powerful, eloquent denunciation of the Soviet Union, seeing what happened there as a betrayal of Communist ideals. Its historical data seemed irrefutable. I trusted the author's commitment and his intelligence. That was the beginning of my own move away from acceptance of the Soviet Union as a socialist or Communist model.

When Khrushchev gave his astounding speech in 1956 acknowledging Stalin's crimes (which involved, although Khrushchev did not stress this, the complicity of so many other members of the Soviet hierarchy), he was affirming what Koestler and other critics of the Soviet Union had been saying for a long time. When Soviet troops invaded Hungary and then Czechoslovakia to crush rebellions, it was clear to me that the Soviet Union was violating a fundamental Marxist value—really, a universal principle, beyond Marxism—of international solidarity.

My faith in the ideal of an egalitarian society, a cooperative commonwealth, in a world without national boundaries, remained secure. My idea that the Soviet Union represented that new world was something I could discard. I had to be willing to call the shots as I saw them in reading the history of the Soviet Union, just as I wanted those who had a romanticized view of the United States to be willing to call the shots as they saw them in the American past. I knew also that it was a temptation to hold onto old beliefs, to ignore uncomfortable facts because one had become attached to ideals, and that I must guard myself against that temptation and be watchful for it in reading other historians.

A historian's strong belief in certain values and goals *can* lead to dishonesty or to distortions of history. But that is avoidable if the historian understands the difference between solidity in ultimate values and openness in regard to historical fact.

There is another kind of dishonesty that often goes unnoticed. That is when historians fail to acknowledge their own values and pretend to "objectivity," deceiving themselves and their readers.

Everyone is biased, whether they know it or not, in possessing fundamental goals, purposes, and ends. If we understand that, we can be properly skeptical of all historians (and journalists and anyone who reports on the world) and check to see if their biases cause them to emphasize certain things in history and omit or give slight consideration to others.

Perhaps the closest we can get to objectivity is a free and honest marketplace of subjectivities, in which we can examine both orthodox accounts of the past and unorthodox ones, commonly known facts and hitherto ignored facts. But we need to try to discover (which is not easy) what items are missing from that marketplace and insist that they be available for scrutiny. We can then decide for ourselves, based on our own values, which accounts are most important and most useful.

Anyone reading history should understand from the start that there is no such thing as impartial history. All written history is partial in two senses. It is partial in that it is only a tiny *part* of what really happened. That is a limitation that can never be overcome. And it is partial in that it inevitably takes sides, by what it includes or omits, what it emphasizes or deemphasizes. It may do this openly or deceptively, consciously or subconsciously.

The chief problem in historical honesty is not outright lying. It is omission or deemphasis of important data. The definition of *important*, of course, depends on one's values.

An example is the Ludlow Massacre.

I was still in college studying history when I heard a song by folksinger Woody Guthrie called "The Ludlow Massacre," a dark, intense ballad, accompanied by slow, haunting chords on his guitar. It told of women and children burned to death in a strike of miners against Rockefeller-owned coal mines in southern Colorado in 1914.

My curiosity was aroused. In none of my classes in American history, in none of the textbooks I had read, was there any mention of the Ludlow Massacre or of the Colorado coal strike. I decided to study the history of the labor movement on my own.

This led me to a book, *American Labor Struggles*, written not by a historian but an English teacher named Samuel Yellen. It contained exciting accounts of some ten labor conflicts in American history, most of which were unmentioned in my courses and my textbooks. One of the chapters was on the Colorado coal strike of 1913–1914.[3]

I was fascinated by the sheer drama of that event. It began with the shooting of a young labor organizer on the streets of Trinidad,

Colorado, in the center of the mining district on a crowded Saturday night, by two detectives in the pay of Rockefeller's Colorado Fuel & Iron Corporation. The miners, mostly immigrants, speaking a dozen different languages, were living in a kind of serfdom in the mining towns where Rockefeller collected their rent, sold them their necessities, hired the police, and watched them carefully for any sign of unionization.

The killing of organizer Gerry Lippiatt sent a wave of anger through the mine towns. At a mass meeting in Trinidad, miners listened to a rousing speech by an eighty-year-old woman named Mary Jones—"Mother Jones"—an organizer for the United Mine Workers: "What would the coal in these mines and in these hills be worth unless you put your strength and muscle in to bring them. . . . You have collected more wealth, created more wealth than they in a thousand years of the Roman Republic, and yet you have not any."[4]

The miners voted to strike. Evicted from their huts by the coal companies, they packed their belongings onto carts and onto their backs and walked through a mountain blizzard to tent colonies set up by the United Mine Workers. It was September 1913. There they lived for the next seven months, enduring hunger and sickness, picketing the mines to prevent strikebreakers from entering, and defending themselves against armed assaults. The Baldwin-Felts Detective Agency, hired by the Rockefellers to break the morale of the strikers, used rifles, shotguns, and a machine gun mounted on an armored car, which roved the countryside and fired into the tents where the miners lived.

They would not give up the strike, however, and the National Guard was called in by the governor. A letter from the vice president of Colorado Fuel & Iron to John D. Rockefeller, Jr., in New York explained,

> You will be interested to know that we have been able to secure the cooperation of all the bankers of the city, who have had three or four interviews with our little cowboy governor, agreeing to back the State and lend it all funds necessary to maintain the militia and afford ample protection so our miners could return to work. . . . Another mighty power has been rounded up on behalf of the operators by the getting together of fourteen of the editors of the most important newspapers in the state.[5]

The National Guard was innocently welcomed to town by miners and their families, waving American flags, thinking that men in the uniform of the United States would protect them. But the guard went to work for the operators. They beat miners, jailed them, and escorted strikebreakers into the mines.[6]

The strikers responded. One strikebreaker was murdered, another brutally beaten, four mine guards killed while escorting a scab. And Baldwin-Felts detective George Belcher, the killer of Lippiatt, who had been freed by a coroner's jury composed of Trinidad businessmen ("justifiable homicide"), was killed with a single rifle shot by an unseen gunman as he left a Trinidad drugstore and stopped to light a cigar.

The miners held out through the hard winter, and the mine owners decided on more drastic action. In the spring, two companies of National Guardsmen stationed themselves in the hills above the largest tent colony, housing a thousand men, women, and children, near a tiny depot called Ludlow. On the morning of April 20, 1914, they began firing machine guns into the tents. The men crawled away to draw fire and shoot back, while the women and children crouched in pits dug into the tent floors. At dusk, the soldiers came down from the hills with torches, and set fire to the tents. The countryside was ablaze. The occupants fled.

The next morning, a telephone linesman, going through the charred ruins of the Ludlow colony, lifted an iron cot that covered a pit dug in the floor of one tent, and found the mangled, burned bodies of two women and eleven children. This became known as the Ludlow Massacre.

As I read about this, I wondered why this extraordinary event, so full of drama, so peopled by remarkable personalities, was never mentioned in the history books. Why was this strike, which cast a dark shadow on the Rockefeller interests and on corporate America generally, considered less important than the building by John D. Rockefeller of the Standard Oil Company, which was looked on as an important and positive event in the development of American industry?

I knew that there was no secret meeting of industrialists and historians to agree to emphasize the admirable achievements of the great corporations and ignore the bloody costs of industrialization in America. But I concluded that a certain unspoken understanding lay beneath the writing of textbooks and the teaching of history: that it would be considered bold, radical, perhaps even "communist" to emphasize class struggle in the United States, a country where the dominant ideology emphasized the oneness of the nation "We the People, in order to . . . etc., etc." and the glories of the American system.

Not long ago, a news commentator on a small radio station in Madison, Wisconsin, brought to my attention a textbook used in high schools all over the nation, published in 1986, titled *Legacy of Freedom,* written by two high-school teachers and one university professor of history and published by a division of Doubleday and Company, one of the giant publishers in the United States. In a foreword "To the Student" we find,

> *Legacy of Freedom* will aid you in understanding the economic growth and development of our country. The book presents the developments and benefits of our country's free enterprise economic system. You will read about the various ways that American business, industry, and agriculture have used scientific and technological advances to further the American free market system. This system allows businesses to generate profits while providing consumers with a variety of quality products from which to choose in the marketplace, thus enabling our people to enjoy a high standard of living.[7]

In this overview one gets the impression of a wonderful, peaceful development, which is the result of "our country's free enterprise economic system." Where is the long, complex history of labor conflict? Where is the human cost of this industrial development, in the thousands of deaths each year in industrial accidents, the hundreds of thousands of injuries, the short lives of the workers (textile mill girls in New England dying in their twenties, after starting work at twelve and thirteen)?

The Colorado coal strike does not fit neatly into the pleasant picture created by most high-school textbooks of the development of the American economy. Perhaps a detailed account of that event would raise questions in the minds of young people as it raised in mine, questions that would be threatening to the dominant powers in this country, that would clash with the dominant orthodoxy. The questioners—whether teachers or principals, or school boards—might get into trouble.

For one thing, would the event not undermine faith in the neutrality of government, the cherished belief (which I possessed through my childhood) that whatever conflicts there were in American society, it was the role of government to mediate them as a neutral referee, trying its best to dispense, in the words of the Pledge of Allegiance, "liberty and justice for all"? Would the Colorado strike not suggest that governors, that perhaps all political leaders, were subject to the power of wealth, and would do the bidding of corporations rather than protect the lives of poor, powerless workers?

A close look at the Colorado coal strike would reveal that not only the state government of Colorado, but the national government in Washington—under the presidency of a presumed liberal, Woodrow Wilson—was on the side of the corporations. While miners were being beaten, jailed, and killed by Rockefeller's detectives or by his National Guard, the federal government did nothing to protect the constitutional rights of its people. (There is a federal statute—Title 10, Section 333—which gives the national government the power to defend the constitutional rights of citizens when local authorities fail to do so.)

It was only after the massacre, when the miners armed themselves and went on a rampage of violence against the mine properties and mine guards, that President Wilson called out the federal troops to end the turmoil in southern Colorado.

And then there was an odd coincidence. On the same day that the bodies were discovered in the pit at Ludlow, Woodrow Wilson, responding to the jailing of a few American sailors in Mexico, ordered the bombardment of the Mexican port of Vera Cruz, landed ten boatloads of marines, occupied the city, and killed more than a hundred Mexicans.

In that same textbook the foreword "To the Student" says: "*Legacy of Freedom* will aid you in understanding our country's involvement in foreign affairs, including our role in international conflicts and in peaceful and cooperative efforts of many kinds in many places." Is that not a benign, misleading, papering over of the history of American foreign policy?

A study of the Ludlow Massacre, alongside the Mexican incident, would also tell students something about our great press, the comfort we feel when picking up, not a scandal sheet or a sensational tabloid, but the sober, dependable *New York Times.* When the U.S. Navy bombarded Vera Cruz, the *Times* wrote in an editorial:

> We may trust the just mind, the sound judgment, and the peaceful temper of President Wilson. There is not the slightest occasion for popular excitement over the Mexican affair; there is no reason why anybody should get nervous either about the stock market or about his business.[8]

There is no *objective* way to deal with the Ludlow Massacre. There is the subjective (biased, opinionated) decision to omit it from history, based on a value system that doesn't consider it important enough. That value system may include a fundamental belief in the beneficence of the American industrial system (as represented by the passage quoted above from the textbook *Legacy of Freedom*) or it may just involve a complacency about class struggle and the intrusion of government on the side of corporations. In any case, a certain set of values has dictated the ignoring of an important historical event.

It is also a subjective (biased, opinionated) decision to tell the story of the Ludlow Massacre in some detail (as I do, in a chapter in my book *The Politics of History*,[9] or in several pages in *A People's History of the United States*). My decision was based on my belief that it is important for people to know the extent of class conflict in our history, to know something about how working people had to struggle to change their conditions, and to understand the role of the government and the mainstream press in the class struggles of our past.

One must inevitably omit large chunks of what is available in historical information. But *what* is omitted is critical in the kind of historical education people get; it may move them one way or another or leave them motionless—passive passengers on a train that is already moving in a certain direction, which they by their passivity seem to accept. My own intention is to select subjects and emphasize aspects of those subjects that will help move citizens into activity on behalf of basic human rights: equality, democracy, peace, and a world without national boundaries. Not by hiding factors from them, but by adding to the orthodox store of knowledge, opening wider the marketplace of information.

The problem of selection in history is strikingly shown in the story of Christopher Columbus, which appears in every textbook of American history on every level from elementary school through college.[10] It is a story, always, of skill and courage, leading to the discovery of the Western Hemisphere.

Something is omitted from that story, in almost every textbook in every school in the United States. What is omitted is that Columbus, in his greed for gold, mutilated, enslaved, and murdered the Indians who greeted him in friendly innocence, and that this was done on such a scale as to deserve the term "genocide"—the destruction of an entire people.[11]

This information was available to historians. In Columbus's own log he shows his attitude from the beginning. After telling how he and his men landed on that first island in the Bahamas and were greeted peaceably by the Arawak Indians, who seemed to have no knowledge of weapons and gave the strangers gifts, Columbus says, "They would make fine servants. . . . With fifty men we could subjugate them all and make them do whatever we want."

The closest we have to a contemporary source on what happened after that first landing is the account by Bartolomeo de las Casas, who as a young priest participated in the conquest of Cuba. In his *History of the Indies*, las Casas wrote, "Endless testimonies . . . prove the mild and pacific temperament of the natives. . . . But our work was to exasperate, ravage, kill, mangle, and destroy. . . . The admiral . . . was so anxious to please the King that he committed irreparable crimes against the Indians."[12]

The "admiral" was Columbus. One of the few historians even to mention the atrocities committed by Columbus against the Indians was Samuel Eliot Morison, who wrote the two-volume biography of Columbus, *Admiral of the Ocean Sea.* [13] In his shorter book, written for a wider audience in 1954, *Christopher Columbus, Mariner*, Morison says, "The cruel policy initiated by Columbus and pursued by his successors resulted in complete genocide."[14] But this statement is on one page, buried in a book that is mostly a glowing tribute to Columbus.

In my book *A People's History of the United States* I commented on Morison's quick mention of Columbus's brutality:

> Outright lying or quiet omission takes the risk of discovery which, when made, might arouse the reader to rebel against the writer. To state the facts, however, and then to bury them in a mass of other information is to say to the reader with a certain infectious calm: yes, mass murder took place, but it's not that important—it should weigh very little in our final judgements; it should affect very little what we do in the world.[15]

Is my own emphasis on Columbus's treatment of the Indians biased? No doubt. I won't deny or conceal that Columbus had courage and skill, was an extraordinary sailor. But I want to reveal something about him that was omitted from the historical education of most Americans.

My bias is this: I want my readers to think twice about our traditional heroes, to reexamine what we cherish (technical competence) and what we ignore (human consequences). I want them to think about how easily we accept conquest and murder because it furthers "progress." Mass murder for "a good cause" is one of the sicknesses of our time. There were those who defended Stalin's murders by saying, "Well, he made Russia a major power." As we have seen, there were those who justified the atom bombing of Hiroshima and Nagasaki by saying "We had to win the war."

* * *

There is still another kind of historical bias that can mislead us, and that is the tendency of the culture to emphasize historical trivia, to learn facts for their own sake. The result of this is to encourage a flat, valueless interest in past facts that have no great significance in the betterment of the human condition, but that are simply "interesting." The interest served, however, is that of diverting us from the truly important uses of history, thus making history, literally, a diversion.

For instance, in the fall of 1986 the *Boston Globe* carried a front-page story about how a scholar for the National Geographic Society had concluded that Columbus landed sixty-five miles to the south of where it had always been assumed he landed, on Samana Cay rather than on Watling Island.[16] The *Globe*, I am quite sure has never carried a front-page story, and probably not any kind of a story (nor has any other newspaper in the United States I suspect), on the revelations that Columbus had murdered countless Indians. The celebrants of Columbus Day would find that story embarrassing. But readers of such news might find it much more important, much more thought provoking, than the exact route of Columbus's voyage.

What is important is not Columbus. To defend or attack Columbus is pointless. What is important is how closely we look today at what is done to human beings, what criteria we use for "progress." We are accustomed to measuring the state of the nation by the numbers on the stock market (the Dow-Jones average), rather than by how many children died of malnutrition.

The very labels we give to eras accustom us to overlooking some events and highlighting others. The Ludlow Massacre took place in that period labeled in so many American history books as "the Progressive Era," a phrase based on the fact that certain pieces of reform legislation were passed by Congress in the early years of the twentieth century. But the Progressive Era was also the period when the greatest numbers of black people, thousands of them, were lynched—hanged, burned, shot by mobs—in the United States.

In all the standard treatments of the twenties, this was the "Jazz Age," a time of fun and prosperity for Americans. But when sociologists Robert and Helen Lynd studied Muncie, Indiana, in the twenties, they found two classes: "the Working Class and the Business Class." They reported that for two-thirds of the city's families "the father gets up in the dark in winter, eats hastily in the kitchen in the gray dawn, and is at work from an hour to two and a quarter hours before his children have to be at school."[17] Historian Merle Curti remarked on how the plight of the majority in the twenties was ignored: "It was, in fact, only the upper ten percent of the population that enjoyed a marked increase in real income."[18]

Fiorello LaGuardia represented a district of poor people in East Harlem during the Jazz Age, and he kept rising to his feet in the House of Representatives, reminding his colleagues that there were people in his district and all over the country who could not pay their rent, could not pay their gas bills, and could not buy adequate food for their families. In 1928 LaGuardia toured the poorer districts of New York and reported; "I confess I was not prepared for what I actually saw. It seemed almost incredible that such conditions of poverty could really exist."[19]

If Americans received a better historical education, if they learned to look beneath the surface of easy labels ("The Era of Good Will," "The Age of Prosperity," etc.), if they understood that our national orthodoxy prefers to conceal certain disturbing facts about our society, they might, in the 1980s and 1990s, look beneath the glitter and luxury and react with anger to the homelessness, poverty, and despair that plague millions of people in this country.

Historians, like journalists, select what they think is important or what they think the publisher will deem important or what they both think the public will consider important. Often they will report on something because everyone else who has written before has reported on it. And they will omit something because it has always been omitted. In other words, there is a conservative bias to history and a tendency to emphasize what previous generations have emphasized. The motive for that is safety, because the historian who breaks the pattern causes stares and suspicions.

This conservative bias, this tendency to repeat the thinking of the past, is true even of people who think themselves revolutionaries. Karl Marx recognized this in his history of counterrevolution in France after the 1848 revolution:

> The tradition of all the dead generations weighs like a nightmare on the brain of the living. And just when they seem engaged in revolutionising themselves and things, in creating something entirely new, precisely in such

> epochs of revolutionary crisis they anxiously conjure up the spirits of the past to their service and borrow from them names, battle slogans and costumes in order to present the new scene of world history in this time-honored disguise and this borrowed language.[20]

In the United States historical education has emphasized the doings of the rich and powerful—the political leaders, the industrial entrepreneurs. The classroom education of the young has often centered on the presidents. One widely used book for teachers (*Push Back the Desks*), spoke admiringly of a classroom where the portraits of all the presidents filled the walls and the history lessons were based on that. We often poke fun in the United States at other countries where political leaders are treated like gods, their portraits and statues everywhere. But in our culture, the most trivial activities of the presidents are considered of great significance, while the life-and-death struggles of ordinary people are ignored.

For instance, on September 17, 1972, the *New York Times* carried a front-page story about Chester A. Arthur, who became president in 1881 and whose administration was hardly noteworthy for any achievements on behalf of human freedom. The headline to the story was "President Arthur Kept Illness a Secret." The story is about a conference of historians in Tarrytown, New York: "President Arthur's tightly held secret (that he had a rare kidney disease), withheld not only from his time but also from history, was made known publicly for the first time at the conference." The *Times* story quoted one of the historians at the conference: "The factual record is substantially corrected, updated and enlarged, and our inherited assumptions about a bygone era receive a sharp jolt."

What should really give us a sharp jolt is that such a piece of trivia should become a front-page story for the nation's major metropolitan newspaper.

The National Historical Papers Commission has spent millions of dollars, given by Congress and by the Ford Foundation, to publish sixty volumes of Thomas Jefferson's papers, sixty volumes of James Madison, seventy-five volumes of George Washington, a hundred volumes on the Adams family. There are plans for sixty-five volumes on Benjamin Franklin (plus, as the editor noted, "several volumes of addenda and errata").[21] One historian, Jesse Lemisch, whose own work dealt with ordinary seamen of the Revolutionary Period and who lamented the lack of historical attention to the working people of the country, referred to this project as "the papers of Great White Men."

What sorts of values and ideals are encouraged in the young people of the coming generation by the enormous emphasis on the Founding Fathers and the presidents? It seems to me that the result is the creation of dependency on powerful political figures to solve our problems.

We were being exploited by England? Well, the Founding Fathers took care of that in leading the struggle for independence. Was the nation morally blighted by the existence of 4 million black slaves? Abraham Lincoln solved that with the Emancipation Proclamation. Did we have a terrible economic crisis in the early 1930s? Franklin Roosevelt got us out of that one. Do we face enormous problems today? Well, the solution is to find the right president, to go to the polls and choose either the Republican or Democratic candidate.

Such a view, embedded in the minds of the American public by an education that focuses on elites, ignores an important part of the historical record. It does not pay sufficient attention to the "crowds" of the Revolutionary period, the grass-roots organizations, rioters, demonstrators, and boycotters who brought the Revolution to a boil.[22]

Not enough credit is given to the great Abolitionist movement of tens of thousands of black and white people, risking their lives and their freedom to demand the end of slavery. It was this movement that galvanized antislavery sentiment in the country between 1830 and 1860 and pressured Lincoln into his first actions against slavery and pushed Congress into passing the Thirteenth, Fourteenth, and Fifteenth Amendments, which made slavery and racial discrimination at last illegal (even if they still existed in different forms).[23]

The New Deal reforms, although presided over by Roosevelt, were given their momentum by the mass movements of that time: the Bonus March of 1932; the general strikes of 1934; the waves of strikes in the auto, rubber, and steel industries in 1936 and 1937; the organizations of tenants and the unemployed; and by turmoil in the cities and the countryside.[24]

Consider how much attention is given in historical writing to military affairs—to wars and battles—and how many of our heroes are military heroes. And consider also how little attention is given to antiwar movements and to those who struggled against the idiocy of war. Everybody who goes to an American school learns about Theodore Roosevelt's charge up San Juan Hill in the Spanish-American war. But how many learn about the Anti-Imperialist League, which criticized the nation's actions in Cuba and the military conquest of the Philippines.[25]

As a result of omitting, or downplaying, the importance of social movements of the people in our history—the actions of abolitionists, labor leaders, radicals, feminists, and pacifists—a fundamental principle of democracy is undermined: the principle that it is the citizenry, rather than the government, that is the ultimate source of power and the locomotive that pulls the train of government in the direction of equality and justice. Such histories create a passive and subordinate citizenry.

This is not the deliberate intention of the historian, but it comes from a desire to avoid controversy, to go along with what has always been done, to stress what has always been stressed, to keep one's job, to stay out of trouble, and to get published. The pollution of history comes about like the pollution of the air and the water. No one *plans* the poisoning of air and water; each one acts for personal gain in a system where "private enterprise" operates in education as in industry to keep society along the old tracks and to keep minds thinking the old way. It is profitable. It is safe.

The pretense of objectivity conceals the fact that all history, while recalling the past, serves some present interest. One university professor, in a review of a book on the imprisonment of Japanese families on the West Coast during World War II (Richard Drinnon's *Keeper of Concentration Camps*), was unhappy with the moral indignation shown by the author. He referred to it as "the fatal flaw . . . of hindsight." The reviewer advised, "Write history as if you were there at the time it occurred and were seeing it through the eyes of the participants. . . . That is difficult, but it must be done to avoid bias."[26]

This is a common argument. Avoid "presentism," it says. Put yourself back then. But "back then" there were different views. Who were "the participants" in the cruel detention of the Japanese: the bureaucrats who planned and supervised it or the Japanese men, women, and children who endured it? To put oneself back in the past does not eliminate the necessity of choosing a viewpoint, a side, or a value. A bias is inevitable. But you can declare it honestly. To pretend "to avoid bias" too often means exonerating (by the *lack* of indignation) the moral crimes of the past.[27]

The claim of historians to objectivity has been examined very closely by Peter Novick.[28] He finds the claim especially false in wartime. For instance, in April 1917, just after the United States had entered the European war, a group of eminent historians met in Washington to discuss "what History men can do for their country now." They set up the National Board for Historical Service to "aid in supplying the public with trustworthy information of historical or similar character."[29]

One result was a huge outpouring of pamphlets written by historians with the purpose of instilling patriotism in the public; 23 million copies of such pamphlets were distributed. Most of them, according to a recent study of the role of historians in World War I propaganda, "reduced war issues to black and white, infused idealism and righteousness into America's role, and established German guilt with finality."[30]

During World War II, Samuel Eliot Morison declared his commitment to not instruct the present but to "simply explain the event exactly as it happened." Yet, in the same essay ("Faith of a Historian"), he criticized those historians who had expressed disillusionment with World War I, saying they "rendered the generation of youth which came to maturity around 1940 spiritually unprepared for the war they had to fight. . . . Historians . . . are the ones who should have pointed out that war does accomplish something, that war is better than servitude."[31]

In the cold war atmosphere of the 1950s, a number of historians selected their facts to conform to the government's position. Two of them wrote a two-volume history of the U.S. entry into World War II, to show, as they put it, "the tortured emergence of the United States of America as leader of the forces of light in a world struggle which even today has scarcely abated."

An honest declaration of their bias would have been refreshing. But, although they had access to official documents not available to others, they said in their preface, "No one, in the State Department or elsewhere, has made the slightest effort to influence our views." Perhaps not. But one of them, William Langer, was director of research for the CIA at one time, and the other, S. Everett Gleason, was deputy executive secretary of the National Security Council.[32]

Langer was also a president of the American Historical Association (AHA). Another president of the AHA, Samuel Flagg Bemis, in his address to that group in 1961, was very clear about what he wanted historians to do:

> Too much . . . self-criticism is weakening to a people. A great people's culture begins to decay when it commences to examine itself. . . . We have been losing sight of our national purpose . . . our military preparedness held back by insidious strikes for less work and more pay. . . . Massive self-indulgence and massive responsibility do not go together. . . . How can our

lazy dalliance and crooning softness compare with the stern discipline and tyrannical compulsion of subject peoples that strengthen the aggressive sinews of our malignant antagonist.[33]

Historian Daniel Boorstin testified before the House Committee on Un-American Activities in 1953. He agreed with the committee that Communists should not be permitted to teach in American universities—presumably because they would be biased. Boorstin told the committee that he expressed his own opposition to communism in two ways. First, by participation in religious activities at the University of Chicago, and "the second form of my opposition has been an attempt to discover and explain to students, in my teaching and in my writing, the unique virtues of American democracy." No bias there.[34]

After studying the objectivity of American historians, and noting how many slanted their work toward support for the United States, Peter Novick wondered if that kind of "hubris," the arrogance of national power, played a part in the ugly American intervention in Vietnam and the cold war itself. He put it this way:

> If ill-considered American global interventionism had landed us in this bloodiest manifestation of the cold war, was it not at least worth considering whether the same hubris had been responsible for the larger conflict of which it was a part? Manifestly by the sixties, the United States was overseeing an empire. Could scholars comfortably argue that it had been acquired as had been said of the British Empire, "in a fit of absence of mind"?[35]

In the sixties, there was a series of tumultuous social movements against racial segregation and against the Vietnam War and for equality between the sexes. This caused a reappraisal of the orthodox histories. More and more books began to appear (or old books were brought to light) on the struggles of black people, on the attempts of women throughout history to declare their equality with men, on movements against war, and on the strikes and protests of working people against their conditions—books that, while sticking close to confirmed information, openly took sides for equality, against war, and for the working classes.

The unapologetic activism of the sixties (making history in the street as well as writing it in the study) was startling to many professional historians. And in the seventies and eighties, it was accused by some scholars and some organs of public opinion of hurting the proper historical education of young people by its insistence on "relevance." As part of the attack, a demand grew for more emphasis on facts, on dates, and on the sheer accumulation of historical information.[36]

In May of 1976 the *New York Times* published a series of articles in which it lamented the ignorance of American students about their own history.[37] The *Times* was pained. Four leading historians whom it consulted were also pained. It seemed students did not know that James Polk was president during the Mexican War, that James Madison was president during the War of 1812, that the Homestead Act was passed earlier than Civil Service reform, or that the Constitution authorizes Congress to regulate interstate commerce but says nothing about the cabinet.

We might wonder if the *Times,* or its historian-consultants, learned anything from the history of this century. It has been a century of atrocities: the death camps of Hitler, the slave camps of Stalin, and the devastation of Southeast Asia by the United States. All of these were done by powerful leaders and obedient populations in countries that had achieved high levels of literacy and education. It seems that high scores on tests were not the most crucial fact about those leaders and those citizens.

In the case of the United States the killing of a million Vietnamese and the sacrifice of 55,000 Americans were carried out by highly educated men around the White House who scored very well in tests and who undoubtedly would have made impressive grades in the *New York Times* exam. It was a Phi Beta Kappa, McGeorge Bundy, who was one of the chief planners of the bombing of civilians in Southeast Asia. It was a Harvard professor, Henry Kissinger, who was a strategist of the secret bombing of peasant villages in Cambodia.

Going back a bit in history, it was our most educated president, Woodrow Wilson—a historian, a Ph.D., and a former president of Princeton—who bombarded the Mexican coast, killing hundreds of innocent people, because the Mexican government refused to salute the American flag. It was Harvard-educated John Kennedy, author of two books on history, who presided over the American invasion of Cuba and the lies that accompanied it.

What did Kennedy or Wilson learn from all that history they absorbed in the best universities in America? What did the American people learn in their high-school history texts that caused them to submerge their own common sense and listen to these leaders? Surely how "smart" a person is on history tests like the one devised by the

Times, or how "educated" someone is, tells you nothing about whether that person is decent or indecent, violent or peaceful, and whether that person will resist evil or become a consultant to warmakers. It does not tell you who will become a Pastor Niemöller (a German who resisted the Nazis) or an Albert Speer (who worked for them), a Lieutenant Calley (who killed children at My Lai), or a Warrant Officer Thompson (who tried to save them).

One of the two top scorers on the *Times* test was described as follows: "Just short of 20 years old, he lists outdoor activities and the Augustana War Games Club as constituting his favorite leisure-time pursuits, explaining the latter as a group that meets on Fridays to simulate historical battles on a playing board."[38]

We do need to learn history, the kind that does not put its main emphasis on knowing presidents and statutes and Supreme Court decisions, but inspires a new generation to resist the madness of governments trying to carve the world and our minds into their spheres of influence.

CHAPTER
FIVE

Just and Unjust War

There are some people who do not question war.

In 1972, the general who was head of the U.S. Strategic Air Command told an interviewer, "I've been asked often about my moral scruples if I had to send the planes out with hydrogen bombs. My answer is always the same. I would be concerned only with my professional responsibility."[1]

It was a Machiavellian reply. Machiavelli did not ask if making war was right or wrong.[2] He just wrote about the best way to wage it so as to conquer the enemy. One of his books is called *The Art of War.*

That title might make artists uneasy. Indeed, artists—poets, novelists, and playwrights as well as musicians, painters, and actors—have shown a special aversion to war. Perhaps because, as the playwright Arthur Miller once said, "When the guns boom, the arts die." But that would make their interest too self-centered; they have always been sensitive to the fate of the larger society around them. They have questioned war, whether in the fifth century before Christ, with the plays of Euripedes, or in modern times, with the paintings of Goya and Picasso.

Machiavelli was being *realistic.* Wars were going to be fought. The only question was how to win them.

Some people have believed that war is not just inevitable but desirable: It is adventure and excitement, it brings out the best qualities in men—courage, comradeship, and sacrifice. It gives respect and glory to a country. In 1897, Theodore Roosevelt wrote to a friend, "In strict confidence . . . I should welcome almost any war, for I think this country needs one."[3]

In our time, fascist regimes have glorified war as heroic and ennobling. Bombing Ethiopia in 1935, Mussolini's son-in-law Count Ciano described the explosions as an aesthetic thrill, as having the beauty of a flower unfolding.

In the 1980s two writers of a book on war see it as an effective instrument of national policy and say that even nuclear war can, under certain circumstances, be justified. They are contemptuous of "the pacifist passions: self-indulgence and fear," and of "American statesmen, who believe victory is an archaic concept." They say, "The bottom line in war and hence in political warfare is who gets buried and who gets to walk in the sun."[4]

Most people are not that enamored of war. They see it as bad, but also as a possible means to something good. And so they distinguish between wars that are just and those that are unjust. The religions of the West and Middle East—Judaism, Christianity, and Islam—approve of violence and war under certain circumstances. The Catholic church has a specific doctrine of "just" and "unjust" war, worked out in some detail. Political philosophers today argue about which wars, or which actions in wars, may be considered just or unjust.[5]

Beyond both viewpoints—the glorification of war and the weighing of good and bad wars—there is a third: that war is too evil to ever be just. The monk Erasmus, writing in the early sixteenth century, was repelled by war of any kind. One of his pupils was killed in battle and he reacted with anguish:

> Tell me, what had you to do with Mars, the stupidest of all the poet's gods, you who were consecrated to the Muses, nay to Christ? Your youth, your beauty, your gentle nature, your honest mind—what had they to do with the flourishing of trumpets, the bombards, the swords?

Erasmus described war: "There is nothing more wicked, more disastrous, more widely destructive, more deeply tenacious, more loathsome." He said this was repugnant to nature: "Whoever heard of a hundred thousand animals rushing together to butcher each other, as men do everywhere?"

Erasmus saw war as useful to governments, for it enabled them to enhance their power over their subjects; " . . . once war has been declared, then all the affairs of the State are at the mercy of the appetites of a few."[6]

This absolute aversion to war of any kind is outside the orthodoxy of modern thinking. In a series of lectures at Oxford University in the 1970s, English scholar Michael Howard talked disparagingly about Erasmus. He called him simplistic, unsophisticated, and someone who did not see beyond the "surface manifestations" of war. He said,

> With all [Erasmus's] genius he was not a profound political analyst, nor did he ever have to exercise the responsibilities of power. Rather he was the first in that long line of humanitarian thinkers for whom it was enough to chronicle the horrors of war in order to condemn it.

Howard had praise for Thomas More: "Very different was the approach of Erasmus's friend, Thomas More; a man who had exercised political responsibility and, perhaps in consequence, saw the problem in all its complexity." More was a realist; Howard says,

> He accepted, as thinkers for the next two hundred years were to accept, that European society was organized in a system of states in which war was an inescapable process for the settlement of differences in the absence of any higher common jurisdiction. That being the case, it was a requirement of humanity, of religion and of common sense alike that those wars should be fought in such a manner as to cause as little damage as possible. . . . For better or worse war was an institution which could not be eliminated from the international system. All that could be done about it was, so far as possible, to codify its rationale and to civilize its means.

Thus, Machiavelli said: Don't question the ends of the prince, just tell him how best to do what he wants to do, make the means more *efficient*. Thomas More said: You can't do anything about the ends, but try to make the means more *moral*.

In the 400 years following the era of Machiavelli and More, making war more humane became the preoccupation of certain liberal "realists." Hugo Grotius, writing a century after More, proposed laws to govern the waging of war *(Concerning the Law of War and Peace)*. The beginning of the twentieth century saw international conferences at The Hague in the Netherlands and at Geneva in Switzerland, which drew up agreements on how to wage war.

These realistic approaches, however, had little effect on the reality of war. Rather than becoming more controlled, war became more uncontrolled and more deadly, using more horrible means and killing more noncombatants than ever before in the history of mankind. We note the use of poison gas in World War I, the bombardment of cities in World War II, the atomic destruction of Hiroshima and Nagasaki near the end of that war, the use of napalm in Vietnam, and the chemical warfare in the Iran-Iraq war of the early 1980s.

Albert Einstein, observing the effects of attempts to "humanize" wars, became more and more anguished. In 1932 he attended a conference of sixty nations in Geneva and listened to the lengthy discussions of which weapons were acceptable and which were not, which forms of killing were legitimate and which were not.

Einstein was a shy, private person, but he did something extraordinary for him: he called a press conference in Geneva. The international press turned out in force to hear Einstein, already world famous for his theories of relativity. Einstein told the assembled reporters, "One does not make wars less likely by formulating rules of warfare. . . . War cannot be humanized. It can only be abolished."[7] But the Geneva conference went on, working out rules for "humane" warfare, rules that were repeatedly ignored in the world war soon to come, a war of endless atrocities.

In early 1990 President George Bush, while approving new weapons systems for nuclear warheads (of which the United States had about 30,000) and refusing to join the Soviet Union in stopping nuclear testing, was willing to agree to destroy chemical weapons, but only over a ten-year period. Such are the absurdities of "humanizing" war.[8]

Liberal States and Just Wars: Athens

The argument that there are just wars often rests on the social system of the nation engaging in war. It is supposed that if a "liberal" state is at war with a "totalitarian" state, then the war is justified. The beneficent nature of a government is assumed to give rightness to the wars it wages.

Ancient Athens has been one of the most admired of all societies, praised for its democratic institutions and its magnificent cultural achievements. It had enlightened statesmen (Solon and Pericles), pioneer historians (Herodotus and Thucydides), great philosophers (Plato and Aristotle), and an extraordinary quartet of playwrights (Aeschylus, Sophocles, Euripides, and Aristophanes). When it went to war in 431 B.C. against its rival power, the city-state of Sparta, the war seemed to be between a democratic society and a military dictatorship.

The great qualities of Athens were described early in that war by the Athenian leader Pericles at a public celebration for the warriors, dead or alive. The bones of the dead were placed in chests; there was an empty litter for the missing. There was a procession, a burial, and then Pericles spoke. Thucydides recorded Pericles's speech in his *History of the Peloponnesian War:*

> Before I praise the dead, I should like to point out by what principles of action we rose to power, and under what institutions and through what manner of life our empire became great. Our form of government does not enter into rivalry with the institutions of others. . . . It is true that we are called a democracy, for the administration is in the hands of the many and not of the few. . . . The law secures equal justice to all alike. . . . Neither is poverty a bar. . . . There is no exclusiveness in our public life. . . . At home the style of our life is refined. . . . Because of the greatness of our city the fruits of the whole earth flow in upon us. . . . And although our opponents are fighting for their homes and we on a foreign soil, we seldom have any difficulty in overcoming them. . . . I have dwelt upon the greatness of Athens because I want to show you that we are contending for a higher prize than those who enjoy none of these privileges.

Similarly, American presidents in time of war have pointed to the qualities of the American system as evidence for the justness of the cause. Woodrow Wilson and Franklin Roosevelt were liberals, which gave credence to their words exalting the two world wars, just as the liberalism of Truman made going into Korea more acceptable and the idealism of Kennedy's New Frontier and Johnson's Great Society gave an early glow of righteousness to the war in Vietnam.

But we should take a closer look at the claim that liberalism at home carries over into military actions abroad.

The tendency, especially in time of war, is to exaggerate the difference between oneself and the opponent, to assume the conflict is between total good and total evil. It was true that Athens had certain features of political democracy. Each of ten tribes selected 50 representatives, by lot, to make a governing council of 500. Trial juries were large, from 100 to 1,000 people, with no judge and no professional lawyers; the cases were handled by the people involved.

Yet, these democratic institutions only applied to a minority of the population. A majority of the people—125,000 out of 225,000—were slaves. Even among the free people, only males were considered citizens with the right to participate in the political process.

Of the slaves, 50,000 worked in industry (this is as if, in the United States in 1990, 50 million people worked in industry as slaves) and 10,000 worked in the mines. H. D. Kitto, a leading scholar on Greek civilization and a great admirer of Athens, wrote; "The treatment of the miners was callous in the extreme, the only serious blot on the general humanity of the Athenians. . . . Slaves were often worked until they died."[9] (To Kitto and others, slavery was only a "blot" on an otherwise wonderful society.)

The jury system in Athens was certainly preferable to summary executions by tyrants. Nevertheless, it put Socrates to death for speaking his mind to young people.

Athens was more democratic than Sparta, but this did not affect its addiction to warfare, to expansion into other territories, to the ruthless conduct of war against helpless peoples.[10] In modern times we have seen the ease with which parliamentary democracies and constitutional republics have been among the most ferocious of imperialists. We recall the British and French empires of the nineteenth century and the United States as a world imperial power in this century.

Throughout the long war with Sparta, Athens's democratic institutions and artistic achievements continued. But the death toll was enormous. Pericles, on the eve of war, refused to make concessions that might have prevented it. In the second year of war, with the casualties mounting quickly, Pericles urged his fellow citizens not to weaken: "You have a great polis, and a great reputation; you must be worthy of them. Half the world is yours—the sea. For you the alternative to empire is slavery."[11]

Pericles's kind of argument ("Ours is a great nation. It is worth dying for.") has persisted and been admired down to the present. Kitto, commenting on that speech by Pericles, again overcome by admiration, wrote,

> When we reflect that this plague was as awful as the Plague of London, and that the Athenians had the additional horror of being cooped up inside their fortifications by the enemy without, we must admire the greatness of the man who could talk to his fellow citizens like this, and the greatness of the people who could not only listen to such a speech at such a time but actually be substantially persuaded by it.

They were enough persuaded by it so that the war with Sparta lasted twenty-seven years. Athens lost through plague and war (according to Kitto's own estimate) perhaps one-fourth of its population.

However liberal it was (for its free male citizens) at home, Athens became more and more cruel to its victims in war, not just to its enemy Sparta, but to every one caught in the crossfire of the two antagonists. As the war went on, Kitto himself says, "a certain irresponsibility grew."

Could the treatment of the inhabitants of the island of Melos be best described as "a certain irresponsibility"? Athens demanded that the Melians submit to its rule. The Melians, however, argued (as reported by Thucydides), "It may be to your interest to be our masters, but how can it be ours to be your slaves?" The Melians would not submit. They fought and were defeated. Thucydides wrote, "The Athenians thereupon put to death all who were of military age, and made slaves of the women and children." (It was shortly after this event that Euripides wrote his great antiwar play, *The Trojan Women.*)

What the experience of Athens suggests is that a nation may be relatively liberal at home and yet totally ruthless abroad. Indeed, it may more easily enlist its population in cruelty to others by pointing to the advantages at home. An entire nation is made into mercenaries, being paid with a bit of democracy at home for participating in the destruction of life abroad.

Liberalism at War

Liberalism at home, however, seems to become corrupted by war waged abroad. French philosopher Jean Jacques Rousseau noted that conquering nations "make war at least as much on their subjects as on their enemies."[12] Tom Paine, in America, saw war as the creature of governments, serving their own interests, not the interests of justice for their citizens. "Man is not the enemy of man but through the medium of a false system of government."[13] In our time, George Orwell has written that wars are mainly "internal."

One certain effect of war is to diminish freedom of expression. Patriotism becomes the order of the day, and those who question the war are seen as traitors, to be silenced and imprisoned.

* * *

Mark Twain, observing the United States at the turn of the century, its wars in Cuba and the Philippines, described in *The Mysterious Stranger* the process by which wars that are at first seen as unnecessary by the mass of the people become converted into "just" wars:

> The loud little handful will shout for war. The pulpit will warily and cautiously protest at first. . . . The great mass of the nation will rub its sleepy eyes, and will try to make out why there should be a war, and they will say earnestly and indignantly: "It is unjust and dishonorable and there is no need for war."
>
> Then the few will shout even louder. . . . Before long you will see a curious thing: anti-war speakers will be stoned from the platform, and free speech will be strangled by hordes of furious men who still agree with the speakers but dare not admit it. . . .
>
> Next, the statesmen will invent cheap lies . . . and each man will be glad of these lies and will study them because they soothe his conscience; and thus he will bye and bye convince himself that the war is just and he will thank God for the better sleep he enjoys by his self-deception.

Mark Twain died in 1910. In 1917, the United States entered the slaughterhouse of the European war, and the process of silencing dissent and converting a butchery into a just war took place as he had predicted.

President Woodrow Wilson tried to rouse the nation, using the language of a crusade. It was a war, he said, "to end all wars." But large numbers of Americans were reluctant to join. A million men were needed, yet in the first six weeks after the declaration of war only 73,000 volunteered. It seemed that men would have to be compelled to fight by fear of prison, so Congress enacted a draft law.

The Socialist party at that time was a formidable influence in the country. It had perhaps 100,000 members, and more than a thousand Socialists had been elected to office in 340 towns and cities. Probably a million Americans read Socialist newspapers. There were fifty-five weekly Socialist newspapers in Oklahoma, Texas, Louisiana, and Arkansas alone; over a hundred Socialists were elected to office in Oklahoma. The Socialist party candidate for president, Eugene Debs, got 900,000 votes in 1912 (Wilson won with 6 million).

A year before the United States entered the European war, Helen

Keller, blind and deaf and a committed Socialist, told an audience at Carnegie Hall:

> Strike against war, for without you no battles can be fought! Strike against manufacturing shrapnel and gas bombs and all other tools of murder! Strike against preparedness that means death and misery to millions of human beings! Be not dumb, obedient slaves in an army of destruction! Be heroes in an army of construction![14]

The day after Congress declared war, the Socialist party met in an emergency convention and called the declaration "a crime against the American people." Antiwar meetings took place all over the country. In the local elections of 1917, Socialists made great gains. Ten Socialists were elected to the New York State legislature. In Chicago the Socialist party had won 3.6 percent of the vote in 1915 and it got 34.7 percent in 1917. But with the advent of war, speaking against it became a crime; Debs and hundreds of other Socialists were imprisoned.

When that war ended, 10 million men of various countries had died on the battlefields of Europe, and millions more had been blinded, maimed, gassed, shell-shocked, and driven mad. It was hard to find in that war any gain for the human race to justify that suffering, that death.

Indeed, when the war was studied years later, it was clear that no rational decision based on any moral principle had led the nations into war. Rather, there were imperial rivalries, greed for more territory, a lusting for national prestige, and the stupidity of revenge. And at the last moment, there was a reckless plunge by governments caught up in a series of threats and counterthreats, mobilizations and countermobilizations, ultimatums and counterultimatums, creating a momentum that mediocre leaders had neither the courage nor the will to stop. As described by Barbara Tuchman in her book *The Guns of August:*

> War pressed against every frontier. Suddenly dismayed, governments struggled and twisted to fend it off. It was no use. Agents at frontiers were reporting every cavalry patrol as a deployment to beat the mobilization gun. General staffs, goaded by their relentless timetables, were pounding the table for the signal to move lest their opponents gain an hour's head start. Appalled upon the brink, the chiefs of state who would be ultimately responsible for their country's fate attempted to back away, but the pull of military schedules dragged them forward.[15]

Bitterness and disillusion followed the end of the war, and this was reflected in the literature of those years: Ernest Hemingway's *A Farewell to Arms*, John Dos Passos's *U.S.A.*, and Ford Madox Ford's *No More Parades.* In Europe, German war veteran Erich Maria Remarque wrote the bitter antiwar novel *All Quiet on the Western Front.*

In 1935 French playwright Jean Giraudoux wrote *La guerre de Troie n'aura pas lieu (The Trojan War Will Not Take Place;* the English translation was retitled *Tiger at the Gates*). The war of the Greeks against Troy, more than a thousand years before Christ, was provoked, according to legend, by the kidnapping of the beautiful Helen by the Trojans. Giraudoux at one point uses Hecuba, an old woman, and Demokos, a Trojan soldier, to show how the ugliness of war is masked by attractive causes, as in this case, the recapture of Helen.

> DEMOKOS: Tell us before you go, Hecuba, what it is you think war looks like.
> HECUBA: Like the bottom of a baboon. When the baboon is up in a tree, with its hind end facing us, there is the face of war exactly: scarlet, scaly, glazed, framed in a clotted, filthy wig.
> DEMOKOS: So war has two faces: this you describe, and Helen's.

An Eager Bombardier

My own first impressions of something called war had come at the age of ten, when I read with excitement a series of books about "the boy allies"—a French boy, an English boy, an American boy, and a Russian boy, who became friends, united in the wonderful cause to defeat Germany in World War I. It was an adventure, a romance, told in a group of stories about comradeship and heroism. It was war cleansed of death and suffering.

If anything was left of that romantic view of war, it was totally extinguished when, at eighteen, I read a book by a Hollywood screenwriter named Dalton Trumbo (jailed in the 1950s for refusing to talk to the House Committee on Un-American Activities about his political affiliations). The book was called *Johnny Got His Gun.* It is, perhaps, the most powerful antiwar novel ever written.

Here was war in its ultimate horror. A slab of flesh in an American uniform had been found on the battlefield, still alive, with no legs, no arms, no face, blind, deaf, unable to speak, but the heart still beating, the brain still functioning, able to think about his past, ponder his present condition, and wonder if he will ever be able to communicate with the world outside.

For him, the oratory of the politicians who sent him off to war—the language of freedom, democracy, and justice—is now seen as the ultimate hypocrisy. A mute, thinking torso on a hospital bed, he finds a way to communicate with a kindly nurse, and when a visiting delegation of military brass comes by to pin a medal on his body, he taps out a message. He says: Take me into the workplaces, into the schools, show me to the little children and to the college students, let them see what war is like.

> Take me wherever there are parliaments and diets and congresses and chambers of statesmen. I want to be there when they talk about honor and justice and making the world safe for democracy and fourteen points and the self determination of peoples. . . . Put my glass case upon the speaker's desk and every time the gavel descends let me feel its vibration. . . . Then let them speak of trade policies and embargoes and new colonies and old grudges. Let them debate the menace of the yellow race and the white man's burden and the course of empire and why should we take all this crap off Germany or whoever the next Germany is. . . . Let them talk more munitions and airplanes and battleships and tanks and gases and why of course we've got to have them we can't get along without them how in the world could we protect the peace if we didn't have them. . . .
>
> But before they vote on them before they give the order for all the little guys to start killing each other let the main guy rap his gavel on my case and point down at me and say here gentlemen is the only issue before this house and that is are you for this thing here or are you against it.[16]

Johnny Got His Gun had a shattering effect on me when I read it. It left me with a bone-deep hatred of war.

Around the same time I read a book by Walter Mills, *The Road to War,* which was an account of how the United States had been led into World War I by a series of lies and deceptions. Afterward I would learn more about those lies. For instance, the sinking of the ship *Lusitania* by German submarines was presented as a brutal, unprovoked act against a harmless passenger vessel. It was later revealed that the *Lusitania* was loaded with munitions intended for use against Germany; the ship's manifest had been falsified to hide that. This didn't lessen the brutality of the sinking, but did show something about the ways in which nations are lured into war.

Class consciousness accounted for some of my feeling about war. I agreed with the judgment of the Roman biographer Plutarch, who said, "The poor go to war, to fight and die for the delights, riches, and superfluities of others."

And yet, in early 1943, at the age of twenty-one, I enlisted in the U.S. Army Air Force. American troops were already in North Africa, Italy, and England; there was fierce fighting on the Russian front and the United States and Britain were preparing for the invasion of Western Europe. Bombing raids were taking place daily on the Continent, U.S. planes bombing during the day, British planes bombing at night. I was so anxious to get overseas and start dropping bombs that after my training in gunnery school and bombing school I traded places with another man who was scheduled to go overseas sooner than me.

I had learned to hate war. But this war was different. It was not for profit or empire, it was a people's war, a war against the unspeakable brutality of fascism. I had been reading about Italian fascism in a book about Mussolini by journalist George Seldes called *Sawdust Caesar.* I was inspired by his account of the Socialist Matteotti, who stood up in the Italian Chamber of Deputies to denounce the establishment of a dictatorship. The black-shirted thugs of Mussolini's party picked up Matteotti outside his home one morning and shot him to death. That was fascism.

Mussolini's Italy, deciding to restore the glory of the old Roman Empire, invaded the East African country of Ethiopia, a pitifully poor country. Its people, armed with spears and muskets, tried to fight off an Italian army equipped with the most modern weapons and with an air force that, unopposed, dropped bombs on the civilian populations of Ethiopian towns and villages. The Ethiopians resisted, were slaughtered, and finally surrendered.

American black poet Langston Hughes wrote,

> The little fox is still—
> The dogs of war have made their kill.[17]

I was thirteen when this happened and was only vaguely aware of headlines: "Italian Planes Bomb Addis Ababa." But later I read about it and also about German Nazism. John Gunther's *Inside Europe* introduced me to the rise of Hitler, the SA, the SS, the attacks on the Jews, the shrill oratory of the little man with the mustache, and the monster rallies of Germans in sports stadia who shouted in unison: "Heil Hitler!

Heil Hitler!" Opponents were beaten and murdered. I learned the phrase *concentration camp.*

I came across a book called *The Brown Book of the Nazi Terror.* It told in detail about the burning of the German Reichstag shortly after Hitler came to power and the arrest of Communists accused of setting the fire, clearly a frame-up. It told also of the extraordinary courage of the defendants, led by the remarkable Bulgarian Communist George Dimitrov, who rose in the courtroom to point an accusing finger at Hermann Goering, Hitler's lieutenant. Dimitrov tore the prosecution's case to shreds and denounced the Nazi regime. The defendants were acquitted by the court. It was an amazing moment, which would never be repeated under Hitler.

In 1936 Hitler and Mussolini sent their troops and planes to support the Spanish Fascist Franco, who had plunged his country into civil war to overthrow the mildly socialist Spanish government. The Spanish Civil War became the symbol all over the world of resistance to fascism, and young men—many of them socialists, Communists, and anarchists—volunteered from a dozen countries, forming brigades (from the United States, the Abraham Lincoln Brigade), going immediately into battle against the better-equipped army of Franco. They fought heroically and died in great numbers. The Fascists won.

Then came the Hitler onslaught in Europe—Austria, Czechoslovakia, and Poland. France and England entered the war, and, a year after the quick defeat of France, 3 million German soldiers supported by tanks, artillery, and dive bombers turned eastward to attack the Soviet Union ("Operation Barbarossa") along a thousand-mile front.

Fascism had to be resisted and defeated. I had no doubts. *This* was a just war.

I was stationed at an airfield out in the countryside of East Anglia (between the towns of Diss and Eye), that part of England that bulges eastward toward the Continent. East Anglia was crowded with military airfields, from which hundreds of bombers went out every day across the Channel.

Our little airfield housed the 490th Bomb Group. Its job was to make sure that every morning twelve B-17s—splendid-looking, low-winged, four-engined heavy bombers—each with a crew of nine—wearing sheepskin jackets and fur-lined boots over electrically heated suits and equipped with oxygen masks and throat mikes—were ready to fly. We would take off around dawn and assemble with other groups of twelve, and then these huge flotillas would make their way east. Our bomb bay was full; our fifty-caliber machine guns (four in the nose, one in the upper turret, one in the ball turret, two in the waist, and one in the tail) were loaded and ready for attacking fighter planes.

I remember one morning standing out on that airfield, arguing with another bombardier over who was scheduled to fly that morning's mission. The target was Regensburg, and Intelligence reported that it was heavily defended by antiaircraft guns, but the two of us argued heatedly over who was going to fly that mission. I wonder today, was his motive like mine—wanting to fly another mission to bring closer the defeat of fascism. Or was it because we had all been awakened at one A.M. in the cold dark of England in March, loaded onto trucks, taken to hours of briefings and breakfast, weighed down with equipment, and after going through all that, he did not want to be deprived of another step toward his air medal, another mission. Even though he might be killed.

Maybe that was partly my motive too, I can't be sure. But for me, it was also a war of high principle, and each bombing mission was a mission of high principle. The moral issue could hardly be clearer. The enemy could not be more obviously evil—openly espousing the superiority of the white Aryan, fanatically violent and murderous toward other nations, herding its own people into concentration camps, executing them if they dared dissent. The Nazis were pathological killers. They had to be stopped, and there seemed no other way but by force.

If there was such a thing as a just war, this was it. Even Dalton Trumbo, who had written *Johnny Got His Gun,* did not want his book to be reprinted, did not want that overpowering antiwar message to reach the American public, when a war had to be fought against fascism.[18]

If, therefore, anyone wants to argue (as I am about to do) that there is no such thing as a just war, then World War II is the supreme test.

I flew the last bombing missions of the war, got my Air Medal and my battle stars. I was quietly proud of my participation in the great war to defeat fascism. But when I packed up my things at the end of the war and put my old navigation logs and snapshots and other mementos in a folder, I marked that folder, almost without thinking, "Never Again."

I'm still not sure why I did that, because it was not until years later that I began consciously to question the motives, the conduct, and the consequences of that crusade against fascism. The point was not that my abhorrence of fascism was in any way diminished. I still believed *something* had to be done to stop fascism. But that clear certainty of moral rightness that propelled me into the Air Force as an enthusiastic bombardier was now clouded over by many thoughts.

Perhaps my conversations with that gunner on the other crew, the one who loaned me *The Yogi and the Commissar*, gave me the first flickers of doubt. He spoke of the war as "an imperialist war," fought on both sides for national power. Britain and the United States opposed fascism only because it threatened their own control over resources and people. Yes, Hitler was a maniacal dictator and invader of other countries. But what of the British Empire and its long history of wars against native peoples to subdue them for the profit and glory of the empire? And the Soviet Union—was it not also a brutal dictatorship, concerned not with the working classes of the world but with its own national power?

I was puzzled. "Why," I asked my friend, "are you flying missions, risking your life, in a war you don't believe in?" His answer astonished me. "I'm here to speak to people like you."

I found out later he was a member of the Socialist Workers party; they opposed the war but believed that instead of evading military service they should enter it and propagandize against the war every moment they could. I couldn't understand this, but I was impressed by it. Two weeks after that conversation with him, he was killed on a mission over Germany.

After the war, my doubts grew. I was reading history. Had the United States fought in World War II for the rights of nations to independence and self-determination? What of its own history of expansion through war and conquest? It had waged a hundred-year war against the native Americans, driving them off their ancestral lands. The United States had instigated a war with Mexico and taken almost half its land, had sent marines at least twenty times into the countries of the Caribbean for power and profit, had seized Hawaii, had fought a brutal war to subjugate the Filipinos, and had sent 5,000 marines into Nicaragua in 1926. Our nation could hardly claim it believed in the right of self-determination unless it believed in it selectively.

Indeed, the United States had observed Fascist expansion without any strong reactions. When Italy invaded Ethiopia, the United States, while declaring an embargo on munitions, allowed American businesses to send oil to Italy, which was crucial for carrying on the war against Ethiopia. An official of the U.S. State Department, James E. Miller, reviewing a book on the relations between the United States and Mussolini, acknowledged that "American aid certainly reinforced the hold of Fascism."[19]

During the Spanish Civil War, while the Fascist side was receiving arms from Hitler and Mussolini, Roosevelt's administration sponsored a Neutrality Act that shut off help to the Spanish government fighting Franco.

Neither the invasion of Austria nor Czechoslovakia nor Poland brought the United States into armed collision with fascism. We went to war only when our possession Hawaii was attacked and when our navy was disabled by Japanese bombs. There was no reason to think that it was Japan's bombing of civilians at Pearl Harbor that caused us to declare war. Japan's attack on China in 1937, her massacre of civilians at Nanking, and her bombardments of helpless Chinese cities had not provoked the United States to war.

The sudden indignation against Japan contained a good deal of hypocrisy. The United States, along with Japan and the great European powers, had participated in the exploitation of China. Our Open Door Policy of 1901 accepted that ganging up of the great powers on China. The United States had exchanged notes with Japan in 1917 saying, "the Government of the United States recognizes that Japan has special interests in China," and in 1928, American consuls in China supported the coming of Japanese troops.[20]

It was only when Japan threatened potential U.S. markets by its attempted takeover of China, but especially as it moved toward the tin, rubber, and oil of Southeast Asia, that the United States became alarmed and took those measures that led to the Japanese attack: a total embargo on scrap iron and a total embargo on oil in the summer of 1941.[21]

A State Department memorandum on Japanese expansion, a year before Pearl Harbor, did not talk of the independence of China or the principle of self-determination. It said,

> Our general diplomatic and strategic position would be considerably weakened—by our loss of Chinese, Indian and South Seas markets (and by our loss of much of the Japanese market for our goods, as Japan would become more and more self-sufficient) as well as by insurmountable restrictions upon our access to the rubber, tin, jute, and other vital materials of the Asian and Oceanic regions.

A War to Save the Jews?

Did the United States enter the war because of its indignation at Hitler's treatment of the Jews? Hitler had been in power a year, and his campaign against the Jews had already begun when, in January 1934, a resolution was introduced into the Senate expressing "surprise and pain" at what the Germans were doing and asking a restoration of Jewish rights. The State Department used its influence to get the resolution buried in committee.[22]

Even after we were in the war against Germany (it should be noted that after Pearl Harbor Germany declared war on the United States, not vice versa) and reports began to arrive that Hitler was planning the annihilation of the Jews, Roosevelt's administration failed to take steps that might have saved thousands of lives.

Goebbels, minister of propaganda for Hitler's Germany, wrote in his diary on December 13, 1942: "At bottom, however, I believe both the English and the Americans are happy we are exterminating the Jewish riffraff." Goebbels was undoubtedly engaging in wishful thinking, but, in fact, the English and American governments had not shown by their actions that they were terribly concerned about the Jews. As for Roosevelt, he shunted the problem to the State Department, where it did not become a matter of high priority.

As an example of this failure to treat the situation as an emergency, Raul Hilberg, a leading scholar of the Holocaust, points to an event that took place in 1942. Early in August of that year, with 1,500,000 Jews already dead, the Jewish leader Stephen Wise was informed indirectly through a German industrialist that there was a plan in Hitler's headquarters for the extermination of all Jews; Wise brought the information to Under Secretary of State Sumner Welles. Welles asked him not to release the story until it was investigated for confirmation. Three months were spent checking the report. During that time a million Jews were killed in Europe.[23]

It is doubtful that all those Jews could have been saved. But thousands could have been rescued. All the entrenched governments and organizations were negligent.[24]

The British were slow and cautious. In March 1943, in the presence of Franklin D. Roosevelt, Secretary of State Hull pressed British Foreign Minister Anthony Eden on plans to rescue the 60,000 Jews in Bulgaria threatened with death. According to a memo by Roosevelt aide

Harry Hopkins who was at that meeting, Eden worried that Polish and German Jews might then also ask to be rescued. "Hitler might well take us up on any such offer and there simply are not enough ships and means of transportation in the world to handle them."[25] When there was a possibility of bombing the railroad lines leading into the murder chambers of Auschwitz, to stop further transportation of Jews there, the opportunity was ignored.

It should be noted that the Jewish organizations themselves behaved shamefully. In 1984, the American Jewish Commission on the Holocaust reviewed the historical record. It found that the American Jewish Joint Distribution Committee, a relief agency set up during World War II by the various Jewish groups, "was dominated by the wealthier and more 'American' elements of U.S. Jewry. . . . Thus, its policy was to do nothing in wartime that the U.S. government would not officially countenance."[26]

Raul Hilberg points out that the Hungarian Jews might have been saved by a bargain: the Allies would not make air raids on Hungary if the Jews would be kept in the cities and not sent away. But "the Jews could not think in terms of interfering with the war effort, and the Allies on their part could not conceive of such a promise. . . . The Allied bombers roared over Hungary at will, killing Hungarians and Jews alike."[27]

As I read this I recalled that one of the bombing raids I had done was on a town in Hungary.

Not only did waging war against Hitler fail to save the Jews, it may be that the war itself brought on the Final Solution of genocide. This is not to remove the responsibility from Hitler and the Nazis, but there is much evidence that Germany's anti-Semitic actions, cruel as they were, would not have turned to mass murder were it not for the psychic distortions of war, acting on already distorted minds. Hitler's early aim was forced emigration, not extermination, but the frenzy of war created an atmosphere in which the policy turned to genocide. This is the view of Princeton historian Arno Mayer, in his book *Why Did the Heavens Not Darken*, and it is supported by the chronology—that not until Germany was at war was the Final Solution adopted.[28]

Hilberg, in his classic work on the Holocaust, says, "From 1938 to 1940, Hitler made extraordinary and unusual attempts to bring about a vast emigration scheme. . . . The Jews were not killed before the emigration policy was literally exhausted." The Nazis found that the Western powers were not anxious to cooperate in emigration and that no one wanted the Jews.[29]

A War for Self-Determination?

We should examine another claim, that World War II was fought for the right of nations to determine their own destiny. This was declared with great fanfare by Winston Churchill and Franklin Roosevelt when they met off the coast of Newfoundland in August 1941 and announced the Atlantic Charter, saying their countries, looking to the postwar world, respected "the right of all peoples to choose the form of government under which they will live." This was a direct appeal to the dependent countries of the world, especially the colonies of Britain, France, Holland, and Belgium, that their rights of self-determination would be upheld after the war. The support of the nonwhite colonial world was seen as crucial to the defeat of fascism.

However, two weeks before the Atlantic Charter, with the longtime French colony of Indochina very much in mind, acting Secretary of State of the United States Sumner Welles had given quiet assurances to the French: "This Government, mindful of its traditional friendship for France, has deeply sympathized with the desire of the French people to maintain their territories and to preserve them intact." And in late 1942, Roosevelt's personal representative told French General Henri Giraud, "It is thoroughly understood that French sovereignty will be reestablished as soon as possible throughout all the territory, metropolitan or colonial, over which flew the French flag in 1939."[30] (These assurances of the United States are especially interesting in view of the claims of the United States during the Vietnam War, that the United States was fighting for the right of the Vietnamese to rule themselves.)

If neither saving the Jews nor guaranteeing the principle of self-determination was the war aim of the United States (and there is no evidence that either was the aim of Britain or the Soviet Union), then what *were* the principal motives? Overthrowing the governments of Hitler, Mussolini, and Tojo was certainly one of them. But was this desired on humanitarian grounds or because these regimes threatened the *positions* of the Allies in the world?

The rhetoric of morality—the language of freedom and democracy—had some substance to it, in that it represented the war aims of many ordinary citizens. However, it was not the citizenry but the governments who decided how the war was fought and who had the power to shape the world afterward.

Behind the halo of righteousness that surrounded the war against fascism, the usual motives of governments, repeatedly shown in history, were operating: the aggrandizement of the nation, more profit for its wealthy elite, and more power to its political leaders.

One of the most distinguished of British historians, A. J. P. Taylor, commented on World War II that "the British and American governments wanted no change in Europe except that Hitler should disappear."[31] At the end of the war, novelist George Orwell, always conscious of *class,* wrote, "I see the railings [which enclosed the parks and had been torn up so the metal could be used in war production] are returning in one London park after another, so the lawful denizens of the squares can make use of their keys again, and the children of the poor can be kept out."[32]

World War II was an opportunity for United States business to penetrate areas that up to that time had been dominated by England. Secretary of State Hull said early in the war,

> Leadership toward a new system of international relationships in trade and other economic affairs will devolve very largely upon the United States because of our great economic strength. We should assume this leadership, and the responsibility that goes with it, primarily for reasons of pure national self-interest.[33]

Henry Luce, who owned three of the most influential magazines in the United States—*Life, Time,* and *Fortune*—and had powerful connections in Washington, wrote a famous editorial for *Life* in 1941 called "The American Century." This was the time, he said, "to accept wholeheartedly our duty and our opportunity as the most powerful and vital nation in the world and in consequence to exert upon the world the full impact of our influence, for such purposes as we see fit and by such means as we see fit."[34]

The British, weakened by war, clearly could not maintain their old empire. In 1944 England and the United States signed a pact on oil agreeing on "the principle of equal opportunity." This meant the United States was muscling in on England's traditional domination of Middle East oil.[35] A study of the international oil business by the English writer Anthony Sampson concluded,

By the end of the war the dominant influence in Saudi Arabia was unqu[illegible]tionably the United States. King Ibn Saud was regarded no longer as a wild desert warrior, but as a key piece in the power-game, to be wooed by the West. Roosevelt, on his way back from Yalta in February, 1945, entertained the King on the cruiser *Quincy*, together with his entourage of fifty, including two sons, a prime minister, an astrologer and flocks of sheep for slaughter.[36]

There was a critic inside the American government, not a politician but poet Archibald MacLeish, who briefly served as assistant secretary of state. He worried about the postwar world: "As things are now going the peace we will make, the peace we seem to be making, will be a peace of oil, a peace of gold, a peace of shipping, a peace, in brief . . . without moral purpose or human interest."[37]

A War Against Racism?

If the war was truly a war of moral purpose, against the Nazi idea of superior and inferior races, then we might have seen action by the U.S. government to eliminate racial segregation. Such segregation had been declared lawful by the Supreme Court in 1896 and existed in both South and North, accepted by both state and national governments.

The armed forces were segregated by race. When I was in basic training at Jefferson Barracks, Missouri, in 1943, it did not occur to me, so typical an American white was I, that there were no black men in training with us. But it was a huge base, and one day, taking a long walk to the other end of it, I was suddenly aware that all the GIs around me were black. There was a squad of blacks taking a ten-minute break from hiking in the sun, lying on a small grassy incline, and singing a hymn that surprised me at the moment, but that I realized later was quite appropriate to their situation: "Ain't Gonna Study War No More."

My air crew sailed to England on the *Queen Mary*. That elegant passenger liner had been converted into a troop ship. There were 16,000 men aboard, and 4,000 of them were black. The whites had quarters on deck and just below deck. The blacks were housed separately, deep in the hold of the ship, around the engine room, in the darkest, dirtiest sections. Meals were taken in four shifts (except for the officers, who ate in prewar *Queen Mary* style in a chandeliered ballroom—the war was not being fought to disturb class privilege), and the blacks had to wait ts of whites had finished eating.

e front, racial discrimination in employment continued, until A. Philip Randolph, head of the Brotherhood of Porters, a union of black workers, threatened to organize a march on Washington during the war and embarrass the Roosevelt administration before the world that the president signed an order setting up a Fair Employment Practices Commission. But its orders were not enforced and job discrimination continued. A spokesman for a West Coast aviation plant said, "The Negro will be considered only as janitors and in other similar capacities. . . . Regardless of their training as aircraft workers, we will not employ them."[38]

There was no organized black opposition to the war, but there were many signs of bitterness at the hypocrisy of a war against fascism that did nothing about American racism. One black journalist wrote: "The Negro . . . is angry, resentful, and utterly apathetic about the war. 'Fight for what?' he is asking. 'This war doesn't mean a thing to me. If we win I lose, so what?' "[39]

A student at a black college told his teacher: "The Army jim-crows us. The Navy lets us serve only as messmen. The Red Cross refuses our blood. Employers and labor unions shut us out. Lynchings continue. We are disenfranchised, jim-crowed, spat upon. What more could Hitler do than that?" That student's statement was repeated by Walter White, a leader of the National Association for the Advancement of Colored People (NAACP), to an audience of several thousand black people in the Midwest, expecting that they would disapprove. Instead, as he recalled, "To my surprise and dismay the audience burst into such applause that it took me some thirty or forty seconds to quiet it."[40]

In January 1943, there appeared in a Negro newspaper a "Draftee's Prayer":

Dear Lord, today
I go to war:
To fight, to die.
Tell me, what for?
Dear Lord, I'll fight,
I do not fear,
Germans or Japs;
My fears are here.
America![41]

In one little-known incident of World War II, two transport ships being loaded with ammunition by U.S. sailors at the Port Chicago naval base in California suddenly blew up on the night of July 17, 1944. It was an enormous explosion, and its glare could be seen in San Francisco, thirty-five miles away. More than 300 sailors were killed, two-thirds of them black, because blacks were given the hard jobs of ammunition loaders. "It was the worst home front disaster of World War II," historian Robert Allen writes in his book *The Port Chicago Mutiny.*[42]

Three weeks later 328 of the survivors were asked to load ammunition again; 258 of them refused, citing unsafe conditions. They were immediately jailed. Fifty of them were then court-martialed on a charge of mutiny, and received sentences ranging from eight to fifteen years imprisonment. It took a massive campaign by the NAACP and its counsel, Thurgood Marshall, to get the sentences reduced.[43]

To the Japanese who lived on the West Coast of the United States, it quickly became clear that the war against Hitler was not accompanied by a spirit of racial equality. After the attack by Japan on Pearl Harbor, anger rose against all people of Japanese ancestry. One congressman said, "I'm for catching every Japanese in America, Alaska and Hawaii now and putting them in concentration camps. . . . Damn them! Let's get rid of them now!"[44]

Hysteria grew. Roosevelt, persuaded by racists in the military that the Japanese on the West Coast constituted a threat to the security of the country, signed Executive Order 9066 in February 1942. This empowered the army, without warrants or indictments or hearings, to arrest every Japanese-American on the West Coast—110,000 men, women, and children—to take them from their homes, to transport them to camps far in the interior, and to keep them there under prison conditions.

Three-fourths of the Japanese so removed from their homes were Nisei—children born in the United States of Japanese parents and, therefore, American citizens. The other fourth—the Issei, born in Japan—were barred by law from becoming citizens. In 1944 the United States Supreme Court upheld the forced evacuation on the grounds of military necessity.[45]

Data uncovered in the 1980s by legal historian Peter Irons showed that the army falsified material in its brief to the Supreme Court. When Congress in 1983 was considering financial compensation to the Japanese who had been removed from their homes and their possessions during the war, John J. McCloy wrote an article in the *New York Times* opposing such compensation, defending the action as necessary. As Peter

Irons discovered in his research, it was McCloy, then assistant secretary of war, who had ordered the deletion of a critical footnote in the Justice Department brief to the Supreme Court, a footnote that cast great doubt on the army's assertions that the Japanese living on the West Coast were a threat to American security.[46]

Michi Weglyn was a young girl when her family experienced evacuation and detention. She tells in her book *Years of Infamy* of bungling in the evacuation; of misery, confusion, and anger; but also of Japanese-American dignity and of fighting back. There were strikes, petitions, mass meetings, refusals to sign loyalty oaths, and riots against the camp authorities.[47]

Only a few Americans protested publicly. The press often helped to feed racism. Reporting the bloody battle of Iwo Jima in the Pacific, *Time* magazine said, "The ordinary unreasoning Jap is ignorant. Perhaps he is human. Nothing . . . indicates it."[48]

In the 1970s, Peter Ota, then fifty-seven, was interviewed by Studs Terkel. His parents had come from Japan in 1904, and became respected members of the Los Angeles community. Ota was born in the United States. He remembered what had happened in the war:

> On the evening of December 7, 1941, my father was at a wedding. He was dressed in a tuxedo. When the reception was over, the FBI agents were waiting. They rounded up at least a dozen wedding guests and took 'em to county jail.
>
> For a few days we didn't know what happened. We heard nothing. When we found out, my mother, my sister and myself went to jail. . . . When my father walked through the door my mother was so humiliated. . . . She cried. He was in prisoner's clothing, with a denim jacket and a number on the back. The shame and humiliation just broke her down. . . . Right after that day she got very ill and contracted tuberculosis. She had to be sent to a sanitarium. . . . She was there till she died. . . .
>
> My father was transferred to Missoula, Montana. We got letters from him—censored, of course. . . . It was just my sister and myself. I was fifteen, she was twelve. . . . School in camp was a joke. . . . One of our basic subjects was American history. They talked about freedom all the time. (Laughs.)[49]

In England there was similar hysteria. People with German-sounding names were picked up and interned. In the panic, a number of Jewish refugees who had German names were arrested and thrown into the same camps. There were thousands of Italians who were living in England, and when Italy entered World War II in June of 1940, Winston Churchill gave the order: "Collar the lot." Italians were picked up and interned, the windows of Italian shops and restaurants were smashed by patriotic mobs. A British ship carrying Italian internees to Canada was sunk by a German submarine and everyone drowned.[50]

A War for Democracy?

It was supposed to be a war for freedom. But in the United States, when Trotskyists and members of the Socialist Workers party spoke out in criticism of the war, eighteen of them were prosecuted in 1943 in Minneapolis. The Smith Act, passed in 1940, extended the anti-free-speech provisions of the World War I Espionage Act to peacetime. It prohibited joining any group or publishing any material that advocated revolution or that might lead to refusal of military service. The Trotskyists were sentenced to prison terms, and the Supreme Court refused to review their case.[51]

Fortunes were made during the war, and wealth was concentrated in fewer and fewer hands. By 1941 three-fourths of the value of military contracts were handled by fifty-six large corporations. Pressure was put on the labor unions to pledge they would not strike. But they saw their wages frozen, and profits of corporations rising, and so strikes went on. There were 14,000 strikes during the war, involving over 6 million workers, more than in any comparable period in American history.

An insight into what great profits were made during the war came years later, when the multimillionaire John McCone was nominated by President John F. Kennedy to head the CIA. The Senate Armed Services Committee, considering the nomination, was informed that in World War II, McCone and associates in a shipbuilding company had made $44 million on an investment of $100,000. Reacting indignantly to criticism of McCone, one of his supporters on the Senate committee asked him:

> SEN. SYMINGTON: Now, it is still legal in America, if not to make a profit, at least to try to make a profit, is it not?
> MCCONE: That is my understanding.[52]

Bruce Catton, a writer and historian working in Washington during the war, commented bitingly on the retention of wealth and power in the same hands, despite a war that seemed to promise a new world of social reform. He wrote:

> We were committed to the defeat of the Axis but to nothing else. . . . It was solemnly decided that the war effort must not be used to bring about social or economic reform and to him that hath shall be given. . . .
>
> And through it all . . . the people were not trusted with the facts or relied on to display that intelligence, sanity, and innate decency of spirit, upon which democracy . . . finally rests. In a very real sense, our government spent the war years looking desperately for some safe middle ground between Hitler and Abraham Lincoln.[53]

Dresden, Hiroshima, and Royan

It becomes difficult to sustain the claim that a war is just when both sides commit atrocities, unless one wants to argue that *their* atrocities are worse than *ours.* True, nothing done by the Allied Powers in World War II matches in utter viciousness the deliberate gassing, shooting, and burning of 6 million Jews and 4 million others by the Nazis. The deaths caused by the Allies were less, but still so massive as to throw doubt on the justice of a war that includes such acts.

Early in the war, various world leaders condemned the indiscriminate bombing of city populations. Italy had bombed civilians in Ethiopia; Japan, in China; Germany and Italy, in the Spanish Civil War. Germany had dropped bombs on Rotterdam in Holland, on Coventry in England, and other places. Roosevelt described these bombings as "inhuman barbarism that has profoundly shocked the conscience of humanity."[54]

But very soon, the United States and Britain were doing the same thing and on a far larger scale. When the Allied leaders met at Casablanca in January 1943, they agreed on massive air attacks to achieve "the destruction and dislocation of the German military, industrial and economic system and the undermining of the morale of the German people to the point where their capacity for armed resistance is fatally weakened."[55] Churchill and his advisers had decided that bombing the working-class districts of German cities would accomplish just that, "the undermining of the morale of the German people."

The saturation bombing of the German cities began. There were raids of a thousand planes on Cologne, Essen, Frankfurt, and Hamburg.

The British flew at night and did "area bombing" with no pretense of aiming at specific military targets.

The Americans flew in the daytime, pretending to precision, but bombing from high altitudes made that impossible. When I was doing my practice bombing in Deming, New Mexico, before going overseas, our egos were built up by having us fly at 4,000 feet and drop a bomb within twenty feet of the target. But at 11,000 feet, we were more likely to be 200 feet away. And when we flew combat missions, we did it from 30,000 feet, and might miss by a quarter of a mile. Hardly "precision bombing."

There was huge self-deception. We had been angered when the Germans bombed cities and killed several hundred or a thousand people. But now the British and Americans were killing tens of thousands in a single air strike. Michael Sherry, in his study of aerial bombing, notes that "so few in the air force asked questions."[56] Sherry says there was no clear thinking about the effects of the bombing. Some generals objected, but were overruled by civilians. The technology crowded out moral considerations. Once the planes existed, targets had to be found.

It was terror bombing, and the German city of Dresden was the extreme example. (The city and the event are immortalized in fiction by Kurt Vonnegut's comic, bitter novel, *Slaughterhouse Five.*) It was February 1945, the Red Army was eighty miles to the east and it was clear that Germany was on the way to defeat. In one day and one night of bombing, by American and British planes, the tremendous heat generated by the bombs created a vacuum, and an enormous firestorm swept the city, which was full of refugees at the time, increasing the population to a million. More than 100,000 people died.[57]

The British pilot of a Lancaster bomber recalled, "There was a sea of fire covering in my estimation some 40 square miles. We were so aghast at the awesome blaze that although alone over the city we flew around in a stand-off position for many minutes before turning for home, quite subdued by our imagination of the horror that must be below."

One incident remembered by survivors is that on the afternoon of February 14, 1945, American fighter planes machine-gunned clusters of refugees on the banks of the Elbe. A German woman told of this years later: "We ran along the Elbe stepping over the bodies."[58]

Winston Churchill, who seemed to have no moral qualms about his policy of indiscriminate bombing, described the annihilation of Dresden in his wartime memoirs with a simple statement: "We made a heavy raid in the latter month on Dresden, then a centre of communication of Germany's Eastern Front."[59]

At one point in the war Churchill ordered thousands of anthrax bombs from a plant that was secretly producing them in the United States. His chief science adviser, Lord Cherwell, had informed him in February 1944: "Any animal breathing in minute quantities of these N [anthrax] spores is extremely likely to die suddenly but peacefully within the week. There is no known cure and no effective prophylaxis. There is little doubt that it is equally lethal to human beings." He told Churchill that a half dozen bombers could carry enough four-pound anthrax bombs to kill everyone within a square mile. However, production delays got in the way of this plan.[60]

The actor Richard Burton once wrote an article for the *New York Times* about his experience playing the role of Winston Churchill in a television drama:

> In the course of preparing myself . . . I realized afresh that I hate Churchill and all of his kind. I hate them virulently. They have stalked down the corridors of endless power all through history. . . . What man of sanity would say on hearing of the atrocities committed by the Japanese against British and Anzac prisoners of war, 'We shall wipe them out, everyone of them, men, women, and children. There shall not be a Japanese left on the face of the earth'? Such simple-minded cravings for revenge leave me with a horrified but reluctant awe for such single-minded and merciless ferocity.[61]

When Burton's statement appeared in the "Arts and Leisure" section of the *New York Times,* he was banned from future BBC productions. The supervisor of drama productions for BBC said, "As far as I am concerned, he will never work for us again. . . . Burton acted in an unprofessional way."[62]

It seems that however moral is the cause that initiates a war (in the minds of the public, in the mouths of the politicians), it is in the nature of war to corrupt that morality until the rule becomes "An eye for an eye, a tooth for a tooth," and soon it is not a matter of equivalence, but indiscriminate revenge.

The policy of saturation bombing became even more brutal when B-29s, which carried twice the bombload as the planes we flew in Europe, attacked Japanese cities with incendiaries, turning them into infernos.

In one raid on Tokyo, after midnight on March 10, 1945, 300 B-29s left the city in flames, fanned by a strong northwest wind. The fires could be seen by pilots 150 miles out in the Pacific Ocean. A million people were left homeless. It is estimated that 100,000 people died that night. Many of them attempting to escape leapt into the Sumida River and drowned. A Japanese novelist who was twelve years old at the time, described the scene years later: "The fire was like a living thing. It ran, just like a creature, chasing us."[63]

By the time the atomic bomb was dropped on Hiroshima (August 6, 1945) and another on Nagasaki (three days later), the moral line had been crossed psychologically by the massive bombings in Europe and by the fire bombings of Tokyo and other cities.

The bomb on Hiroshima left perhaps 140,000 dead; the one on Nagasaki, 70,000 dead. Another 130,000 died in the next five years. Hundreds of thousands of others were left radiated and maimed. These numbers are based on the most detailed report that exists on the effects of the bombings; it was compiled by thirty-four Japanese specialists and was published in 1981.[64]

The deception and self-deception that accompanied these atrocities was remarkable. Truman told the public, "The world will note that the first atomic bomb was dropped on Hiroshima, a military base. That was because we wished in this first attack to avoid, insofar as possible, the killing of civilians."[65]

Even the possibility that American prisoners of war would be killed in these bombings did not have any effect on the plans. On July 31, nine days before Nagasaki was bombed, the headquarters of the U.S. Army Strategic Air Forces on Guam (the take-off airfield for the atomic bombings) sent a message to the War Department:

> Reports prisoner of war sources not verified by photo give location of Allied prisoner-of-war camp, one mile north of center of city of Nagasaki. Does this influence the choice of this target for initial Centerboard operation? Request immediate reply.

The reply came, "Targets previously assigned for Centerboard remain unchanged."[66]

The terrible momentum of war continued even after the bombings of Hiroshima and Nagasaki. The end of the war was a few days away, yet B-29s continued their missions. On August 14, five days after the Nagasaki bombing and the day before the actual acceptance of surrender terms, 449 B-29s went out from the Marianas for a daylight strike and

372 more went out that night. Altogether, more than 1,000 planes were sent to bomb Japanese cities. There were no American losses. The last plane had not yet returned when Truman announced the Japanese had surrendered.

Japanese writer Oda Makoto describes that August 14 in Osaka, where he lived. He was a boy. He went out into the streets and found in the midst of the corpses American leaflets written in Japanese, which had been dropped with the bombs: "Your government has surrendered; the war is over."[67]

The American public, already conditioned to massive bombing, accepted the atomic bombings with equanimity, indeed with joy. I remember my own reaction. When the war ended in Europe, my crew flew our plane back to the United States. We were given a thirty-day furlough and then had to report for duty to be sent to Japan to continue bombing. My wife and I decided to spend that time in the countryside. Waiting for the bus to take us, I picked up the morning newspaper, August 7, 1945. The headline was "Atomic Bomb Dropped on Hiroshima." My immediate reaction was elation: "The war will end. I won't have to go to the Pacific."

I had no idea what the explosion of the atomic bomb had done to the men, women, and children of Hiroshima. It was abstract and distant, as were the deaths of the people from the bombs I had dropped in Europe from a height of six miles; I was unable to see anything below, there was no visible blood, and there were no audible screams. And I knew nothing of the imminence of a Japanese surrender. It was only later when I read John Hersey's *Hiroshima,* when I read the testimony of Japanese survivors, and when I studied the history of the decision to drop the bomb that I was outraged by what had been done.

It seems that once an initial judgment has been made that a war is just, there is a tendency to stop thinking, to assume then that everything done on behalf of victory is morally acceptable. I had myself participated in the bombing of cities, without even considering whether there was any relationship between what I was doing and the elimination of fascism in the world. One of my bombing missions had been on the city of Pilsen (now Plzeň) in Czechoslovakia. The inhabitants were Czechs—the very people who had been among the first victims of Nazi expansion—yet we were dropping bombs on them. I don't remember being conscious of that irony, or questioning our mission.

After the war I looked up the official Air Force history and found this description of the Pilsen bombing:

> The last attack on an industrial target by the Eighth Air Force occurred on 25 April, when the famous Skoda works at Pilsen, Czechoslovakia, received 500 well-placed tons. Because of a warning sent out ahead of time the workers were able to escape, except for five persons."[68]

In 1966, I encountered two Czech citizens who had lived in Pilsen at that time, and they told me that several hundred people died in that bombing raid.

There was another mission I flew, again unthinking and unfeeling, like a programmed robot. This was the bombing of the little French town of Royan, on the Atlantic coast near Bordeaux. The Allies were well into Germany, and it was clear that the war was almost over (it ended three weeks later). There was no reason for bombing Royan. True, there were several thousand German soldiers stationed outside the town, left behind by the Nazi retreat from France, but they were just waiting for the war to end.

Our raid was reported in the *New York Times:*

> More than 1300 Flying Fortresses and Liberators of the U.S. Eighth Air Force prepared the way for today's successful assault by drenching the enemy's positions on both sides of the Gironde controlling the route to Bordeaux with about 460,000 gallons of liquid fire that bathed in flames the German positions and strong points.[69]

This was one of the earliest uses of napalm in modern warfare. It may well be that one of the reasons for the raid was to try out this new weapon. Also, there were all these planes and all these well-trained crews, and here was something for them to do—not an unusual motive in war. Still another reason: the French military leaders on the ground were aching for some glory before the war ended.

A ground assault by the French followed up on the bombing. The *Times* reported,

> French troops mopped up most of Royan, on the north side of the river's mouth. . . . Royan, a town of 20,000, once was a vacation spot. About 350 civilians, dazed or bruised by two terrific air bombings in forty-eight hours, crawled from the ruins and said the air attacks had been "such hell as we never believed possible."[70]

General de Larminat, in charge of the French forces in that region and much criticized for the attack, was silent for a long time, but several years after the war he said,

> All wars carry these painful errors. . . . This is the painful ransom, the inevitable ransom of war. . . . We do not linger on the causes of these unfortunate events because, in truth, there is only a single cause: War, and the only ones truly responsible are those who wanted war."[71]

A similar statement was made by British Air Marshal Sir Robert Saundby after the bombing of Dresden:

> It was one of those terrible things that sometimes happen in wartime, brought about by an unfortunate combination of circumstances. . . . It is not so much this or the other means of making war that is immoral or inhumane. What is immoral is war itself. Once full-scale war has broken out it can never be humanized or civilized. . . . So long as we resort to war to settle differences between nations, so long will we have to endure the horrors, the barbarities and excesses that war brings with it. That, to me, is the lesson of Dresden.[72]

Dissident Voices

What is remarkable is how close these statements, by two military men, come to the one made by Albert Einstein, on the occasion of the Geneva Disarmament Conference in 1932. All of them were suggesting that once war is made, an atmosphere is created and a momentum begins in which the worst horrors become inevitable. Thus a war that apparently begins with a "good" cause—stopping aggression, helping victims, or punishing brutality—ends with its own aggression, creates more victims than before, and brings out more brutality than before, on both sides. The Holocaust, a plan made and executed in the ferocious atmosphere of war, and the saturation bombings, also created in the frenzy of war, are evidence of this.

The good cause in World War II was the defeat of fascism. And, in fact, it ended with that defeat: the corpse of Mussolini hanging in the public square in Milan; Hitler burned to death in his underground bunker; Tojo, captured and sentenced to death by an international tribunal. But 40 million people were dead, and the *elements* of fascism—militarism, racism, imperialism, dictatorship, ferocious nationalism, and war—were still at large in the postwar world.

Two of those 40 million were my closest air force friends, Joe Perry and Ed Plotkin. We had suffered through basic training and rode horses and flew Piper Cubs in Burlington, Vermont, and played basketball at Santa Ana before going our own ways to different combat zones. Both were killed in the final weeks of the war. For years afterward, they appeared in my dreams. In my waking hours, the question grew: What did they really die for?

We were victorious over fascism, but this left two superpowers dominating the world, vying for control of other nations, carving out new spheres of influence, on a scale even larger than that attempted by the Fascist powers. Both superpowers supported dictatorships all over the world: the Soviet Union in Eastern Europe and the United States in Latin America, Korea, and the Philippines.

The war machines of the Axis powers were destroyed, but the Soviet Union and the United States were building military machines greater than the world had ever seen, piling up frightful numbers of nuclear weapons, soon equivalent to *a million* Hiroshima-type bombs. They were preparing for a war to keep the peace, they said (this had also been said before World War I) but those preparations were such that if war took place (by accident? by miscalculation?) it would make the Holocaust look puny.

Hitler's aggression was over but wars continued, which the superpowers either initiated or fed with military aid or observed without attempting to halt them. Two million people died in Korea; 2 to 5 million in Vietnam, Cambodia, and Laos; 1 million in Indonesia; perhaps 2 million in the Nigerian civil war; 1 million in the Iran-Iraq War; and many more in Latin America, Africa, and the Middle East. It is estimated that, in the forty years after 1945, there were 150 wars, with 20 million casualties.[73]

The victorious and morally righteous superpowers stood by in the postwar world while millions—more than had died in Hitler's Holocaust—starved to death. They made gestures, but allowed national ambitions and interpower rivalries to stand in the way of saving the hungry. A United Nations official reported, with great bitterness, that

> in pursuit of political objectives in the Nigerian Civil War, a number of great and small nations, including Britain and the United States, worked to prevent supplies of food and medicine from reaching the starving children of rebel Biafra.[74]

Swept up in the obvious rightness of a crusade to rid the world of fascism, most people supported or participated in that crusade, to the point of risking their lives. But there were skeptics, especially among the nonwhite peoples of the world—blacks in the United States and the colonized millions of the British Empire (Gandhi withheld his support).

The extraordinary black writer Zora Neale Hurston wrote her memoir, *Dust Tracks on a Road,* at the start of World War II. Just before it was to come out the Japanese attacked Pearl Harbor, and her publisher, Lippincott, removed a section of the book in which she wrote bitterly about the "democracies" of the West and their hypocrisy. She said:

> All around me, bitter tears are being shed over the fate of Holland, Belgium, France and England. I must confess to being a little dry around the eyes. I hear people shaking with shudders at the thought of Germany collecting taxes in Holland. I have not heard a word against Holland collecting one twelfth of poor people's wages in Asia. Hitler's crime is that he is actually doing a thing like that to his own kind. . . .
>
> As I see it, the doctrines of democracy deal with the aspirations of men's souls, but the application deals with things. One hand in somebody else's pocket and one on your gun, and you are highly civilized. . . . Desire enough for your own use only, and you are a heathen. Civilized people have things to show to their neighbors.[75]

The editor at Lippincott wrote on her manuscript, "Suggest eliminating international opinions as irrelevant to autobiography."[76] Only when the book was reissued in 1984 did the censored passages appear.[77]

Hurston, in a letter she wrote to a journalist friend in 1946, showed her indignation at the hypocrisy that accompanied the war:

> I am amazed at the complacency of Negro press and public. Truman is a monster. I can think of him as nothing else but the Butcher of Asia. Of his grin of triumph on giving the order to drop the Atom bombs on Japan. Of his maintaining troops in China who are shooting the starving Chinese for stealing a handful of food.[78]

Some white writers were resistant to the fanaticism of war. After it was over, Joseph Heller wrote his biting, brilliant satire *Catch-22* and Kurt Vonnegut wrote *Slaughterhouse Five.* In the 1957 film *Bridge on the River Kwai,* the Japanese military is obsessed with building a bridge, and the British are obsessed with destroying it. At the end it is blown up and a British lieutenant, barely surviving, looks around at the river strewn with corpses and mutters: "Madness. Madness."

There were pacifists in the United States who went to prison rather than participate in World War II. There were 350,000 draft evaders in the United States. Six thousand men went to prison as conscientious objectors; one out of every six inmates in U.S. federal prisons was a conscientious objector to the war.[79]

But the general mood in the United States was support. Liberals, conservatives, and Communists agreed that it was a just war. Only a few voices were raised publicly in Europe and the United States to question the motives of the participants, the means by which the war was being conducted, and the *ends* that would be achieved. Very few tried to stand back from the battle and take a long view. One was the French worker-philosopher Simone Weil. Early in 1945 she wrote in a new magazine called *Politics*,

> Whether the mask is labelled Fascism, Democracy, or Dictatorship of the Proletariat, our great adversary remains the Apparatus—the bureaucracy, the police, the military. . . . No matter what the circumstances, the worst betrayal will always be to subordinate ourselves to this Apparatus, and to trample underfoot, in its service, all human values in ourselves and in others.[80]

The editor of *Politics* was an extraordinary American intellectual named Dwight MacDonald, who with his wife, Nancy, produced the magazine as an outlet for unorthodox points of view. After the bombing of Hiroshima, MacDonald refused to join in the general jubilation. He wrote with a fury:

> The CONCEPTS "WAR" AND "PROGRESS" ARE NOW OBSOLETE: . . . THE FUTILITY OF MODERN WARFARE SHOULD NOW BE CLEAR. Must we not now conclude, with Simone Weil, that the technical aspect of war today is the evil, regardless of political factors? Can one imagine that the atomic bomb could ever be used "in a good cause"?[81]

But what was the alternative to war, with Germany on the march in Europe, Japan on its rampage through Asia, and Italy looking for empire? This is the toughest possible question. Once the history of an epoch has run its course, it is very difficult to imagine an alternate set of events, to imagine that some act or acts might set in motion a whole new train of circumstances, leading in a different direction.[82]

Would it have been possible to trade time and territory for human life? Was there an alternative preferable to using the most modern weapons of destruction for mass annihilation? Can we try to imagine instead of a six-year war a ten-year or twenty-year period of resistance; of guerrilla warfare, strikes, and noncooperation; of underground movements, sabotage, and paralysis of vital communication and transportation; and of clandestine propaganda for the organization of a larger and larger opposition?

Even in the midst of war, some nations occupied by the Nazis were able to resist: the Danes, the Norwegians, and the Bulgarians refused to give up their Jews.[83] Gene Sharp, on the basis of his study of resistance movements in World War II, writes:

> During the second World War—in such occupied countries as the Netherlands, Norway and Denmark—patriots resisted their Nazi overlords and internal puppets by such weapons as underground newspapers, labor slowdowns, general strikes, refusal of collaboration, special boycotts of German troops and quislings, and noncooperation with fascist controls and efforts to restructure their societies' institutions.[84]

Guerrilla warfare is more selective, its violence more limited and more discriminate, than conventional war. It is less centralized and more democratic by nature, requiring the commitment, the initiative, and the cooperation of ordinary people who do not need to be conscripted, but who are motivated by their desire for freedom and justice.

History is full of instances of successful resistance (although we are not informed very much about this) without violence and against tyranny, by people using strikes, boycotts, propaganda, and a dozen different ingenious forms of struggle. Gene Sharp, in his book *The Politics of Non-Violent Action*,[85] records hundreds of instances and dozens of methods of action.

Since the end of World War II, we have seen dictatorships overthrown by mass movements that mobilized so much popular opposition that the tyrant finally had to flee: in Iran, in Nicaragua, in the Philippines, and in Haiti. Granted, the Nazi machine was formidable, efficient, and ruthless. But there are limits to conquest. A point is reached where the conquerer has swallowed too much territory, has to control too many people. Great empires have fallen when it was thought they would last forever.

We have seen, in the eighties, mass movements of protest arise in the tightly controlled Communist countries of Eastern Europe, forcing dramatic changes in Hungary, Czechoslovakia, Poland, Bulgaria, Rumania, and East Germany. The Spanish people, having lost a million lives in their civil war, waited out Franco. He died, as all men do, and the dictatorship was over. For Portugal, the resistance in its outlying African Empire weakened control; corruption grew and the long dictatorship of Salazar was overthrown—without a bloodbath.

There is a fable written by German playwright Bertolt Brecht that goes roughly like this: A man living alone answers a knock at the door. When he opens it, he sees in the doorway the powerful body, the cruel face, of The Tyrant. The Tyrant asks, "Will you submit?" The man does not reply. He steps aside. The Tyrant enters and establishes himself in the man's house. The man serves him for years. Then The Tyrant becomes sick from food poisoning. He dies. The man wraps the body, opens the door, gets rid of the body, comes back to his house, closes the door behind him, and says, firmly, "No."

Violence is not the only form of power. Sometimes it is the least effective. Always it is the most vicious, for the perpetrator as well as for the victim. And it is corrupting.

Immediately after the war, Albert Camus, the great French writer who fought in the underground against the Nazis, wrote in *Combat,* the daily newspaper of the French Resistance. In his essay called "Neither Victims Nor Executioners," he considered the tens of millions of dead caused by the war and asked that the world reconsider fanaticism and violence:

> All I ask is that, in the midst of a murderous world, we agree to reflect on murder and to make a choice. . . . Over the expanse of five continents throughout the coming years an endless struggle is going to be pursued between violence and friendly persuasion, a struggle in which, granted, the former has a thousand times the chances of success than has the latter. But I have always held that, if he who bases his hopes on human nature is a fool, he who gives up in the face of circumstances is a coward. And henceforth, the only honorable course will be to stake everything on a formidable gamble: that words are more powerful than munitions.[86]

Whatever alternative scenarios we can imagine to replace World War II and its mountain of corpses, it really doesn't matter any more. That war is over. The practical effect of declaring World War II *just* is not for that war, but for the wars that follow. And that effect has been a dangerous one, because the glow of rightness that accompanied that war has been transferred, by false analogy and emotional carryover, to other wars. To put it another way, perhaps the worst consequence of World War II is that it kept alive the idea that war could be just.

Looking at World War II in perspective, looking at the world it created and the terror that grips our century, should we not bury for all time the idea of just war?

Some of the participants in that "good war" had second thoughts. Former GI Tommy Bridges, who after the war became a policeman in Michigan, expressed his feelings to Studs Terkel:

> It was a useless war, as every war is. . . . How gaddamn foolish it is, the war. They's no war in the world that's worth fighting for, I don't care where it is. They can't tell me any different. Money, money is the thing that causes it all. I wouldn't be a bit surprised that the people that start wars and promote 'em are the men that make the money, make the ammunition, make the clothing and so forth. Just think of the poor kids that are starvin' to death in Asia and so forth that could be fed with how much you make one big shell out of.[87]

Higher up in the military ranks was Admiral Gene LaRocque, who also spoke to Studs Terkel about the war:

> I had been in thirteen battle engagements, had sunk a submarine, and was the first man ashore in the landing at Roi. In that four years, I thought, What a hell of a waste of a man's life. I lost a lot of friends. I had the task of telling my roommate's parents about our last days together. You lose limbs, sight, part of your life—for what? Old men send young men to war. Flag, banners, and patriotic savings. . . .
>
> We've institutionalized militarism. This came out of World War Two. . . . It gave us the National Security Council. It gave us the CIA, that is able to spy on you and me this very moment. For the first time in the history of man, a country has divided up the world into military districts. . . .
>
> You could argue World War Two had to be fought. Hitler had to be stopped. Unfortunately, we translate it unchanged to the situation today. . . .

> I hate it when they say, "He gave his life for his country." Nobody gives their life for anything. We steal the lives of these kids. We take it away from them. They don't die for the honor and glory of their country. We kill them.[88]

Granted that we have started in this century with the notion of just war, we don't have to keep it. Perhaps the change in our thinking can be as dramatic, as clear, as that in the life of a French general, whose obituary in 1986 was headed: "Gen. Jacques Paris de Bollardiere, War Hero Who Became a Pacifist; Dead at the age of 78."

He had served in the Free French Forces in Africa during World War II, later parachuted into France and Holland to organize the Resistance, and commanded an airborne unit in Indochina from 1946 to 1953. But in 1957, according to the obituary, he "caused an uproar in the French army when he asked to be relieved of his command in Algeria to protest the torture of Algerian rebels." In 1961 he began to speak out against militarism and nuclear weapons. He created an organization called The Alternative Movement for Non-Violence and in 1973 participated in a protest expedition to France's South Pacific nuclear testing site.

It remains to be seen how many people in our time will make that journey from war to nonviolent action against war. It is the great challenge of our time: How to achieve justice, with struggle, but without war.

CHAPTER

SIX

Law and Justice

In 1978 I was teaching a class called "Law and Justice in America," and on the first day I handed out the course outline. At the end of the hour one of the students came up to the desk. He was a little older than the others. He said, "I notice in your course outline you will be discussing the case of *U.S. vs. O'Brien.* When we come to that I would like to say something about it."

I was a bit surprised, but glad that a student would take such initiative. I said, "Sure. What's your name?"

He said, "O'Brien. David O'Brien."

It was, indeed, his case. On the morning of March 31, 1966, while American troops were pouring into Vietnam and U.S. planes were bombing day and night, David O'Brien and three friends climbed the steps of the courthouse in South Boston where they lived—a mostly Irish, working-class neighborhood—held up their draft registration cards before a crowd that had assembled, and set the cards afire.

According to Chief Justice Earl Warren, who rendered the Supreme Court decision in the case: "Immediately after the burning, members of the crowd began attacking O'Brien," and he was ushered to safety by an FBI agent. As O'Brien told the story to my class, FBI agents pulled him into the courthouse, threw him into a closet, and gave him a few blows as they arrested him.

Chief Justice Warren's decision said, "O'Brien stated to FBI agents that he had burned his registration certificate because of his beliefs, knowing that he was violating federal law." His intention was clear. He wanted to express to the community his strong feelings about the war in Vietnam, trying to call attention, by a dramatic act, to the mass killing our government was engaged in there. The burning of his draft card would get special attention precisely because it was against the law, and so he would risk imprisonment to make his statement.

O'Brien claimed in court that his act, although in violation of the draft law, was protected by the free speech provision of the Constitution. But the Supreme Court decided that the government's need to regulate the draft overcame his right to free expression, and he went to prison.[1]

O'Brien had engaged in an act of civil disobedience—the deliberate violation of a law for a social purpose.[2] To violate a law for individual gain, for a private purpose, is an ordinary criminal act; it is not civil disobedience. Some acts fall in both categories, as in the case of a mother stealing bread to feed her children, or neighbors stopping the eviction of a family that hadn't been able to pay the rent. Although limited to one family's need, they carry a larger message to the society about its failures.

In either instance, the law is being disobeyed, which sets up strong emotional currents in a population that has been taught obedience from childhood.

Obedience and Disobedience

"Obey the law." That is a powerful teaching, often powerful enough to overcome deep feelings of right and wrong, even to override the fundamental instinct for personal survival. We *learn* very early (it's not in our genes) that we must obey "the law of the land." Tommy Trantino, a poet and artist, sitting on death row in Trenton State Prison, wrote (in his book *Lock the Lock*) a short piece called "The Lore of the Lamb":

> i was in prison long ago and it was the first grade and i have to take a shit and . . . the law says you must first raise your hand and ask the teacher for permission so i obeyer of the lore of the lamb am therefore busy raising my hand to the fuhrer who says yes thomas what is it? and i thomas say I have to take a i mean may i go to the bathroom please? didn't you go to the bathroom yesterday thomas she says and i say yes ma'am mrs parsley sir but i have to go again today but she says NO . . . And I say eh . . . I GOTTA TAKE A SHIT DAMMIT and again she says NO but I go anyway except

> that it was not out but in my pants that is to say right in my corduroy knickers goddamm. . . .
>
> i was about six years old at the time and yet i guess that even then i knew without cerebration that if one obeys and follows orders and adheres to all the rules and regulations of the lore of the lamb one is going to shit in one's pants and one's mother is going to have to clean up afterwards ya see?[3]

Surely not all rules and regulations are wrong. One must have complicated feelings about the obligation to obey the law. Obeying the law when it sends you to war seems wrong. Obeying the law against murder seems absolutely right. To *really* obey that law, you should refuse to obey the law that sends you to war.

But the dominant ideology leaves no room for making intelligent and humane distinctions about the obligation to obey the law. It is stern and absolute. It is the unbending rule of every government, whether Fascist, Communist, or liberal capitalist. Gertrude Scholtz-Klink, chief of the Women's Bureau under Hitler, explained to an interviewer after the war the Jewish policy of the Nazis, "We always obeyed the law. Isn't that what you do in America? Even if you don't agree with a law personally, you still obey it. Otherwise life would be chaos."[4]

"Life would be chaos." If we allow disobedience to law we will have anarchy. That idea is inculcated in the population of every country. The accepted phrase is "law and order." It is a phrase that sends police and the military to break up demonstrations everywhere, whether in Moscow or Chicago. It was behind the killing of four students at Kent State University in 1970 by National Guardsmen. It was the reason given by Chinese authorities in 1989 when they killed hundreds of demonstrating students in Beijing.

It is a phrase that has appeal for most citizens, who, unless they themselves have a powerful grievance against authority, are afraid of disorder. In the 1960s, a student at Harvard Law School addressed parents and alumni with these words:

> The streets of our country are in turmoil. The universities are filled with students rebelling and rioting. Communists are seeking to destroy our country. Russia is threatening us with her might. And the republic is in danger. Yes! danger from within and without. We need law and order! Without law and order our nation cannot survive.

There was prolonged applause. When the applause died down, the student quietly told his listeners: "These words were spoken in 1932 by Adolf Hitler."[5]

Surely, peace, stability, and order are desirable. Chaos and violence are not. But stability and order are not the only desirable conditions of social life. There is also *justice*, meaning the fair treatment of all human beings, the equal right of all people to freedom and prosperity. Absolute obedience to law may bring order temporarily, but it may not bring justice. And when it does not, those treated unjustly may protest, may rebel, may cause disorder, as the American revolutionaries did in the eighteenth century, as antislavery people did in the nineteenth century, as Chinese students did in this century, and as working people going on strike have done in every country, across the centuries.

Are we not more obligated to achieve justice than to obey the law? The law may serve justice, as when it forbids rape and murder or requires a school to admit all students regardless of race or nationality. But when it sends young men to war, when it protects the rich and punishes the poor, then law and justice are opposed to one another. In that case, where is our greater obligation: to law or to justice?[6]

The answer is given in democratic theory at its best, in the words of Jefferson and his colleagues in the Declaration of Independence. Law is only a means. Government is only a means. "Life, Liberty, and the pursuit of Happiness"—these are the ends. And "whenever any Form of Government becomes destructive of these ends, it is the Right of the People to alter or to abolish it, and to institute new government."

True, the disorder itself may become unjust if it involves indiscriminate violence against people, as the Cultural Revolution in China in the period 1966–1976 started out with the aim of equality but became vengeful and murderous. But that danger should not lead us back to the old injustices to have stability. It should only lead us to seek methods of achieving justice that, although disorderly and upsetting, avoid massive violence to human rights.

Should we worry that disobedience to law will lead to anarchy? The answer is best given by historical experience. Did the mass demonstrations of the black movement in the American South, in the early sixties, lead to anarchy? True, they disrupted the order of racial segregation. They created scenes of disorder in hundreds of towns and cities in the country (although it might be argued that the police, responding to nonviolent protest, were the chief creators of that disorder). But the result of all that tumult was not general lawlessness.[7] Rather the result was a healthy reconstitution of the social order toward greater justice and a healthy new understanding among Americans (not all, of course) about the need for racial equality.

The orthodox notion is that law and order are inseparable. However, absolute obedience to all laws will violate justice and sooner or later lead to enormous disorder. Hitler, calling for law and order, threw Europe into the hellish disorder of war. Every nation uses the power of law to keep its population obedient and to mobilize acquiescent armies, threatening punishment for those who refuse. Thus the law that inside each nation creates conscript armies leads to the unspeakable disorder of war, to the bloody chaos of the battlefield, and to international turmoil.

If law and order are only ways of making injustice legitimate, then the "order" on the surface of everyday life may conceal deep mental and emotional disorder among the victims of injustice. This is also true for the powerful beneficiaries of the system, in the way that slavery distorts the psyches of both slave and master. In such a case, the order will only be temporary; when it is broken, it may be accompanied by a bloodbath of disorder—as in the United States, when the tightly controlled order of slavery ended in civil war and 600,000 men died in a country of 35 million people.

The Modern Era of Law

We take much pride in that phrase of John Adams, second president of the United States, when he spoke of the "rule of law" replacing the "rule of men." In ancient societies, in feudal society, there were no clear rules, written in statute books, accompanied by constitutions. Everyone was subject to the whims of powerful men, whether the feudal lord, the tribal chief, or the king.

But as societies evolved modern times brought big cities, international trade, widespread literacy, and parliamentary government. With all that came the rule of law, no longer personal and arbitrary, but written down. It claimed to be impersonal, neutral, apply equally to all, and, therefore, democratic.

We profess great reverence for certain symbols of the modern rule of law: the Magna Carta, which set forth what are men's rights as against the king; the American Constitution, which is supposed to limit the powers of government and provide a Bill of Rights; the Napoleonic

Code, which introduced uniformity into the French legal system. But we might get uneasy about the connection between law and democracy when we read the comment of two historians (Robert Palmer and Joel Colton) on Napoleon: "Man on horseback though he was, he believed firmly in the rule of law."[8]

I don't want to deny the benefits of the modern era: the advance of science, the improvements in health, the spread of literacy and art beyond tiny elites, and the value of even an imperfect representative system over a monarchy. But those advantages lead us to overlook the fact that the modern era, replacing the arbitrary rule of men with the impartial rule of law, has not brought any fundamental change in the facts of unequal wealth and unequal power. What was done before—exploiting the poor, sending the young to war, and putting troublesome people in dungeons—is still done, except that this no longer seems to be the arbitrary action of the feudal lord or the king; it now has the authority of neutral, impersonal law.

The law appears impersonal. It is on paper, and who can trace it back to what men? And because it has the look of neutrality, its injustices are made legitimate. It was not easy to hold onto the "divine right" of kings—everyone could see that kings and queens were human beings. A code of law is more easily deified than a flesh-and-blood ruler.

Under the rule of men, the oppressor was identifiable, and so peasant rebels hunted down the lords, slaves killed plantation owners, and revolutionaries assassinated monarchs. In the era of the corporate bureaucracies, representative assemblies, and the rule of law, the enemy is elusive and unidentifiable. In John Steinbeck's depression-era novel *The Grapes of Wrath* a farmer having his land taken away from him confronts the tractor driver who is knocking down his house. He aims a gun at him, but is confused when the driver tells him that he takes his orders from a banker in Oklahoma City, who takes his orders from a banker in New York. The farmer cries out: "Then who can I shoot?"

The rule of law does not do away with the unequal distribution of wealth and power, but reinforces that inequality with the authority of law. It allocates wealth and poverty (through taxes and appropriations) but in such complicated and indirect ways as to leave the victim bewildered.

Exploitation was obvious when the peasant gave half his produce to the lord. It still exists, but inside the complexity of a market society and enforced by a library of statutes. A mine owner in Appalachia was asked, some years ago, why the coal companies paid so little taxes and kept so much of the wealth from the coal fields, while local people starved. The owner replied: "I pay exactly what the law asks me to pay."

There is a huge interest in the United States in crime and corruption as ways of acquiring wealth. But the greatest wealth, the largest fortunes, are acquired legally, aided by the laws of contract and property, enforced in the courts by friendly judges, handled by shrewd corporation lawyers, figured out by well-paid accountants. When our history books get to the 1920s, they dwell on the Teapot Dome scandals of the Harding administration, while ignoring the far greater reallocations of wealth that took place legally, through the tax laws proposed by Secretary of the Treasury Andrew Mellon (a very rich man, through oil and aluminum), and passed by Congress in the Coolidge Administration.

How can this be? Didn't the modern era bring us democracy? Who drew up the Constitution? Wasn't it all of us, getting together to draw up the rules by which we would live, a "social contract"? Doesn't the Preamble to the Constitution start with the words: "We the People, in order to . . . etc., etc."?

In fact, while the Constitution was certainly an improvement over the royal charters of England, it was still a document drawn up by rich men, merchants, and slaveowners who wanted a bit of political democracy, but had no sympathy for economic democracy. It was designed to set up a "rule of law," which would efficiently prevent rebellion by dissatisfied elements in the population. As the Founding Fathers assembled in Philadelphia, they still had in mind farmers who had recently taken up arms in western Massachusetts (Shays' Rebellion) against unjust treatment by the wealth-controlled legislature.[9]

It is a deception of the citizenry to claim that the "rule of law" has replaced the "rule of men." It is still men (women are mostly kept out of the process) who enact the laws, who sit on the bench and interpret them, who occupy the White House or the Governor's mansion, and have the job of enforcing them.

These men have enormous powers of discretion. The legislators decide which laws to put on the books. The president and his attorney-general decide which laws to enforce. The judges decide who has a right to sue in court, what instructions to give to juries, what rules of law apply, and what evidence should not be allowed in the courtroom.

The lawyers, to whom ordinary people must turn for help in making their way through the court system, are trained and selected in such a way as to ensure their conservatism. The exceptions, when they appear, are noble and welcome, but too many lawyers are more concerned about being "good professionals" than achieving justice. As one student of the world of lawyers put it: "It is of the essence of the professionalization process to divorce law from politics, to elevate technique and craft over power, to search for 'neutral principles,' and to deny ideological purpose."[10]

Equal Justice Under Law is the slogan one sees on the marble pillars of the courthouse. And there is nothing in the words of the Constitution or the laws to indicate that anyone gets special treatment. They look as if they apply to everyone. But in the actual administration of the laws are rich and poor treated equally? Blacks and whites? Foreign born and natives? Conservatives and radicals? Private citizens and government officials?

There is a mountain of evidence on this: a CIA official (Richard Helms) commits perjury and gets off with a fine (Alger Hiss spent four years in jail for perjury), a president (Nixon) is pardoned in advance of prosecution for acts against the law, and Oliver North and other Reagan administration officials are found guilty of violating the law in the Iran-Contra affair, but none go to prison.

Still, the system of laws, to maintain its standing in the eyes of the citizenry and to provide safety valves by which the discontented can let off steam, must keep up the appearance of fairness. And so the law itself provides for change. When the pressure of discontentment becomes great, laws are passed to satisfy some part of the grievance. Presidents, when pushed by social movements, may enforce good laws. Judges, observing a changing temper in the society, may come forth with humane decisions.

Thus we have alternating currents of progress and paralysis. Periods of war alternate with periods of peace. There are times of witch-hunts for dissenters and times of apologies for the witch-hunts. We have "conservative" presidents giving way to liberal presidents and back again. The Supreme Court makes decisions one week on behalf of civil liberties and the next week curtails them. No one can get a clear fix on the system that way.

The modern system of the rule of law is something like roulette. Sometimes you win and sometimes you lose. No one can predict in any one instance whether the little ball will fall into the red or the black, and no one is really responsible. You win, you lose. But as in roulette, in the end you almost always lose. In roulette the results are fixed by the structure of the wheel, the laws of mathematical probability, and the rules of "the house." In society, the rich and strong get what they want by the law of contract, the rules of the market, and the power of the authorities to change the rules or violate them at will.

What is the structure of society's roulette wheel that ensures you will, in the end, lose? It is, first of all, the great disparities in wealth that give a tremendous advantage to those who can buy and sell industries, buy and sell people's labor and services, buy and sell the means of communication, subsidize the educational system, and buy and sell the political candidates themselves. Second, it is the system of "checks and balances," in which bold new reforms (try free medical care for all or sweeping protections of the environment) can be buried in committee, vetoed by one legislative chamber or by the president, interpreted to death by the Supreme Court, or passed by Congress and unenforced by the president.

In this system, the occasional victories may ease some of the pain of economic injustice. They also reveal the usefulness of protest and pressure, suggest even greater possibilities for the future. And they keep you in the game, giving you the feeling of fairness, preventing you from getting angry and upsetting the wheel. It is a system ingeniously devised for maintaining things as they are, while allowing for limited reform.

Obligation to the State

Despite all I have said about the gap between law and justice and despite the fact that this gap is visible to many people in the society, the idea of obligation to law, obligation to government, remains powerful. President Jimmy Carter reinstated the draft of young men for military service in 1979, and when television reporters asked the men why they were complying with the law (about 10 percent were not), the most common answer was "I owe it to my country."

The obligation that people feel to one another goes back to the very beginning of human history, as a natural, spontaneous act in human relations.[11] Obligation to government, however, is not natural. It must be taught to every generation.

Who can teach this lesson of obligation with more authority than the great Plato? Plato has long been one of the gods of modern culture, his reputation that of an awesome mind and a brilliant writer of dialogue, his work the greatest of the Great Books. Shrewdly, Plato puts his ideas about obligation in the mouth of Socrates. Socrates left no writings that we know of, so he can be used to say whatever Plato wants. And Plato could have no better spokesman than a wise, gentle old man who was put to death by the government of Athens in 399 B.C. for speaking his mind. Any words coming from such a man will be especially persuasive.

But they are Plato's words, Plato's ideas. All we know of Socrates is what Plato tells us. Or, what we read in the recollections of another contemporary, Xenophon. Or what we can believe about him from reading Aristophanes's spoof on his friend Socrates, in his play *The Clouds.*

So we can't know for sure what Socrates really said to his friend Crito, who visited him in jail, after he had been condemned to death. But we do know what Plato has him say in the dialogue *Crito*[12] (written many years after Socrates's execution), which has been impressed on the minds of countless generations, down to the present day, with deadly effect. Plato's ideas have become part of the orthodoxy of the nation, absorbed into the national bloodstream and reproduced in ordinary conversations and on bumper stickers. ("Love it or leave it"—summing up Plato's idea of obligation.)

Plato's message is presented appealingly by a man calmly facing death, whose courage disarms any possible skepticism. It is made even more appealing by the fact that it follows another dialogue, the *Apology,* in which (according to Plato), Socrates addresses the jury in an eloquent defense of free speech, saying those famous words: "The unexamined life is not worth living."

Plato then unashamedly (lesson one in intellectual bullying: speak with utter confidence) presents us with some unexamined ideas. Having established Socrates's credentials as a martyr for independent thought, he proceeds in the *Crito* to put on Socrates's tongue an argument for blind obedience to government.

It is hardly a dialogue, although Plato is famous for dialogue and the "Socratic method" is based on teaching through dialogue. Poor Crito, who visits Socrates in prison to persuade him to let his friends plan his escape, is virtually tongue-tied. He is reduced to saying, to every one of Socrates's little speeches: "Yes . . . of course . . . clearly . . . I agree. . . . Yes . . . I think that you are right. . . . True." And Socrates is going on and on, like the good trouper that he is, saying Plato's lines, making Plato's argument. We know the ideas are Plato's because in his well-known and much bigger dialogue the *Republic* he makes an even more extended case for a totalitarian state.

To Crito's offer of escape, Socrates replies: I must obey the law. True, he says, Athens has committed an injustice by ordering him to die for speaking his mind (he seems slightly annoyed at this!), but if he complained about this injustice, Athens could rightly say: "We brought you into the world, we raised you, we educated you, we gave you and every other citizen a share of all the good things we could."

Socrates accepts this argument of the state. He tells Crito that by not leaving Athens he agreed to obey its laws. So he must go to his death. Yes, it is Plato's own bumper sticker: "Love it or leave it."

If Plato had lived another 2,000 years or so he would have encountered the argument of Henry David Thoreau, the quiet hermit of Walden Pond who wrote a famous essay on civil disobedience. Thoreau said that whatever good things we have were not given us by the state, but by the energies and talents of the people of the country. And he would be damned if he would pay taxes to support a war against Mexico based on such a paltry argument.

Plato, the Western world's star intellectual, makes a number of paltry arguments in this so-called dialogue. He has Socrates imagining the authorities addressing him: "What complaint have you against us and the state, that you are trying to destroy us? Are we not, first of all, your parents? Through us your father took your mother and brought you into the world."[12]

What complaint? Only that they are putting him to death! The state as parents? Now we understand those words: the Motherland, the Fatherland, the Founding Fathers, Uncle Sam. What neat spades for planting the idea of obligation. It's not some little junta of military men and politicians who are sending you to die in some muddy field in Asia or Central America, it's your mother, your father, or your father's favorite brother. How can you say no? "Through us your father took your mother and brought you into the world." What stately arrogance! To give the state credit for marriage and children, as if without government men and women would remain apart and celibate. Socrates listens meekly to the words of the law:

> Are you too wise to see your country is worthier, more to be revered, more sacred, and held in higher honor both by the gods and by all men of understanding, than your father and your mother and all your other ancestors; that you ought to reverence it and to submit to it . . . and to obey in silence if it orders you to endure flogging or imprisonment or if it sends you to battle to be wounded or to die?[13]

In the face of this seductive argument, Crito is virtually mute, a sad sack of a debater. You would think that Plato, just to maintain his reputation for good dialogue, would give Crito some better lines. But he took no chances.

Plato says (again, through Socrates bullying Crito): "In war, and in the court of justice, and everywhere, you must do whatever your state and your country tell you to do, or you must persuade them that their commands are unjust."

Why not insist that the *state* persuade *us* to do its bidding? There is no equality in Plato's scheme: the citizen may use persuasion, but no more; the state may use force.

It is curious that Socrates (according to Plato) was willing to disobey the authorities by preaching as he chose, by telling the young what he saw as the truth, even if that meant going against the laws of Athens. Yet, when he was sentenced to death, and by a divided jury (the vote was 281 to 220), he meekly accepted the verdict, saying he owed Athens obedience to its laws, giving that puny 56 percent majority vote an absolute right to take his life.

And so it is that the admirable obligation human beings feel to one's neighbors, one's loved ones, even to a stranger needing water or shelter, becomes confused with blind obedience to that deadly artifact called government. And in that confusion, young men, going off to war in some part of the world they never heard of, for some cause that cannot be rationally explained, then say: "I owe it to my country."

It seems that the idea of *owing*, of obligation, is strongly felt by almost everyone. But what does one owe the government? Granted, the government may do useful things for its citizens: help farmers, administer old-age pensions and health benefits, regulate the use of drugs, apprehend criminals, etc. But because the government administers these programs (for which the citizens pay taxes, and for which the government officials draw salaries), does this mean that you owe the government your life?

Plato is enticing us to confuse the *country* with the *government*. The Declaration of Independence tried to make clear that the people of the country *set up* the government, to achieve the aims of equality and justice, and when a government no longer pursues those aims it loses its legitimacy, it has violated *its* obligation to the citizens, and deserves no more respect or obedience.

We are intimidated by the word *patriotism*, afraid to be called unpatriotic. Early in the twentieth century, the Russian-American anarchist and feminist Emma Goldman lectured on patriotism. She said,

> Conceit, arrogance and egotism are the essentials of patriotism. . . . Patriotism assumes that our globe is divided into little spots, each one surrounded by an iron gate. Those who had the fortune of being born on some particular spot, consider themselves better, nobler, grander, more intelligent than the living beings inhabiting any other spot. It is, therefore, the duty of everyone living on that chosen spot to fight, kill, and die in the attempt to impose his superiority upon all the others.[14]

Even the symbols of patriotism—the flag, the national anthem—become objects of worship, and those who refuse to worship are treated as heretics. When in 1989 the U.S. Supreme Court decided that a citizen has a right to express himself or herself by burning the American flag, there was an uproar in the White House and in Congress. President Bush, almost in tears, began speaking of a Constitutional amendment to make flag burning a crime. Congress, with its customary sheepishness, rushed to pass a law providing a year in prison for anyone hurting the flag.

The humorist Garrison Keillor responded to the president with some seriousness:

> Flag-burning is a minor insult compared to George Bush's cynical use of the flag for political advantage. Any decent law to protect the flag ought to prohibit politicians from wrapping it around themselves! Flag-burning is an impulsive act by a powerless individual—but the cool pinstripe demagoguery of this powerful preppie is a real and present threat to freedom.[15]

If patriotism were defined, not as blind obedience to government, not as submissive worship to flags and anthems, but rather as love of one's country, one's fellow citizens (all over the world), as loyalty to the principles of justice and democracy, then patriotism would require us to disobey our government, when it violated those principles.

Accept Your Punishment!

Socrates's position—that he must accept death for his disobedience—has become one of the cardinal principles in the liberal philosophy of civil disobedience and part of the dominant American orthodoxy in the United States, for both conservatives and liberals. It is usually stated this way: it's your right to break the law when your conscience is offended; but then you must accept your punishment.[16]

Why? Why agree to be punished when you think you have acted rightly and the law, punishing you for that, has acted wrongly? Why is it all right to disobey the law in the first instance, but then, when you are sentenced to prison, start obeying it?[17]

Some people, to support the idea of accepting punishment, like to quote Martin Luther King, Jr., one of the great apostles of civil disobedience in this century. In his "Letter from Birmingham City Jail," written in the spring of 1963, in the midst of tumultuous demonstrations against racial segregation, he said, "I submit that an individual who breaks a law that conscience tells him is unjust, and willingly accepts the penalty by staying in jail to arouse the conscience of the community over its injustices is in reality expressing the very highest respect for law."[18]

King was writing in answer to pleas by some white church leaders that he stop the demonstrations. They urged him to take his cause to the courts but "not in the streets." I believe King's reply has been seriously misinterpreted. It was an impassioned defense of nonviolent direct action, but it is obvious that he wanted to persuade those conservative church leaders of his moderation. He was anxious to show that, while committing civil disobedience he was "expressing the very highest respect for law."

The "law" that King respected, we know unquestionably from his life, his work, and his philosophy, was not man-made law, neither segregation laws nor even laws approved by the Supreme Court nor decisions of the courts nor sentences meted out by judges. He meant respect for the higher law, the law of morality, of justice.

To be "one who willingly accepts" punishment is not the same as thinking it *right* to be punished for an act of conscience. If this were so, why would King agree to be released from jail by behind-the-scenes pressure, as he did in 1960 when a mysterious benefactor in a high position (someone close to President-elect Kennedy) pulled strings to get him out of prison? The meaning of "willingly accepts" is that you know you are risking jail and are willing to take that risk, but it doesn't mean it is morally right for you to be punished.

King talks about "staying in jail to arouse the conscience of the community over its injustice." He does not speak of staying in jail because he *owes* that to the government and that (as Plato argues) he has a duty to obey whatever the government tells him to do. Not at all. He remains in jail not for philosophical or moral reasons, but for a practical purpose, to continue his struggle "to arouse the conscience of the community over its injustice."

Knowing King's life and thought, we can safely say that if the circumstances had been different, he might well have agreed (unlike Socrates) to escape from jail. What if he had been sentenced, not to six months in a Georgia prison, but to death? Would he have "accepted" this?

Would King have condemned those black slaves who were tried under the Fugitive Slave Act of 1850 and ordered to return to slavery, but who refused to give themselves up? Would he have criticized Angela Davis, the black militant who, accused of abetting the escape of a black prisoner from a courtroom, and fearing a police attempt on her life, refused to stand trial and went underground?

We can imagine another test of King's attitude toward "accepting" punishment. During the Vietnam War, which King powerfully opposed ("The long night of war must be stopped," he said in 1965), the Catholic priest-poet Daniel Berrigan committed an act of civil disobedience. He and other men and women of the "Catonsville Nine," entered a draft board in Catonsville, Maryland, removed draft records, and set them afire in a public "ceremony." Father Berrigan delivered a meditation:

> Our apologies, good friends, for the fracture of good order, the burning of paper instead of children. . . . We could not, so help us God, do otherwise. . . . We say: killing is disorder, life and gentleness and community and unselfishness is the only order we recognize. For the sake of that order we risk our liberty, our good name. The time is past when good men can remain silent, when obedience can segregate men from public risk, when the poor can die without defense.[19]

Although he used the term *men*, one of the Catonsville Nine was a woman, Mary Moylan. When the Nine were found guilty, sentenced to jail terms, and lost their appeals, she and Daniel Berrigan refused to turn themselves in, going "underground." Berrigan was found after four months, Mary Moylan was never apprehended. She wrote from underground: "I don't want to see people marching off to jail with smiles on their faces. I just don't want them going. . . . I don't want to waste the sisters and brothers we have by marching them off to jail."

Berrigan and Moylan thought the war was wrong and thought their going to jail for opposing it was wrong. If, like King, they felt it would serve some practical use, they probably would have "accepted it." Going to jail can make a certain kind of statement to the public: "Yes, I feel so strongly about what is happening in the world that I am willing to risk jail to express my feelings."

Refusing to go to jail makes a different kind of statement: "The system that sentenced me is the same foul system that is carrying on this war. I will defy it to the end. It does not deserve my allegiance." As Daniel Berrigan said, yes, we respect the order of "gentleness and community" but not the "order" of making war on children.

Daniel Berrigan and I had traveled together in early 1968 to Hanoi to pick up three American pilots released from prison by the North Vietnamese. We became good friends, and I was soon in close contact with the extraordinary Catholic resistance movement against the Vietnam War.

In early 1970 his last appeal was turned down; facing several years in prison, he "disappeared," sending the FBI into a frantic effort to find him. They had caught sight of him at a huge student rally in the Cornell University gymnasium, then the lights went out and before they could make their way through the crowd he was spirited away inside a huge puppet, to a nearby farmhouse.

A few days after his disappearance, I received a phone call at my home in Boston. I was being invited to speak at a Catholic church on the Upper West Side of Manhattan, on the issues of the war and the Berrigans. Philip Berrigan, Daniel's brother, a priest and one of the Catonsville Nine, was also living underground and had just been found by the FBI in the tiny apartment of the church's pastor.

The church was packed with perhaps 500 people. FBI agents mingled with the crowd, alerted that Daniel Berrigan might show up. I made a brief speech. Another friend of Daniel's spoke. As the two of us sat on the platform, a note was passed to us, to meet two nuns at a Spanish-Chinese restaurant farther up Broadway, near Columbia University. There we were given directions to New Jersey, to the house where Daniel was hiding out.

The next morning we rented a car, drove to New Jersey, and met him. The house he was staying in was not secure (in fact, an FBI agent lived across the street!). We arranged a trip to Boston, a car, a driver, and a destination. From that point on, for the next four months, he eluded and exasperated the FBI, staying underground, but surfacing from time to time, to deliver a sermon at a church in Philadelphia, to be interviewed on national television, to make public statements about the war, to make a film (*The Holy Outlaw*) about his actions against the war, both overt and underground.[20]

During those four months, while helping take care of Dan Berrigan, I was teaching my course at Boston University in political theory. My students were reading the *Crito,* and I asked them to analyze reasons for not escaping punishment and also to consider Daniel Berrigan's reasons for going underground. They did not know, of course, that Berrigan was right there in Boston, living out his ideas.

I think it is a good guess, despite those often-quoted words of his on "accepting" punishment, that Martin Luther King, Jr., would have supported Berrigan's actions. The principle is clear. If it is right to disobey unjust laws, it is right to disobey unjust punishment for breaking those laws.

The idea behind "accept your punishment" (advanced often by "liberals" sympathetic with dissent) is that whatever your disagreement with some specific law or some particular policy, you should not spread disrespect for the law *in general,* because we need respect for the law to keep society intact.[21]

This is like saying because apples are good for children, we must insist that they not refuse the rotten ones, because that might lead them to reject all apples. Well, good apples are good for your health, and rotten apples are bad. Bad laws and bad policies endanger our lives and our freedoms. Why can't we trust human intelligence to make the proper distinctions—among laws as among apples?

The domino theory is in people's minds: Let one domino fall and they will all go. It is a psychology of absolute control, in which the need for total security brings an end to freedom. Let anyone evade punishment and the whole social structure will come down.

We must ask, however: Can a decent society exist (*that* is our concern, not the *state*), if people humbly obey all laws, even those that violate human rights? And when unjust laws and unjust policies become the rule, should not the state (in Plato's words) "be overthrown"?

Most people quickly accept the idea of disobedience in a totalitarian society or in a blatantly undemocratic situation as in the American South with its racial segregation. But they look differently on breaking the law in a liberal society, where parties compete for the votes of citizens, where laws are passed by bodies of elected representatives, and where people have some opportunities for free expression of their ideas.[22]

What this argument misses is that civil disobedience gives an *intensity* to expression by its dramatic violation of law, which other means—voting, speaking, and writing—do not possess. If we are to avoid majority tyranny over oppressed minorities, we must give a dissident minority a way of expressing the fullness of its grievance.

The fiery editor of the abolitionist newspaper in Boston, William Lloyd Garrison, understood the need. Criticized by another antislavery person for his strong language ("I will not hesitate, I will not equivocate, I will not retreat a single inch, and I will be heard") and his dramatic actions (he set a copy of the United States Constitution afire at a public gathering, to call attention to the Constitution's support of slavery), Garrison replied, "Sir, slavery will not be overthrown without excitement, a most tremendous excitement."

Several of Garrison's contemporaries understood his role. One said that Garrison had roused the country from a sleep so deep "nothing but a rude and almost ruffian-like shake could rouse her."[23] Another said, "he will shake our nation to its center, but he will shake slavery out of it."[24]

Protest beyond the law is not a departure from democracy; it is absolutely essential to it. It is a corrective to the sluggishness of "the proper channels," a way of breaking through passages blocked by tradition and prejudice. It is disruptive and troublesome, but it is a necessary disruption, a healthy troublesomeness.

Disobedience and Foreign Policy

In a little book he wrote in the 1960s, Supreme Court justice Abe Fortas worried about all the civil disobedience taking place and spoke of "the all-important access to the ballot box."[25]

In later chapters I discuss the insufficiency of the ballot box to deal with racial discrimination or with economic justice. But probably the most clear-cut illustration of the inadequacy of that "all-important access to the ballot box" is in the area of foreign policy.

In foreign policy access to the ballot box means very little. Foreign policy is made by the president and a small circle of people around him, his appointed advisers. Again and again, Americans have voted for a president to keep them out of a war, only to see the "peace" candidate elected who then brings the nation into war.

Woodrow Wilson was elected in 1916 on a peace platform: "There is such a thing as a nation being too proud to fight." The next year he asked Congress to declare war. Franklin Roosevelt was elected in 1940 with a pledge to keep the United States out of the war, yet his policies were more and more designed to bring the United States into the war.

In 1964 the situation in Vietnam was tense. Lyndon Johnson ran for president on a platform opposing military intervention in Southeast Asia, while his opponent, Barry Goldwater, urged such action. The voters chose Johnson, but they got Goldwater's policy: escalation and intervention.

The Constitution says it is up to Congress to declare war. James Madison, who presided over the Constitutional Convention in 1787, explained the reasoning of the Founding Fathers in a letter to Thomas Jefferson written years later: "The constitution supposes, what the history of all Govts demonstrates, that the Executive is the branch of power most interested in war and most prone to it. It has accordingly with studied care vested the question of war in the legislature."[26]

However, again and again, the president has made the decision to go to war, and Congress has obsequiously gone along. In the two most recent American wars, the Korean War and the Vietnam War, Congress, while ignored, nevertheless appropriated the money asked by the president to carry on the war. When it comes to making war, we might just as well have a monarchy as a constitutional government.

It seems that the closer we get to matters of life and death—war and peace—the more undemocratic is our so-called democratic system.[27] Once the government, ignoring democratic procedures, gets the nation into war, it creates an atmosphere in which criticism of the war may be punished by imprisonment—as happened in the Civil War and in both world wars. Thus democracy gets a double defeat in matters of war and peace.

The Supreme Court itself, which (we were told back in junior-high-school civics class) is supposed to interpret the Constitution, presumably in the interests of democracy (checks and balances and all that) has interpreted it in such a way as to eliminate democracy in foreign policy. In a decision it made in 1936 (*U.S. v. Curtiss-Wright Export Corp.*), the Court gave the president total power over foreign policy, including the right to ignore the Constitution:

> The broad statement that the federal government can exercise no powers except those specifically enumerated in the Constitution, and such implied

> powers as are necessary and proper to carry into effect the enumerated powers, is categorically true only in respect of our internal affairs.[28]

This is a shocking statement to any American who learned in school that the powers of government are limited to what the Constitution allows. But that decision has never been overturned.[29] And all through the history of the United States we find Congress behaving like a flock of sheep when the president decides on war.

President Polk in 1846 (coveting California and other Mexican land) provoked a war with Mexico by sending troops into a disputed area. A battle took place, and when he asked Congress to declare war, they rushed to comply, the Senate spending just one day on debating the war resolution, the House of Representatives allowing two hours.

A century later in the summer of 1964 President Lyndon Johnson reported attacks on U.S. naval vessels off the coast of Vietnam in the Gulf of Tonkin. Congress took the president's account as truth (it turned out to be full of deceptions) and voted overwhelmingly (unanimously in the House, two dissenting votes in the Senate) to give the president blanket power to take whatever military action he wanted.

There was no declaration of war, as the Constitution required, but when citizens challenged this, the Supreme Court acted as timidly as Congress. The court never decided on the constitutionality of the Vietnam War. It would not even agree to discuss the issue.

For instance, in 1972 a man named Ernest Da Costa brought his case to the Supreme Court. He had been conscripted into the U.S. Army, but when ordered to go to Vietnam he refused, arguing that the American war in Vietnam had not been authorized by Congress, and, therefore, Congress could not draft him for overseas service. The Court refused even to hear his case. It takes the assent of four Supreme Court Justices to bring a case before the Court; only two wanted to hear Da Costa's argument.[30] The Supreme Court's claim was that such questions are "political"—meaning that they are too important to be decided by the nonelected Supreme Court and should be decided by the "political" branches of government, those subject to election, namely the president and Congress.

But we have seen that Congress has never had the boldness to challenge a president's call for war. So much for those checks and balances that, we learned in school, would save us from one-man rule. It turns out that the much-praised "proper channels" are not channels at all, but mazes, into which we are invited, like experimental animals, to get lost.

The concentration of dictatorial power in the hands of the president, in regard to military actions, was underlined when Secretary of State Dean Rusk testified before Congress in 1962. He was explaining the attempt to invade Cuba the year before, an action planned secretly by the CIA and the White House without the involvement of Congress. You shouldn't get upset over being ignored on this, Rusk assured Congress, because it's been done lots of times. He then gave them a list compiled by the State Department called "Instances of the Use of United States Armed Forces Abroad 1798–1945," describing 127 military actions by the United States, carried out by presidential order.[31] A small sample of that list includes (in the language of the State Department):

> 1852–53—Argentina—Marines were landed and maintained in Buenos Aires to protect American interests during a revolution.
>
> 1854—Nicaragua—San Juan del Norte [Greytown was destroyed to avenge an insult to the American Minister to Nicaragua].
>
> 1855—Uruguay—U.S. and European naval forces landed to protect American interests during an attempted revolution in Montevideo.

When U.S. troops were finally withdrawn from Vietnam in 1973, over 50,000 American men were dead after a war begun by the president, aided by a submissive Congress and a hands-off Supreme Court. Now Congress, mustering a bit of courage, passed a War Powers Act, intended to limit the power of the president in sending the American military into warlike situations. The act declared, among other provisions, "The President, in every possible instance, shall consult with Congress before introducing United States Armed Forces into hostilities or into situations where imminent involvement in hostilities is clearly indicated by the circumstances."

This War Powers Act has been ignored again and again, by various presidents. President Ford invaded a Cambodian island and bombed a Cambodian town in the spring of 1975 after the crew of an American merchant ship, the *Mayaguez,* was detained, but not harmed, by Cambodian authorities. According to the War Powers Act, Ford should have consulted with Congress. Senator Mike Mansfield, the Democratic leader of the Senate, said "I was not consulted, but notified after the fact."[32]

President Ronald Reagan in the fall of 1982 sent troops into a dangerous situation in Lebanon, again without following the requirements of the War Powers Act, and soon after that over 200 marines were killed in Lebanon by a bomb that exploded in their barracks. In the spring of 1983 Reagan sent U.S. forces to invade the Caribbean island of Grenada, again only notifying Congress, not consulting them. And in 1986 U.S. planes bombed the capital of Libya, again without consulting Congress. In 1989 President Bush launched an invasion of Panama (he called it Operation Just Cause), again without consulting Congress.

We have been speaking of *open* military actions undertaken by the president, uncontrolled by Congress. But the absence of democracy in foreign policy is even more obvious when you consider how much is done secretly by the president and his advisers, behind the backs of the American public, as well as behind the backs of their elected representatives.

The list of secret actions includes the CIA's overthrow of the government of Iran in 1953, restoring the Shah to the throne; the 1954 invasion of Guatemala and the ousting of its democratically elected president; the invasion of Cuba in 1961; and the wide range of covert operations in Indochina in the 1950s and 1960s, including the secret bombing of Cambodia. More recently, we find the series of attempts to overthrow the Sandinista government in Nicaragua by arming a counterrevolutionary force (the "contras") across the border in Honduras, and mining Nicaragua's harbors, as well as the secret transfer of arms to the contras in violation of a law passed by Congress.

When the "Iran-Contra" scandal became public in 1986–1987, President Reagan feigned innocence—the doctrine of "plausible denial" again. With astounding hypocrisy, Reagan said in his State of the Union Address at the beginning of 1987 (the bicentennial of the Constitution), "In those other constitutions, the government tells the people what they are allowed to do. In our Constitution, we the people tell the government what it can do and that it can do only those things listed in that document and no other."

These actions (the word *covert* is used officially, perhaps it sounds more respectable than secret) are fundamentally undemocratic; they take place behind the backs of the American people. The people who carry them out are, therefore, not accountable to any democratic process. The government has bypassed its own channels. For the citizens to stop this, civil disobedience may be needed.

Is Civil Disobedience Always Right?

There is a common argument against civil disobedience that goes like this: If I approve *your* act of civil disobedience, am I not honor bound to approve *anyone's* civil disobedience? If I approve Martin Luther King's violations of law, must I not also approve the Ku Klux Klan's illegal activities?

This argument comes from a mistaken idea about civil disobedience. The violation of law for the purpose of committing an injustice (like the Governor of Alabama preventing a black student from entering a public school or Colonel Oliver North buying arms for terrorists in Central America) is not defensible. Whether it was *legal* (as it was until 1954) or illegal (after 1954) to prevent black children from entering a school, it would still be wrong. The test of justification for an act is not its legality but its morality.

The principle I am suggesting for civil disobedience is not that we must tolerate all disobedience to law, but that we refuse an absolute *obedience* to law. The ultimate test is not law, but justice.

This troubles many people, because it gives them a heavy responsibility, to weigh social acts by their moral consequences. This can get complicated and requires a never-ending set of judgments about practices and policies. It is much easier to lie back and let the law make our moral judgments for us, whatever the law happens to say at the moment, whatever politicians have made into law on the basis of *their* interests, however the Supreme Court interprets the law at the moment. Yes, easier. But recall Jefferson's words: "Eternal vigilance is the price of liberty."

There is fear that this kind of citizens' judgment about when to obey and when to disobey the law will lead to terrible consequences. In the summer of 1968 four people who called for resistance to the draft as a way of halting the war in Vietnam—Dr. Benjamin Spock, Reverend William Sloane Coffin, writer Mitchell Goodman, and Harvard student Michael Ferber—were sentenced to prison by Judge Francis Ford in Boston, who said, "Where law and order stops, obviously anarchy begins."[33]

That is the same basically conservative impulse that once saw minimum wage laws as leading to Bolshevism, or bus desegregation leading to intermarriage, or communism in Vietnam leading to world communism. It assumes that all actions in a given direction rush toward the extreme, as if all social change takes place at the top of a steep, smooth hill, where the first push ensures a plunge to the bottom.

In fact an act of civil disobedience, like any move for reform, is more like the first push *up* a hill. Society's tendency is to maintain what has been. Rebellion is only an occasional reaction to suffering in human history; we have infinitely more instances of submission to authority than we have examples of revolt. What we should be most concerned about is not some natural tendency toward violent uprising, but rather the inclination of people faced with an overwhelming environment of injustice to submit to it.

Historically, the most terrible things—war, genocide, and slavery—have resulted not from disobedience, but from obedience.

Vietnam and Obedience

There are rare moments in the history of nations when citizens, their indignation overflowing, begin to refuse obedience to the authorities. Such a moment in the history of the United States was the war in Vietnam. When Americans saw their nation, which they had been taught to believe was civilized and humane, killing Vietnamese peasants with napalm, fragmentation bombs, and other horrible instruments of modern war, they refused to stay inside the polite and accepted channels of expression.

Most of the actions taken against the war were not acts of civil disobedience. They were not illegal, but extra legal—outside the regular procedures of government: rallies, petitions, picketing, and lobbying. A national network of educational activities spontaneously grew: alternative newspapers, campus teach-ins, church gatherings, and community meetings.

When the supposed clash between U.S. naval vessels and North Vietnamese patrol boats took place in the Gulf of Tonkin during the summer of 1964, I was teaching in a Freedom School in Jackson, Mississippi. In August the bodies of three missing civil rights workers, shot to death, were found near Philadelphia, Mississippi, and many of us working in the movement drove up to attend a memorial meeting held outdoors not far from where they had been killed.

At the meeting, one of the organizers of the Mississippi movement, Bob Moses, stood up to speak. He held aloft the morning newspaper from Jackson. The headline was "LBJ Says Shoot to Kill in Gulf of

Tonkin." Moses spoke with a quiet bitterness (this is a rough recollection of his words): "The president wants to send soldiers to kill people on the other side of the world, people we know nothing about, while here in Mississippi he refuses to send anyone to protect black people against murderous violence."

That fall as the U.S. involvement in Vietnam began to grow, I was starting to teach at Boston University and became immediately involved in the movement against the war. It was at first a puny movement, which seemed to have no hope of prevailing against the enormous power of the government. But as the war in Vietnam became more vicious and as it became clear that noncombatants were being killed in large numbers; that the Saigon government was corrupt, unpopular, and under the control of our own government; and that the American public was being told lies about the war by our highest officials, the movement grew with amazing speed.

In the spring of 1965 I and some others spoke against the war on the Boston Common to perhaps a hundred people. In October 1969 when antiwar meetings took place in hundreds of towns and cities around the country, there was another rally on the Boston Common, and 100,000 people were there. As the American involvement escalated—to 500,000 troops, to millions of tons of bombs dropped—the antiwar movement also escalated.

Young black civil rights workers connected with Student Nonviolent Coordinating Committee (SNCC) were among the first to resist the war. In mid-1965 in McComb, Mississippi, young blacks who had just learned that a classmate of theirs was killed in Vietnam distributed a leaflet:

> No Mississippi Negroes should be fighting in Viet Nam for the White man's freedom, until all the Negro people are free in Mississippi.
>
> Negro boys should not honor the draft here in Mississippi. Mothers should encourage their sons not to go.[34]

In the summer of 1966 six young black men, members of SNCC, invaded an induction center to protest the war. They were arrested and sentenced to prison. Julian Bond, another SNCC member, who had just been elected to the Georgia House of Representatives, spoke out against the war and the draft, and the House voted that he not be seated. (The Supreme Court later restored his seat, saying his First Amendment right to free speech had been violated.)

Martin Luther King, Jr., spoke out publicly against the war, ignoring the advice of some other civil rights leaders, who feared that criticism might weaken Johnson's program of domestic reform. King refused to be silenced:

> Somehow this madness must cease. We must stop now. I speak as a child of God and brother to the suffering poor of Vietnam. I speak for those whose land is being laid waste, whose homes are being destroyed, whose culture is being subverted. I speak for the poor of America who are paying the double price of smashed hopes at home and death and corruption in Vietnam. I speak as a citizen of the world as it stands aghast at the path we have taken. I speak as an American to the leaders of my own nation. The great initiative in this war is ours. The initiative to stop it must be ours.[35]

Young men began to refuse to register for the draft or to refuse induction if called. Students signed petitions headed We Won't Go. Over a half million men resisted the draft. About 200,000 were prosecuted, 3,000 became fugitives. There were too many cases to pursue and most were dropped. Finally, 8,750 men were convicted of draft evasion.[36]

A student of mine, Philip Supina, wrote to his draft board in Tucson, Arizona, on May 1, 1968: "I am enclosing the order for me to report for my pre-induction physical exam for the armed forces. I have absolutely no intention to report for that exam, or for induction, or to aid in any way the American war effort against the people of Vietnam."[37] He was sentenced to four years in prison.

In previous wars, there had been opposition within the armed forces, but the Vietnam War produced open protests and silent desertions on a scale never seen before. As early as June 1965, West Point graduate Richard Steinke refused to board an aircraft taking him to a remote Vietnamese village. He said, "The Vietnamese war is not worth a single American life."

There were many individual acts of disobedience. A black private in Oakland refused to board a troop plane to Vietnam. A navy nurse was court-martialed for marching in a peace demonstration while in uniform and for dropping antiwar leaflets from a plane onto navy installations. In Norfolk, Virginia, a sailor refused to train fighter pilots because he thought the war was immoral. An army lieutenant was arrested in Washington, D.C., in early 1968 for picketing the White House with a sign that said "120,000 American casualties—Why?" Two black marines were given prison sentences of six and ten years, respectively, for talking to other black marines against the war.

Desertions from the armed forces multiplied. We can't be sure of the exact number, but there may have been 100,000. Thousands went to Western Europe—France, Sweden, and Holland. Most deserters crossed the border into Canada; 34,000 were court-martialed and imprisoned. There were over a half million less-than-honorable discharges.[38]

The GI movement against the war became organized. Antiwar coffeehouses were set up near military bases around the country, where GIs could come to meet others who were opposed to what was going on in Vietnam. Underground newspapers sprang up at military bases across the country—fifty of them by 1970. These newspapers printed antiwar articles, gave news about the harassment of GIs, and gave practical advice on the legal rights of people in the military.

The dissidence spread to the war front itself. When antiwar demonstrations were taking place in October 1969 all over the United States, some GIs in Vietnam wore arm bands to show their support. One soldier stationed at Cu Chi wrote to a friend on October 26, 1970, that separate companies had been set up for men refusing to go into the field to fight. He said, "It's no big thing here anymore to refuse to go." A news dispatch in April 1972 reported that 50 infantrymen of a company of 142 refused for an hour and a half to go out on patrol round Phu Bai. They shouted, "We're not going! This isn't our war." Others commented, "Why the hell are we fighting for something we don't believe in?"[39] One army sergeant, captured by the Vietnamese, told later about his march to the prisoner-of-war camp, "Until we got to the first camp, we didn't see a village intact; they were all destroyed. I sat down and put myself in the middle and asked myself: Is this right or wrong? Is it right to destroy villages? Is it right to kill people en masse? After a while it just got to me."

The French newspaper *Le Monde* reported that in four months, 109 soldiers of the first air cavalry division were charged with refusal to fight. "A common sight," the correspondent for *Le Monde* wrote, "is the black soldier, with his left fist clenched in defiance of a war he has never considered his own."

In the summer of 1970, 28 commissioned officers of the military, including some veterans of Vietnam, said they represented about 250 other officers and announced the formation of the Concerned Officers Movement Against the War. In mid-1973 it was reported there were drop-outs among West Point cadets. A reporter wrote that West Point officials attributed this to "an affluent, less disciplined, skeptical and questioning generation and to the anti-military mood that a small radical minority and the Vietnam war had created."[40]

There is probably no more disciplined, obedient, highly trained element of the armed forces than the fliers of the air force. But when the ferocious bombings of civilians in Hanoi and Haiphong was ordered by the Nixon administration around Christmas 1972, several B-52 pilots refused to fly.

The massive civil disobedience against the Vietnam War—by men in the military, by draftees, and by civilians—cannot be justified simply because it was civil disobedience, but because it was disobedience on behalf of a human right—the right of millions of people in Vietnam not to be killed because the United States saw in Southeast Asia (as President John F. Kennedy put it), "an important piece of real estate."

Actions outside the law or against the law must be judged by their human consequences. That is why the civil disobedience of Colonel Oliver North, illegally sending military aid to the terrorist contras in Central America who committed acts of terrorism against Nicaraguan farmers cannot be justified. But the civil disobedience of those who wanted to stop the killing in Vietnam was necessary and right.

The congressional committee that interrogated Oliver North in 1987 as part of the Iran-Contra hearings did not ask him about the innocent people killed in Nicaragua because of what he had done. They concentrated, as the American court system generally does, on the technical question of *whether* he had violated the law, not on the more important question: for what *purpose* did he violate the law.

It is interesting to note that North did not hold to the rule of law over the rule of men. He was willing to break the law to obey the president. He told the hearing committee, "And if the Commander-in-Chief tells this Lieutenant Colonel to go sit in the corner and stand on his head I will do so."

Justice in the Courts

Those who run the legal system in the United States do not want the public to accept the idea of civil disobedience—even though it rests on the Declaration of Independence, even though it has the approval of some of the great minds of human history, even though some of the great achievements for equality and liberty in the United States have been the result of movements outside of and against the law. They are afraid that the idea will take hold, and they are right, because the commonsense belief of most people, I think, is that justice is more important than law.

During the Vietnam War, not long after I got back from Hanoi, where I had visited villages devastated by American bombs, I was asked to testify at a trial in Milwaukee. Fourteen people, many of them Catholic priests and nuns, had invaded a draft board and destroyed documents to protest the war.

I was to testify as a so-called expert witness, to tell the judge and jury about the history of civil disobedience in the United States, to show its honorable roots in the American Revolution, and its achievements for economic justice and for racial equality.

I started out talking about the Declaration of Independence and then about Thoreau's civil disobedience, and then gave a brief history of civil disobedience in the United States. The judge pounded his gavel and said, "Stop. You can't discuss that. This is getting to the heart of the matter."

The defense attorney asked me, "What is the difference between law and justice?" The prosecution objected, and the judge said, "Sustained." More questions about civil disobedience. More objections, all sustained. I turned to the judge (something a witness is not supposed to do) and asked, in a voice loud enough for the courtroom to hear, "Why can't I say something important? Why can't the jury hear something important?"

The judge was angry. He replied, "You are not permitted to speak out like that. If you do that once more I will have you put in jail for contempt of court." Later I felt I should have been more courageous and joined my act of civil disobedience to that of the defendants.

What the judge wanted to hear about in his courtroom was merely the technical violations of law committed by the defendants—breaking and entering, destroying government documents, and trespassing. "This is a case about arson and theft." He did not want to hear *why* these usually upright and law-abiding citizens were breaking the law. He did not want to hear about the war in Vietnam. He did not want to hear about the tradition of civil disobedience.

To have the mechanical requirements of "due process"—a trial, contending arguments, and decision by a jury of citizens—is insufficient if the arguments are not fully made, if the jury does not know what is at stake, and if it cannot make a decision on the *justice* of the defendants' action, regardless of legality. Supposedly, it is the judge who sees to it that the law is made clear to the jury, but then it is up to the jury to see that justice is done. However, if the judge prevents the jury from hearing testimony about the issues, the jury is being compelled to stay within the narrow, technical confines of the law, and the democratic purpose of a jury trial is extinguished.

The courtroom, one of the supposed bastions of democracy, is essentially a tyranny. The judge is monarch. He is in control of the evidence, the witnesses, the questions, and the interpretation of law. In the mid-1980s I was called as a witness by some people in Providence, Rhode Island, who had done some small symbolic damage at the launching of a nuclear-armed submarine, in protest against the huge expenditure of money for deadly weapons and the escalation of the arms race. I was to tell the jury about the importance of civil disobedience for American democracy.

The judge would not let me speak. From the very first question—"Can you tell us about the history of civil disobedience in the United States?"—as I began to answer, the judge stopped me. "Objection sustained," he said loudly. I had not heard any objection from the prosecuting attorney.

Indeed, at this point the prosecuting attorney, a young man, spoke up, "Your honor, I did not object."

"Well," said the judge, "why didn't you?"

"Because," the prosecutor said, "I thought the question was relevant."

"I disagree," the judge said, with finality.

I was not able to say anything to the jury. It was clear that the judge was furious at these antimilitary protesters and was determined to send them to prison. They were facing a felony charge, calling for ten years in prison, and a misdemeanor, calling for one year in prison. The prosecutor, obviously not convinced that these defendants were dangerous criminals, perhaps a bit sympathetic to their cause, dropped the felony charge, telling the defendants, confidentially, that he did that because he was sure the judge would give the defendants the full ten-year sentence.

The quality of justice in the United States is strained through the sieve of the power and prejudice of judges. Free speech in the courtroom does not exist, because the judge decides what can and cannot be said. In 1980 a New York City judge dropped a case against fifteen people who protested at a research facility for nuclear weapons on the advice of the prosecutor, who told him, "We want to prevent these defendants from using the Criminal Court as a forum for their views."[41]

Judges are, for the most part, creatures of comfort—that is, they come from the affluent classes and tend to be conservative and hostile toward radicals, demonstrators, protesters, and violators of "law and order." They are also creatures of the American environment, subject to the dominant ideology.

But when the national mood changes, when the political atmosphere becomes differently charged, judges may be affected by that. If they then allow juries to hear the reasons why protesters acted, the common sense of juries comes into play. They may vote to acquit the defendants even if they have broken the law. Given the opportunity, when not bullied by judges, juries may choose justice before law.

By 1967 there was a formidable movement all across the country against the war in Vietnam. In Oakland, California, demonstrations that disrupted the normal operations of the Induction Center resulted in the prosecution of the Oakland Seven, charged with conspiracy to trespass, create a public nuisance and resist arrest. The judge permitted the defendants to tell the jury about their belief in the illegality of the war and told the jury they should take that belief into consideration in determining whether there was criminal intent in the defendants' actions. The jury acquitted all of the Seven. One of the jury members said later, "I'm not a puppet. I'm a free thinker."[42]

I was in Camden, New Jersey, in 1973, having been asked to testify in the trial of "The Camden Twenty-eight." The twenty-eight men and women, mostly young and from the Philadelphia area, had broken into a draft board in the night for the purpose of doing away with some of the files as their protest against the Vietnam War. There was a police informer in their midst and they were caught before they got to the files, so here they were in federal court, charged with various counts of trespassing, breaking, and entering and there was no question about the fact that they had violated several laws.

Similar cases of draft board protests had resulted, again and again, in verdicts of guilty. Six years earlier (1967) the "Baltimore Four" were convicted and Philip Berrigan had been sentenced to six years in prison.

But by 1973 public opinion had shifted sharply against the war, and the United States had reached an agreement with North Vietnam to withdraw American troops. The judge in Camden, a conservative man, treated the defendants with respect. They were representing themselves in court, with the advice of several "movement lawyers" from Philadelphia. The judge allowed them to bring whatever witnesses they wanted, to tell the jury whatever they wanted to say about the war.

I had recently returned from testifying in the Pentagon Papers case on the West Coast, where I had told the jury what was in the Pentagon Papers—in effect, giving them a two-hour lecture on the history of American intervention in Vietnam. The judge in Camden allowed me to go into that history. He also allowed the defendants to speak from their hearts to the jury about what had impelled them to do what they did.

The jury returned a verdict of acquittal. The woman who headed the jury then threw a party for the defendants. Juror Samuel Braithwaite, a fifty-three-year-old black taxi driver from Atlantic City who had spent eleven years in the army, wrote a letter to the defendants after the verdict was in:

> To you, the clerical physicians with your God-given talent, I say, well done. Well done for trying to heal the sick irresponsible men, men who were chosen by the people to govern and lead them. These men, who failed the people, by raining death and destruction on a hapless country. . . . You went out to do your part while your brothers remained in their ivory towers watching . . . and hopefully some day in the near future, peace and harmony may reign to people of all nations.[43]

Jury Nullification

The Camden jury had exercised a right that judges never tell juries about: the right to come to a verdict following their conscience rather than the strict requirements of the law—to choose justice over law.

That right of "jury nullification" goes back to eighteenth-century Britain, when jurors, despite being fined and jailed, refused to convict two Englishmen for speaking to a street crowd. A plaque in the famous Old Bailey courthouse in London commemorates the courage of these jurors and records the final opinion of the Chief Justice, "which established the Right of Juries to give their Verdict according to their conviction."[44]

In America the principle of jury nullification was affirmed in 1735 when John Peter Zenger, a New York printer who was charged with seditious libel for printing material not authorized by the British mayor, was acquitted by a jury that ignored the instructions of the judge. The jury apparently followed the advice of the defense attorney to "see with their own eyes, to hear with their own ears and to make use of their consciences."

The antislavery preacher Theodore Parker, after the passage of the Fugitive Slave Act of 1850, spoke in New England about what he would do if a slave escaped from South Carolina to Massachusetts and "a Mr. Greatheart" helped her to escape, harbored and concealed her, and was then prosecuted, and he, Parker, was on the jury. He declared:

> I may take the juror's oath to give a verdict according to the law and the testimony. The law is plain, let us suppose and the testimony conclusive. . . . If I have extinguished my manhood by my juror's oath, then I shall do my official business and find Greatheart guilty, and I shall seem to be a true man; but if I value my manhood, I shall answer after my natural duty to love a man and not hate him, to do him justice, not injustice, to allow him the natural rights he has not alienated, and shall say, "Not guilty."[45]

Around the middle of the nineteenth century, however, the courts began to rule that juries did not have the right to decide the law, only the facts, that they had to obey the judge's instructions as to the law. This does not really settle the matter. The jury may not have the right to rule on questions of law, but they don't have to write legal opinions when they give their verdict; they can vote their consciences, regardless of the law explained to them by the judge. A distinguished legal scholar, Wigmore, wrote in 1929 about the importance of jury nullification to achieve justice.

> Law and Justice are from time to time inevitably in conflict. That is because law is a general rule; . . . while justice is the fairness of this precise case under all its circumstances. . . . The jury, in the privacy of its retirement, adjusts the general rule of law to the justice of the particular case. . . . The jury, and the secrecy of the jury room, are the indispensable elements in popular justice.[46]

Another famous legal scholar, Roscoe Pound, had written back in 1910 that "jury lawlessness is the great corrective" in legal proceedings.[47]

In other words, the jury must match the defendants' civil disobedience with its own disobedience of law, if, as a matter of conscience, it believes the defendants did the right thing. When it is submissive before the overbearing authority of a judge, it surrenders its own conscience.

In the case of Dr. Spock and his other antiwar defendants who were found guilty by the jury, one of the jury members said later, "I was in full agreement with the defendants until we were charged by the judge. That was the kiss of death!"[48]

Another juror in the *Spock* case, Frank Tarbi, wrote in the *Boston Globe* about his anguish:

> How and why did I find four men guilty? All men of courage and individuals whom I grew to admire as the trial developed. . . . As the father of three teen-aged sons, two eligible for draft, and a veteran myself, my abhorrence of war is understandable. . . . Was I ready to commit my sons? . . . Rev. Coffin's thought-provoking argument struck home—"Isn't the Cross higher than the flag? Must we not obey God before we obey man?" . . . The paradox was that I agreed wholeheartedly with these defendants, but . . . I felt that technically they did break the law. . . .
>
> I departed to the waiting car and then to home. There I was embraced by my loved ones and I began to think and try to explain. . . . These four men were trying to save my sons whom I love dearly. Yet I found them guilty. To hell with my ulcer. After four or five stiff hookers (I lost count) I began to cry bitterly.[49]

In the case of the Catonsville Nine draft board invaders, the Circuit Court of Appeals, while affirming their convictions, made a remarkable statement in support of jury nullification:

> We recognize . . . the undisputed power of the jury to acquit, even if its verdict is contrary to the law as given by the judge and contrary to the evidence. . . . If the jury feels that the law under which the defendant is accused is unjust, or that exigent circumstances justified the actions of the accused, or for any reason which appeals to their logic or passion, the jury has the power to acquit, and the courts must abide by that decision.[50]

Nevertheless, it is always a struggle in the courtroom to get the judge to agree to admit into evidence those things that will allow the jury to vote its conscience. In the period since the Vietnam War, political protesters against the arms race, or against military intervention in Central America, have tried to introduce the defense of "necessity," or "justification." This defense is based on the idea that while a technical violation of law has taken place, it was necessary to prevent a greater harm to the community.

In 1980 the "Plowshares Eight" invaded a General Electric plant in King of Prussia, Pennsylvania, and did some minor damage to nuclear nose cones, as a protest against the arms race. They were charged with trespassing and destroying property. The judge would not allow a necessity defense, and when the jury was out for eight hours, the judge speeded up their decision by threatening to sequester them overnight. The jury then came in with a verdict of guilty. Juror Michael de Rosa said later, "I didn't think they really went to commit a crime. They went to protest. . . . We really didn't want to convict them on anything. But we had to because of the way the judge said the only thing you can use is what you get under the law."

When juries have been allowed to hear the evidence of "necessity," the results may be startling. In Burlington, Vermont, in 1984 "The Winooski Forty-four" were arrested for refusing to leave the hallway outside of a senator's office. They were protesting his votes to give arms to the contras across the Nicaraguan border. The judge accepted the defendants' right to a necessity defense. He allowed them to call various expert witnesses: a refugee from Central America, who told the jury about the terror caused by American military intervention; a former leader of the contras, who explained that he had left their ranks after he realized they were organized and financed by the CIA and were committing atrocities against the people of Nicaragua; and I testified about the history of civil disobedience in the United States and its usefulness in bringing about healthy social change.

The prosecuting attorney told the jury to disregard all that testimony. He pointed to a large chart on a stand facing the jury—one of the exhibits, which was a map of the senator's offices where the defendants had crowded into the corridor and refused to leave. He said, "The issue is not Nicaragua, not American foreign policy. *This* is the issue—trespassing."

When he had finished, a woman lawyer for the defendants rose for her summation. She walked over to the chart of the senator's office and folded it back, to reveal something underneath—a large map of Central America. She pointed and said, "*This* is the issue." The jury voted to acquit.

At another trial shortly after, in western Massachusetts, a number of people (including activist Abbie Hoffman and Amy Carter, daughter of an ex-president) were charged with blocking recruiters for the Central Intelligence Agency who had shown up at the University of Massachusetts in Amherst. Witnesses were called, including ex-CIA agents who told the jury that the CIA had engaged in illegal and murderous activities all around the world. The jury listened and voted to acquit.

One juror, a hospital worker named Ann Gaffney, said later, "I was not that familiar with the CIA's activities. I was surprised. I was shocked. . . . I was kind of proud of the students." Another juror, Donna Moody, said, "All the expert testimony against the CIA was alarming. It was very educational." The county district attorney himself, Michael Ryan, had this reaction: "If there is a message, it was that this jury was composed of middle America. . . . Middle America doesn't want the C.I.A. doing what they are doing."[51]

In this case the judge allowed the defense of necessity and gave the green light to the jury in considering human rights more important than a technical violation of law. But the courts will continue to remain barricades against change, stiff upholders of the prevailing order, unless juries defy conservative judges and vote their consciences, commit their own civil disobedience in the courtroom, and ignore the law to achieve justice.

Or perhaps we should say "ignore man-made law, the law of the politicians" to obey the higher law—what Reverend Coffin and Father Berrigan would call "the law of God" and what others might call the law of human rights, the principles of peace, freedom, and justice. (Daniel Berrigan's elderly mother was asked by a reporter, when Dan went underground, how she felt about her son defying the law; she responded quietly, "It's not God's law.")[52]

The truth is so often the total reverse of what has been told us by our culture that we cannot turn our heads far enough around to see it. Surely, it is *obedience* to governments, in their appeals to patriotism, their calls for war, that is responsible for the terrible violence of our century. The disobedience of conscientious citizens, for the most part nonviolent, has been directed to stopping the violence of war. The psychologist Erich Fromm, thinking about nuclear war, once referred to the biblical Genesis of the human race and the bite into the forbidden apple: "Human history began with an act of disobedience and it is not unlikely that it will be terminated by an act of obedience."

Violence

It should be stressed that where protesters, rebels, and radicals have gone outside the law, they have been, for the most part nonviolent. Where they have been guilty of "violence," it is usually violence to property and not to human beings.

The issue came up in a 1985 trial in Missouri of a group of people who had gone into a nuclear missile silo and did some minor damage to (as the judge described it) "the concrete, the handle of the access hatch, the antenna and transmission boxes." This exchange took place between Judge Hunter and defendant Martin Holladay:

> JUDGE: I don't agree with your definition of nonviolence. Violence includes injury to property.
>
> HOLLADAY: The question also is—can a nuclear weapon be considered the same kind of property as a desk, or a stove? As long as this country sees nuclear weapons as property to defend and protect, more sacred than the lives they will destroy—what is proper property? The gas ovens in Germany?

Holladay was sentenced to eight years in prison for doing "violence" to the most atrocious instruments of violence ever developed. It is a part of the dominant ideology of our culture to treat damage to property—especially certain kinds of property—as terrible crimes of violence because they have been committed illegally by private citizens protesting government policy, while accepting large-scale murder because it is legal and official.

In 1974 I was asked to write an article on violence for Scribner's *Dictionary of American History.* In my six-page article, I began by defining violence as "that which inflicts injury or death on human beings." I said that damage to property "is excluded here as less worthy of concern among people who claim to put supreme value on human life and health." I also said that violence by individuals and groups in American history had received much attention, but that "the greatest amount of violence by far has been done by government itself, through armies and police force, while expanding across the continent, extending national power overseas, and suppressing rebellion and protest at home and abroad."

The editors left all that in, despite my unorthodox point of view, but there was one paragraph in my article that they omitted completely:

> It should be kept in mind that our definition omits an enormous amount of damage—physical and mental—caused by industrial and highway accidents, by economic exploitation, racial humiliation, and imprisonment, and by

> those conditions of poor housing, health and sanitation which cause infant mortality, malnutrition, sickness, and early death. For instance, "black lung" disease among miners, and the inhalation of deadly fibers among asbestos workers, have caused untold death and suffering. Thus, any moral assessment of the violence caused by race and class rebellion must weigh that against the wrongs of everyday life for millions of people—conditions which injure and kill but are not usually defined as violence.

I had already stretched the editors' tolerance to its limits, I realized. That paragraph was going too far.

The movement against the Vietnam War reveals the double standard of government, treating the burning of pieces of paper (draft cards and draft records) as violent acts, while dropping 7 million tons of bombs on Southeast Asia (twice as many as were dropped in all theaters of operation in World War II).

It was a remarkably nonviolent movement. There was one instance, so rare that it must be noted, where antiwar protesters in Madison, Wisconsin, planted a bomb in a military research building, timed to go off in the middle of the night, when no one would be in the building. But one man was working there, and he was killed.

The movement, while sometimes involving illegal acts of civil disobedience, mostly consisted of *extralegal* actions, that is, actions done outside the regular channels of government, aimed directly at informing and arousing the public. Here is a short list that suggests the variety of actions.

1. Eight hundred Peace Corps volunteers protested the war to President Johnson.
2. Thousands of people refused to pay taxes.
3. Hunger strikes for peace.
4. Three men and two women, calling themselves "The East Coast Conspiracy to Save Lives," sabotaged rail equipment near a factory making bomb casings for Vietnam.
5. In the Pacific Ocean, two young seamen hijacked an American munitions ship carrying bombs into the war zone and diverted it to neutral Cambodia.
6. Various demonstrations at college commencements. At Brown University, two-thirds of the graduates turned their backs on Henry Kissinger during the 1969 commencement ceremony.
7. Fifty writers and publishers, at a National Book Award ceremony,

walked out on a speech by Vice President Hubert Humphrey. One of them, a novelist, called out, "Mr. Vice President, we are burning children in Vietnam, and you and we are all responsible."

8. Distinguished writers, invited to the White House, refused to go: the poet Robert Lowell and the playwright Arthur Miller. On the other hand, the singer Eartha Kitt attended a lawn party given by Mrs. Johnson at the White House and used the occasion to make a public statement against the war.
9. Young Americans in London crashed a party at the American ambassador's elegant Fourth of July reception, called for everyone's attention, and proposed a toast: "To the dead and dying of Vietnam."
10. Teenagers called to the White House to accept 4-H Club prizes shook hands with the president and asked him to stop the war.

Does Protest Matter?

It is not easy to prove that protest changes policy. But in the case of the Vietnam War, there is powerful evidence. In the government's own top-secret documents, the "Pentagon Papers," we find anxious government memos about "public opinion . . . increasing pressure to stop the bombing . . . the breadth and intensity of public unrest and dissatisfaction with the war . . . especially with young people, the underprivileged, the intelligentsia and the women . . . a limit beyond which many Americans and much of the world will not permit the United States to go."

And in the spring of 1968, with over half a million troops in Vietnam and General Westmoreland asking President Johnson for 200,000 more, he was advised by a small study group in the Pentagon not to escalate the war further. There would be more U.S. casualties, the group said, more taxes needed. And

> the growing disaffection accompanied as it certainly will be, by increased defiance of the draft and growing unrest in the cities because of the belief that we are neglecting domestic problems, runs great risks of provoking a domestic crisis of unprecedented proportions.[53]

Johnson, right after this report, refused Westmoreland's request, announced a limitation on the bombing of North Vietnam, and agreed to go to the peace table in Paris to negotiate with the North Vietnamese.

Even President Nixon, who had said of the growing antiwar activity that "under no circumstance will I be affected whatever by it," confessed in his memoirs, nine years later,

> Although publicly I continued to ignore the raging antiwar controversy, . . . I knew, however, that after all the protests and the Moratorium [the nationwide protests of October 1969], American public opinion would be seriously divided by any military escalation of the war.[54]

Thoreau, Jefferson, and Tolstoy

The great artists and writers of the world, from Sophocles in the fifth century B.C. to Tolstoy in the modern era, have understood the difference between law and justice. They have known that, just as imagination is necessary to go outside the traditional boundaries to find and to create beauty and to touch human sensibility, so it is necessary to go outside the rules and regulations of the state to achieve happiness for oneself and others.

Henry David Thoreau, in his famous essay "Civil Disobedience," wrote,

> A common and natural result of an undue respect for law is, that you may see a file of soldiers, colonels, captains, corporals, privates, powder-monkeys, and all, marching in admirable order over hill and dale to the wars, against their wills, ay, against their common sense and consciences, which makes it very steep marching indeed, and produces a palpitation of the heart."

When farmers rebelled in western Massachusetts in 1786 (Shays' Rebellion), Thomas Jefferson was not sympathetic to their action. But he hoped the government would pardon them. He wrote to Abigail Adams:

> The spirit of resistance to government is so valuable on certain occasions that I wish it to be always kept alive. It will often be exercised when wrong, but better so than not to be exercised at all. I like a little rebellion now and then. It is like a storm in the atmosphere.[55]

What kind of person can we admire, can we ask young people of the next generation to emulate—the strict follower of law or the dissident who struggles, sometimes within, sometimes outside, sometimes against the law, but always for justice? What life is best worth living—the life of the proper, obedient, dutiful follower of law and order or the life of the independent thinker, the rebel?

Leo Tolstoy, in his story "The Death of Ivan Ilyich," tells of a proper, successful magistrate, who on his deathbed wonders why he suddenly feels that his life has been horrible and senseless. " 'Maybe I did not live as I ought to have done. . . . But how can that be, when I did everything properly?' . . . and he remembered all the legality, correctitude and propriety of his life."

CHAPTER

SEVEN

Economic Justice: The American Class System

In the summer of 1989, on the twentieth anniversary of the first human landing on the moon, I listened to a television discussion on space exploration. I heard a black woman poet, Maya Angelou, struggle, politely but with obvious frustration, against three famous male writers who spoke enthusiastically about spending more billions to send men to the moon and to Mars.

Against them, she seemed to be climbing the steepest mountain. She kept saying, Yes, I am excited, too, about exploring space, but where will we get money to help the poor people, black and white and Asian, here at home? The three men were perplexed by her stubborn refusal to join all the self-congratulation about the conquest of space.

In the summer of 1969 as preparations were made for landing men on the moon, a *New York Times* reporter wrote from Florida:

> Within the shadow of the John F. Kennedy Space Center, the hungry people sit and watch. . . .
>
> They sit and watch the early morning crush of cars filled with engineers and technicians move toward 'the Cape,' 18 miles north, in the feverish days before the moon launching on Wednesday morning.

> "The irony is so apparent here," said Dr. Henry Jerkins, the county's only Negro doctor. "We're spending all this money to go to the moon and here, right here in Brevard, I treat malnourished children with prominent ribs and pot bellies."[1]

In 1987 a full-page advertisement for Tiffany's, the famous jewelry store, appeared in the *New York Times,* with a photo of "the definitive sports watch in eighteen karat gold. Men's, $9,800. Women's, $7,800." Several months before, the *Times* carried a story, datelined East Hartford, Connecticut, with the following lead paragraph: "A 28-year-old man, described as despondent after a long period of unemployment, shot and killed his three young children today, then committed suicide with the same pistol."[2]

The unemployed man from East Hartford was not an oddity. In the year 1987 (according to a report of the Ways and Means Committee of the House of Representatives) one-fifth of the population in the United States—more than 50 million people—lived in families whose annual income averaged $5,000 a year.

The success and failure of the United States of America lies in those stories. Staggering technological advance alongside poverty and hunger. A class of extremely rich people; another class of quite prosperous people (but nervous about the security of their situation); another class of men, women, and children living in desperation and misery within sight of colossal wealth. Who could be surprised that crime, violence, and drug addiction would accompany such contrasts? Or that psychic disorder, broken families, and alcoholism would accompany such insecurity?

We have a class system, unmistakably, in a country that promises "liberty and justice for all." Where is the *justice* of a society that has such extremes of luxury for some, misery for others? Or does the middle-class comfort with which most of us live in the United States prevent us from asking that question with genuine indignation?

Part of me holds on to the class anger that grew in me as a teenager when, watching the belongings of hardworking people put out on the street because they could not pay their rent, their eviction overseen by club-carrying policemen, I became conscious that something was terribly wrong.

My father had a fourth-grade education, my mother got as far as the seventh grade. They both worked very hard, but we lived in dirty buildings, roach-infested cold rooms, with no refrigerator, no shower, no phone. I would come back from school in the winter's early dusk and find the house dark, no electricity, and the gas oven not working because the bills had not been paid. I did not believe it was the work of God, and after some thought I concluded that it was not an accident; it was systematic, recurring, man-made, and approved by the law.

The anger melts repeatedly as I enjoy the good things that this rich country supplies to two-thirds of its population, but the anger returns when I read that the U.S. Defense Department proposes to spend $70 *billion* for still another war plane (a moral monster, called the Stealth Bomber) while the government cuts subsidies for public housing and 2 million Americans, including hundreds of thousands of children, have no place to live.

During the Reagan administration of the 1980s, the country's rich became richer and the poor, poorer. Reagan's attorney-general, Edwin Meese, said cheerfully he was not aware of people being hungry. Around the time he was saying this a Physicians Task Force reported that 15 million American families had an income of under $10,000 a year, received no food stamps, and were chronically unable to get adequate food.[3] A report by the Harvard School of Public Health in 1984 said that its researchers found that over 30,000 people had to beg for food to avoid starvation.

Since the end of World War II there has been a fanatic, almost insane willingness to spend billions on weapons, while millions of American families lack the basic necessities of life. The following story appeared in the *New York Times* in the summer of 1984:

> An investigation of the Navy's newest and most technically advanced cruiser by the staff of the House Appropriations Committee has found the ship overweight, sluggish, and in possible danger of capsizing. . . . The *Ticonderoga* . . . cost $1 billion. . . . The Reagan Administration plans to order between 18 and 24 of the ships in coming years.[4]

A few months before that report, a United Press International dispatch appeared in the press:

> The Reagan Administration's budget includes welfare cuts for pregnant women and those who get aid under the program for the aged, the blind and the disabled, government officials said today. . . . The change could save $1.5 million to $3.5 million.[5]

In early 1990, as dramatic changes in the Soviet Union and in Eastern Europe made a "Soviet threat" extremely unlikely, President George Bush, in a speech to a local chamber of commerce, "warned that he would not support new domestic programs paid for with money taken from the military budget."[6]

When we read about people standing in long lines in Moscow hoping to buy food, the old arguments about socialism and capitalism don't seem useful any more. As we near the end of the twentieth century, it seems clear that neither the Soviet system nor the American system has been able to meet the fundamental needs of the entire population—to food, house, educate, and provide medical care. Perhaps we need to put aside that theoretical argument (an argument between two frozen bodies of thought, neither one fitting the complicated human situation of today's world) and just try to answer a few important questions.

What *is* economic justice? What are the proper goals of a good economic system? What is the reality of wealth, poverty, and class distinction in this country? And how do we get from this reality to something close to justice?

Rugged Individualism and Self-Help

Thomas Jefferson wrote in the Declaration of Independence, "We hold these truths to be self-evident, that all men are created equal." (Or as amended by women who gathered in 1848 in Seneca Falls, New York, at a women's rights convention: "that all men and women are created equal." Or as a possible children's convention might say: "that all children are created equal.")

A common reaction to Jefferson's phrase "created equal" is that it is just not so; people are endowed with different physical and mental capacities, and with different talents, drives, and energies. But this is a misreading of the Declaration of Independence. There is no period after the word "equal," but a comma, and the sentence goes on: "that they are endowed by their Creator with certain unalienable Rights, that among these are Life, Liberty, and the pursuit of Happiness." In other words, people are equal not in their natural abilities but in their *rights.*

Jefferson said this was "self-evident," and I would think that most people would agree. But some *selves* do not think it evident at all. We know that Jefferson and the Founding Fathers, almost all of whom were very wealthy, did not really mean for that equality to be established, certainly not between slave and master, not between rich and poor. And when, eleven years after they adopted the Declaration, they wrote a constitution, it was designed to keep the distribution of wealth pretty much as it existed at the time—which was very unequal. But that is no reason for anyone to surrender those rights, any more than the ignoring of the racial equality demanded by the Fourteenth Amendment was reason for discarding that goal.

To say that people have an equal right to life, liberty, the pursuit of happiness, means that if, in fact, there is inequality in those things, society has a responsibility to correct the situation and to ensure that equality.

Not everyone thinks so. One man whose thinking was close to that of the Reagan administration in the eighties (Charles Murray, *Losing Ground*) wrote enthusiastically about doing away with government aid to the poor: "It would leave the working-aged person with no recourse whatsoever except the job market, family members, friends, and public or private locally funded services."

It is a restatement of laissez-faire—let things take their natural course without government interference. If people manage to become prosperous, good. If they starve, or have no place to live, or no money to pay medical bills, they have only themselves to blame; it is not the responsibility of society. We mustn't make people dependent on government—it is bad for them, the argument goes. Better hunger than dependency, better sickness than dependency.

But dependency on government has never been bad for the rich. The pretense of the laissez-faire people is that only the poor are dependent on government, while the rich take care of themselves. This argument manages to ignore all of modern history, which shows a consistent record of laissez-faire for the poor, but enormous government intervention for the rich.

The great fortunes of the first modern millionaires depended on the generosity of governments. In the British colonies of North America, how did certain men obtain millions of acres of land? Certainly not by their own hard work, but by government grants. The British Crown gave one semifeudal proprietor control of all of the land of Maryland. How did Captain John Evans of New York get an area of close to half a million acres? Simply because he was a friend of Governor Fletcher, who granted three-fourths of the land of New York to about thirty people.[7]

After the Revolutionary War, the new Constitution of the United

States was drafted by fifty-five men who were mostly wealthy slaveowners, lawyers, merchants, bondholders, and men of property. Their guiding philosophy was that of Alexander Hamilton, George Washington's closest adviser and the first secretary of the treasury. Hamilton wrote, "All communities divide themselves into the few and the many. The first are the rich and well-born, the other the mass of the people. . . . Give therefore to the first class a distinct permanent share in the government."

The Founding Fathers, whether liberal like James Madison or conservative like Alexander Hamilton, felt the same way about the relationship of government and the wealthy classes. Madison and Hamilton collaborated on a series of articles *(The Federalist Papers)* to persuade voters in New York to ratify the new Constitution. In one of these articles (*Federalist #10*) Madison urged ratification on the grounds that the new government would be able to control class conflict, which came from "the various and unequal distribution of property." By creating a large republic of thirteen states, the Constitution would prevent a "majority faction" from creating trouble. "The influence of factious leaders may kindle a flame within their particular States, but will be unable to spread a general conflagration through the other States."

What kind of trouble was Madison worried about? He was blunt. "A rage for paper money, for an abolition of debts, for an equal distribution of property, or for any other improper or wicked project." Like the other makers of the Constitution, he wanted a government that would be able to control the rebellion of the poor, the kind of rebellion that had just taken place in western Massachusetts when farmers, unable to pay their debts, refused to let the courts take over their farms.

The Constitution set up a government that the rich could depend on to protect their property. The phrase "life, liberty, and the pursuit of happiness," which appeared in the Declaration of Independence, was dropped when the Constitution was adopted, and the new phrase, which became part of the Fifth Amendment and later the Fourteenth Amendment, was "life, liberty, or property."[8]

In 1987 the Mobil Oil Corporation celebrated the adoption of that phrase in ads appearing in eight major newspapers, reaching 50 million people:

> Why was property so important as to be included with life and liberty as a fundamental right? Because the Framers saw it as one of the great natural

rights . . . to keep what one had earned or made—that ought to be forever secure from the tyranny of governments or any covetous majority.

That phrase "covetous majority" goes back to Madison's feared majority wanting "an equal division of property, or . . . any other improper or wicked project."

The new government of the United States began immediately to give aid to the rich. Congress passed a Fugitive Slave Act to enforce the provision in the Constitution that persons "held to Service or Labor in one State" who escaped into another "shall be delivered up" to the owner.

"Why make the slaveowner dependent on the government?" a slave, holding to the conservative idea of rugged individualism, might ask. "You want your slave back? You're on your own."

The first Congress also adopted the economic program of Alexander Hamilton, which provided money for bankers setting up a national bank, subsidies to manufacturers in the form of tariffs, and a government guarantee for bondholders. To pay for all those subsidies to the rich, it began to exact taxes from poor farmers. When farmers in western Pennsylvania rebelled against this in 1794 (Whiskey Rebellion), the army was sent to enforce the laws.

This was only the beginning in the history of the United States of the long dependency of the rich on the government.[9] In the decades before the Civil War, great fortunes were made because state legislatures gave special help to capitalists. The builders of railroads and canals, needing large sums of money, were not told Raise your own capital. They became dependents of the government, using their initial capital not to start construction, but to bribe legislators. In Wisconsin in 1856 the LaCrosse and Milwaukee Railroad got a million acres free, after distributing about $900,000 in stocks and bonds to seventy-two state legislators and the governor.

Altogether, in the decade of the 1850s, state governments gave railroad speculators 25 million acres of public land, free of charge, along with millions of dollars in loans. During the Civil War, the national government gave a gift of over 100 million acres to various railroad capitalists.

The first transcontinental railroad was not built by laissez-faire. The railroad capitalists did it with government land and money. The great romantic story of the American railroads owes everything to government welfare. The Central Pacific, starting on the West Coast, got 9 million acres of free land and $24 million in loans (after spending

$200,000 in Washington for bribes). The Union Pacific, starting in Nebraska and going west, got 12 million acres of free land and $27 million in government loans.[10]

And what did the government do for the 20,000 workers—war veterans and Irish immigrants—who laid five miles of track a day, who died by the hundreds in the heat and the cold? Did it give their families a bit of land as payment for their sacrifice? Did it give loans to the 10,000 Chinese and 3,000 Irish, who worked on the Central Pacific for $1 or $2 a day? No, because *that* would be welfare, a departure from the principle of laissez-faire.

The historical practice in the United States of aid to the rich and laissez-faire for the poor was particularly evident in the 1920s, when the secretary of the treasury was Andrew Mellon. One of the wealthiest men in America, he sat atop a vast empire of coal, coke, gas, oil, and aluminum. Mellon cut taxes for the very rich, whose high living gave the decade its name "The Jazz Age." Meanwhile, many millions of Americans lived in poverty, with no aid from the government.[11]

When the nation's economy collapsed after the stock market crash of 1929, a third of the labor force lost their jobs. Hunger and homelessness spread all over the country, and the historian Charles Beard wrote an essay called "The Myth of Rugged American Individualism."[12] He noted the hypocrisy of those who said the poor should make it on their own. He recounted the ways in which the government had aided the business world: regulation of the railroads and donation of hundreds of millions of dollars to improve rivers and harbors and to build canals. Government also granted subsidies to the shipping business, built highways, and gave huge gifts to manufacturers (at the expense of consumers) through higher and higher tariffs.

Beard pointed to the use of the nation's military force to help business interests around the world, a most crass violation of the laissez-faire philosophy. In our time, the dependence of very rich corporations on the military power of the United States and on its secret interventions in other countries has become very clear. In 1954, the CIA organized the overthrow of the elected president of Guatemala to save the properties of the United Fruit Company. In 1973 the U.S. government worked with the IT&T Corporation to overthrow the elected socialist leader of Chile, Salvador Allende. Allende had not been friendly enough to the foreign corporations that exploited Chile's wealth for so long.

In 1946 a secret air force guideline (which became public knowledge when it was declassified in 1960) said that the aircraft companies would go out of business unless the government made sure they got contracts. Since that time certain major aircraft companies have depended totally for their existence on government contracts: Lockheed, North America, and Aero-jet.

The giant businesses depend on the government to arrange tax schedules that will, in some cases, permit them to pay no taxes, in other cases, to pay a much smaller percentage of income than the average American family. For instance, five of the top twelve American military contractors in 1984, although they made substantial profits from their contracts, paid no federal income taxes. The average tax rate for those twelve contractors, who made $19 billion in profits for 1981, 1982, and 1983, was 1.5 percent. Middle-class Americans paid 15 percent.[13]

All through the nineteenth and twentieth centuries, landlords have depended on the government to suppress the protests of tenants (for instance, the antirent movement of the 1840s in the Hudson River Valley of New York) and to enforce evictions (as in the thousands of evictions during the Depression years). Employers have depended on local government's use of police and the federal government's use of soldiers to break strikes—as in the railway strikes of 1877, the eight-hour-day strikes of 1886, the Pullman rail boycott of 1894, the Lawrence textile strike of 1912, the Colorado coal strike of 1913, the auto and rubber and steel strikes of the 1930s, and hundreds more. If those employers were truly "rugged individualists," as they asked their workers to be, they would have rejected government aid.

Furthermore, employers with the money to hire lawyers and to influence judges have depended on the courts to declare strikes and boycotts illegal, to limit picketing, and to put strike leaders in jail (as when Eugene Debs, the leader of the Pullman strike and boycott of 1894, was jailed for six months because he would not call off the strike).

Through the nineteenth century, according to legal historian Morton Horwitz, the courts made clear their intention to protect the business interests. Mill owners were given the legal right to destroy other people's property by flood to carry on their business. The law of "eminent domain" was used to take farmers' land and give it to canal companies or railroad companies as subsidies. Judgments for damages against businessmen were taken out of the hands of juries, which were unpredictable, and given to judges. Horwitz concludes,

> By the middle of the nineteenth century the legal system had been reshaped to the advantage of men of commerce and industry at the expense of farmers,

> workers, consumers, and other less powerful groups within the society. . . . It actively promoted a legal redistribution of wealth against the weakest groups in the society.[14]

Yet when someone advocates "a legal redistribution of wealth" on behalf of the poor, the cry goes up against "government interference" and for "rugged individualism."

After the Civil War, the Fourteenth Amendment's phrase "life, liberty, or property," which turned out to be useless to protect the liberty of black people, was used in the courts to protect the property of corporations. Between 1890 and 1910, of the cases involving the Fourteenth Amendment that came before the Supreme Court, 19 were concerned with the lives and liberties of blacks and 288 dealt with the property rights of corporations.[15]

The working conditions in American industry during that much-praised time of speedy industrialization were horrible and also legal. (The Senate's Committee on Industrial Relations reported that in the year 1914 alone, 35,000 workers were killed in industrial accidents and 700,000 injured.) This led to thousands of strikes, and to demands for protective legislation.

But when the New York legislature passed a law limiting bakery workers to a ten-hour day, six-day week, the U.S. Supreme Court in 1905 declared this law unconstitutional, saying it violated "freedom of contract."[16] It took the economic crisis of the 1930s and the turmoil it produced to get the Supreme Court to reverse its stand and approve a minimum wage law in Washington, D.C. The Court in 1937 decided that the freedom of contract was not as important as the freedom to be healthy.[17]

However, the Supreme Court has been careful to keep intact the present distribution of wealth and the benefits in health and education that come from that wealth. In 1973 it decided a case where poor people in Texas, seeing that much less money was allocated for the schools in a poor county than in a rich one, sued for the right of poor children to equal funds for their education. The Court turned down their plea, saying that these children (mostly Mexican-American) were not completely denied an education, but just denied an *equal* education, and education was not a fundamental right guaranteed by the Constitution.[18]

Clearly, the same would apply to the right to food and medical care, which, like education, are not specifically mentioned in the Constitution as fundamental rights. One constitutional lawyer, however, has argued that the Fourteenth Amendment's requirement that no state can deprive any person of "life" ("life, liberty, or property") could be used to provide an equal right of the poor to food, medical care, a job. Professor Edward V. Sparer of the University of Pennsylvania Law School has said:

> We guarantee income to farmers for not producing crops. We guarantee subsidies to railroads and to oil companies. It seems to me only reasonable that we should guarantee the subsidy of life to those who are starving and to those without shelter or medicine—reasonable not only on humanitarian grounds, but because there is a 14th Amendment, which guarantees equal protection of the laws.[19]

Most of the accumulation of wealth is strictly legal. And if any question comes up about the legality of corporate behavior, lawyers are available to straighten out any accuser. The columnist Russell Baker once wrote, "There are plenty of rich men who have no yachts and others who have no Picassos. . . . Every last one of them, however has a lawyer. . . . Having a lawyer is the very essence of richness. . . . What we have here is a class structure defined by degree of access to the law."[20]

When the rich commit the truly grand larcenies, which become too flagrant to ignore, their lawyers work out deals with the government and no one goes to jail, as would happen to a petty thief. For instance, in 1977 the Federal Energy Administration found that the Gulf Oil Corporation had overstated by $79 million its costs for crude oil obtained from foreign affiliates. It then passed on these false costs to consumers. The following year the administration announced that to avoid going into a court of law, Gulf would pay back $42 million. Gulf cheerfully informed its stockholders that "the payments will not affect earnings since adequate provision was made in prior years."[21] One wonders if a bank robber would be let off if he were to return half his loot.

Jimmy Carter was president at that time. It seemed that liberal Democrats did not behave terribly different from conservative Republicans where wealthy corporations were involved.

Adam Smith's famous book *The Wealth of Nations,* published around the time of the American Revolution, is considered one of the bibles of capitalism. He spoke candidly on the class character of governments:

> Laws and governments may be considered in this and indeed in every case as a combination of the rich to oppress the poor, and preserve to themselves the inequality of the goods which would otherwise be soon destroyed by the attacks of the poor, who if not hindered by the government would soon reduce the others to an equality with themselves by open violence.[22]

Around the same time, Jean Jacques Rousseau wrote his *Discourse on the Origin of Inequality*, an imaginative account of how government and laws came into existence, and concluded that

> society and laws which gave new fetters to the weak and new forces to the rich, irretrievably destroyed natural liberty, established forever the law of property and of inequality, changed adroit usurpation into an irrevocable right, and for the profit of a few ambitious men henceforth subjected the entire human race to labor, servitude, and misery.

A roughly similar point was made in the 1980s, by a black taxi driver in Los Angeles, who was interviewed by a filmmaker about "democracy." The man laughed and said, "We have government by the dollar, of the dollar, for the dollar."[23]

Surely we need to clear guilt from the air in the poorer districts of our cities (there are enough impurities there already) by asking: Why *shouldn't* people in need be dependent on the government, which presumably was set up exactly for the purpose of ensuring the well-being of its citizens? The words *promote the general welfare* do appear in the Preamble to the Constitution, even if ignored in the rest of it.

Indeed, is there such a thing in this complicated society of the twentieth century as true independence? Are we not all dependent on one another, and is that not a necessity of modern life? We all depend on the government for schools, garbage collection, protection against fire and theft, and many other things. Welfare is only one kind of dependency.

Talent and Need

Playwright and social critic George Bernard Shaw, in his book *The Intelligent Woman's Guide to Socialism*, after pointing out the evil effects of inequality in society (rebellion, resentment, envy, and violent conflict) wrote in favor of a simple equality of wealth.[24]

Against him, there are those who argue that the poor have so little because they deserve so little and that wealth follows talent. That claim goes something like this: "If you have the right stuff, you'll be a success; if you don't have the stuff, you just won't make it, and that's the way it should be."[25]

The economist Milton Friedman argues like that. He insists that in a capitalist system like the United States "the market" sees to it that people are paid "in accordance with product," in other words, according to how much they produce.[26] He seems to be living in a world of his own creation, far removed from this one. In his world, everything works beautifully by the laws of the market, and people get more if they produce more, less if they produce less.

But how do you measure this thing called *product* and decide who deserves more or less? Here is the executive whose corporation produces nuclear bombs or deodorants or plastic toys. Did *he* himself produce it or did a thousand workers produce it? How can you measure *his* contribution to the final product? Indeed, in our complex modern society, where things are produced by the participation of huge numbers of individuals, how can you measure the contribution of each one to the product?

Furthermore, if one corporation produces cigarettes, which are bad for people's health, and another produces antibiotics to cure people with infectious diseases, do you only care about the *quantity* of production in deciding the rightness of reward or should you be concerned with the *kind* of thing that is produced? Indeed, how can you measure contribution to product when these products are so vastly different? Why does the producer of jeweled dog collars deserve a hundred times as much as the producer of a single poem?

The United States is full of talented people, in many different fields, who have a difficult time making enough money to keep alive. Is the skilled machinist who is unemployed as a result of a layoff any less skilled than the one who is kept on? There are large numbers of artists—painters, musicians, writers, and actors—who have loads of talent but, because our government does not give out contracts to finance artistic performances as it gives out contracts to finance the performance of nuclear submarines, they cannot make a living. (In 1979 the average published author earned about $4,775 from writing.)[27]

While there are some talented people who make money, there are others equally talented who do not. It must be clear to anyone living in this country that there are many people with no talent—except the talent to make money—who are very rich. In short, there is no logical relationship between talent and money.[28]

Surely, all human beings deserve the fundamental requirements for living—housing, food, medical care, and education—regardless of their talents. Think of children, whose talents are not yet evident: are they to get food and medical care on the basis of their parents' talents? Is that fair?

And how are we to judge the monetary worth of this talent versus that other talent? Our capitalist society has made its decision on the basis of the *market,* meaning on the basis of what sells, not on the basis of logic or reason or morality. It has decided that the head of an advertising agency deserves ten times as much money as a skilled carpenter, and that a man who juggles papers as vice president of an aerospace company deserves a hundred times as much money as a gifted teacher of elementary school children.

The *market* has decided that housewives produce nothing; therefore, it pays them nothing. The welfare mother taking care of her children is, in fact, working. The wife who stays home with the children is giving her talent and energy to one of the most important jobs in society. But she has no rights to a share of society's wealth on the basis of her work. The market, and its supporters, have decided that.

As for incentives, talented people are not productive on the basis of money incentives. Indeed, when money determines what they do, their art may be distorted by what the market requires; the artists will draw what the lingerie company wants, not what he or she desires to do; the writer will produce advertising copy rather than a poem. Talented people exercise their talents for the pleasure of it and will be able to do this if only they can have their basic economic needs taken care of. Their incentive will not be money, but the satisfaction of doing what they are impelled to do and the respect of others for what they have done.

George Bernard Shaw discussed the issue of merit and talent:

> Between persons of equal income there is no social distinction except the distinction of merit. Money is nothing: character, conduct, and capacity are everything. Instead of all the workers being levelled down to low wage standards and all the rich levelled up to fashionable income standards, everybody under a system of equal incomes would find her and his own natural level. There would be great people and ordinary people and little people; but the great would always be those who had done great things, and never the idiots whose mothers had spoiled them and whose fathers had left them

> a hundred thousand a year; and the little would be persons of small minds and mean characters, and not poor persons who had never had a chance. That is why idiots are always in favor of inequality of income (their only chance of eminence), and the really great in favor of equality.[29]

Shaw was arguing for equal incomes for all. This, of course, is a quite radical proposal. Like other radical proposals, it need not be taken literally; no one expects a program of exactly equal incomes to be put into effect even by the most revolutionary of governments. But the idea is shocking to us because we have been brought up in a society of such vastly different incomes that we cannot comprehend a different situation. There is a tendency to think that what we grew up with, what we have seen all our lives, is natural and inevitable. That any other way would be against human nature.

But we have very simple examples of where human beings have adjusted quickly to the idea of equal incomes. The family is the most obvious example: the members of the family get food, clothing, shelter, and medical care not on the basis of their work, their talent, or their contribution, but simply because they have the same basic needs. And no one would think of questioning that equality.

One might argue that a society of strangers is much different than a family. But we have the example of the Israeli kibbutz. People who have joined the kibbutz, even people who come from very competitive capitalist countries, yes, Americans, have adjusted quickly to the equality of the kibbutz, where people do different kinds of work, but all get the same benefits, their basic needs taken care of in the same way.

And while it may be thought that the jump from a family to a kibbutz is one thing, but the next jump to a nation makes the idea of egalitarianism impossible, one should consider the old factor of patriotism. Think of what power there is in patriotism and how that power has often been used for terrible purposes. Yet at the very time it was being used to carry on murderous wars, it revealed its power to unite the people of the country in egalitarian sacrifice.

During World War II, people left high-paying jobs to do civilian jobs that were necessary for the war. Or to go into the armed forces. If we put aside the class distinctions of officers and enlisted men, we can see that men, imbued with the idea they were fighting in a great moral cause, did not perform on the basis of their pay, but gave what they could, even their lives, for that cause.

In short, the capacity of human beings to give their talents, their energies, and their all, not on the basis of monetary reward, but on the basis of some larger collective purpose, has been demonstrated again and again. That fact is encouraging to the idea of an equal distribution of wealth.[30]

How Should Wealth Be Distributed?

There are variations on the theme of equal income. The Marxian idea was not to give everyone the same, but distribute the wealth of society according to *need*, because there are large families and small families, sick people and healthy people, children and old people, all with different needs to enjoy life. Marx did not see the possibility of doing this in the early stages of a socialist society, because production would not be developed enough, there would not be enough to go around. But:

> In a higher phase of communist society, after the enslaving subordination of individuals under division of labor, and therewith also the antithesis between mental and physical labor, has vanished . . . after the productive forces have also increased with the all-round development of the individual, and all the springs of cooperative wealth flow more abundantly—only then can the narrow horizon of bourgeois right be fully left behind and society inscribe on its banners: from each according to his ability, to each according to his needs![31]

The argument has been made that the "needs" formula is not good for everything. Michael Walzer agrees that the formula is good for medical care and other things. But, he says, "Marx's slogan doesn't help at all with regard to the distribution of political power, honor and fame, leisure time, rare books, and sailboats." And so he argues that there can be no one principle of distribution. "Equality requires a diversity of principles, which mirrors the diversity both of mankind and of social goods."[32]

This makes sense. But we don't have to get into complicated arguments about *exactly* how all things will be distributed. It would be an enormous accomplishment to get agreement that the fundamental requirements of existence—food, housing, medical care, education, and work—be distributed according to need. It is shocking, it is irrational, it is unjust, that in a country as wealthy as the United States, any human being living within its borders should not have these basic things.

As of 1986, 37 million adults and children had no medical insurance. About 88 percent of these were working people or their families. Most of them did not earn enough to pay the high cost of individual medical insurance and many of them were not eligible for Medicaid (state medical payments for the poor) because they made just enough money to put them over the eligibility line.

For instance, a forty-one-year-old nurse's assistant who had injured herself lifting a patient and couldn't work was receiving $500 a month in workmen's compensation, not enough to pay for her own insurance and too much to qualify for Medicaid. Her medical bills piled up unpaid and she was being treated for depression at a community health clinic.[33]

In 1989 the American Cancer Institute presented the results of a study that showed that lack of money results in higher rates of cancer, because it goes undetected for longer periods of time, and poor people do not get as good treatment as the rich.

The philosopher Robert Nozick has argued that *entitlement* should be the key to the way wealth is distributed. If, he says, someone has "holdings" (wealth, land, resources, and skills) that he has acquired "by legitimate means," no one should take them away for any purpose, however desperate the need.

Does he mean that if I have a million dollars, which I got by legitimate means, then the government has no right to tax any of that to raise money to build homes for homeless people, or to provide medical insurance for the aged?

What does *legitimate* mean? Does it mean *legal?* Then the money accumulated by corporations through tax breaks obtained by paying lobbyists to get the right laws passed or paying accountants to make the best use of those laws is legitimate and government has no right to tax that to pay for Medicare. If *legitimate* means moral, we could argue that none of the great fortunes came about morally, certainly not without the exploitation of labor. Then, using Nozick's principle, the way would be wide open, to distribute those fortunes in whatever way they would be useful to people in need.[34]

If we traced the holdings of the rich back in history (Rockefeller, Morgan, Vanderbilt, Carnegie, Mellon, Astor, and Frick) we would encounter shrewdness, managerial ability, luck, ruthlessness, and violence. In Carl Sandburg's poem "*The People, Yes*" we find the following exchange:

"Get off this estate."
"What for?"
"Because it's mine."
"Where did you get it?"
"From my father."
"Where did he get it?"
"From his father."
"And where did he get it?"
"He fought for it."
"Well, I'll fight you for it."[35]

Truth is it shouldn't matter how the rich got that way. If people have fundamental needs that are matters of life and death, why should we not, by taxation, take from people who will not suffer as a result of the taking, to meet those needs?

There seems to be no reasonable relationship between how *hard* someone works and that person's income. There are people who do no work and are poor, and people who do no work and are rich. There are hard workers among people who earn $100,000 a year and hard workers among people who earn $15,000 a year. How is one to measure the work done by a cleaning woman and the work done by the corporate executive whose office she cleans? In our culture, by definition, someone who makes a lot of money is working hard. That's a nice circular definition—it saves us the job of really examining what people do.

Farmers work hard and the market pays them so little they desperately need government help to survive. Housewives work hard and the market pays them nothing. Students can work hard too, but because we measure work by money earned, that doesn't count at all. The thought that people get paid according to their contribution to society does not stand up to more than a moment of examination. Among the lowest paid of workers are teachers, social workers, and nurses. Among the highest paid are the executives of corporations that make weapons.

I once put a question to a class of about 400 students: On what basis do people get paid in American society? What determines their income? Is it intelligence, hard work, or contribution to society? There was a lively, argumentative discussion, in which some students pointed to their parents as proof that hard work or contribution to society or special talent had resulted in prosperity—and others were skeptical. Finally, I asked for a show of hands of those who would claim that, in general, leaving room for exceptions, Americans got paid according to their intelligence. Three or four hands were raised. Hard work? A similar response. Contribution to society? Again, just a few hands raised. Then I asked who believed it was "none of the above." Almost everyone raised his hand. It took only a little reflection, apparently, for these young people to see that the way income is distributed today does not follow any rational or just principle.

Coercion

Those who oppose distributing wealth on the basis of need often say they do not want the government to interfere in people's lives, which they call "coercion." But are people who lack the basic resources of society free from interference with their lives? Do they not face continuous interference from policemen, social workers, landlords, loan sharks, and employers? Are these not the people *most* vulnerable to interference with their lives? Perhaps the concern is only about interference in the lives of the rich, who might be taxed to redistribute wealth.

In fact, certain necessities of life can be provided with virtually no government interference. Retired workers get social security checks in the mail each month. That's hardly an intrusion on their liberty. The Social Security system seems to operate efficiently with a minimum of bureaucracy. If medical care were free, there would be much less interference in peoples' lives then with the present hodgepodge system. If college education were free of all the financial rigmarole connected with federal loans—if there were a civilian bill of rights for education as there was a GI Bill after World War II—there would be less interference than there is now.

The economist Milton Friedman worries about coercion.[35] No one should be compelled to share their greater wealth with someone else, he says. And he gives us examples. It is always interesting to see a scholar choose examples. Friedman chooses two. First, he says, let's imagine four Robinson Crusoes marooned on four neighboring islands, and one of them finds his island full of good stuff and the others are barely surviving. "Would the other three be justified in joining forces and compelling him to share his wealth with them? Many a reader will be tempted to say yes."

Of course. Especially if the survival of the other three is at stake, or they will be living in a miserable state, hungry, sick, and unhappy. Friedman doesn't like the reader's response to this example. So he moves to another one, saying to the reader: "But before yielding to this temptation, consider precisely the same situation in different guise." (Let's see if this will be *precisely* the same situation.) "Suppose you and three friends are walking along the street and you happen to spy and retrieve a $20 bill on the pavement. It would be generous of you, of course, if you were to divide it equally with them, or at least blow them to a drink. But suppose you do not. Would the other three be justified in joining forces and compelling you to share the $20 equally with them? I suspect most readers will be tempted to say no."

Friedman has moved his example from the issue of survival or quality of life to the issue of a few dollars or a drink! And so he entices his reader to say no, such coercion is unjustified. As it is, in that case.

But why doesn't he give an example that will accurately reflect the serious question under discussion? Why doesn't he ask if the lawyer Roy Cohn (who made $100,000 a year in salary and perhaps a million dollars a year in "expenses" and whose contribution to society seems to have been more negative than positive[37]) should have been "coerced" (taxed) into giving up half of his phony expenses and half of his overblown salary? That tax money could then be used so that old people living in Cohn's city could live in their own apartments, with the help of various paid social workers, rather than having to live in an institution where there is not enough help to push their wheelchairs down to the dining room.

Let's grant Roy Cohn was extraordinary, but not alone in his use of the tax system to retain a luxurious life-style. We could ask the same question about more ordinary wealthy taxpayers. A report of the House Ways and Means Committee in 1989 showed that in the period 1979–1987 (the last two years of the Carter administration and the first six years of the Reagan administration), the poorest fifth of the population had their family income decrease by 9.8 percent (the report said this change was affected by the fact that "more jobs now pay poverty level wages or below"—so it is not simply a matter of people on welfare). In the same period, the wealthiest fifth of the population *increased* their family income by 15.6 percent.[38]

Now would it be coercion for the government to tax that rich fifth by the amount of its gain and contribute this to the lowest fifth? Well, all taxation is coercion, so the real question is when is it just and when is it unjust? To pretend that leaving alone the present disparities in wealth increases our freedom hardly makes sense, unless you interpret the "our" to mean only the wealthiest portion of society.

Is the working class free from coercion under the capitalist system? Friedman describes our "complex enterprise and money-exchange economy" as one in which "individuals are effectively free to enter or not to enter into any particular exchange" and the employee "is protected from coercion by the employer because of other employers for whom he can work."[39]

Is this a realistic picture of the work situation for most people in the United States? Perhaps it describes high-level professionals who have a choice of employers. It does not describe secretaries, factory workers, waiters, or even teachers and social workers and certainly not most people in the arts. Most people who work for a living have zero or limited choice of their employers. The permanent body of unemployed people ensures that employers have the upper hand; whatever freedom there is seems to be unevenly distributed.[40]

Incentives and the Profit Motive

One of the ways in which the corporate rich protect their incomes is to warn all of us that if they are taxed more heavily, (to give help to poorer people), they will lose their incentive to produce. The motive of high profit, they say, is necessary to keep the economy going.

It is an interesting argument, because it is usually coupled with the statement that if we help the poor they will have no incentive to work. In other words, monetary incentives will make the rich work harder and the poor work less. Frances Piven and Barbara Ehrenreich, writing about the welfare state, respond to this kind of thinking by asking, Do the poor and the rich have different human natures?[41]

In the early years of the Reagan administration when tax cuts for corporations were proposed, presumably to stimulate business, some enterprising reporter dug up Department of Commerce figures showing that from 1970 to 1983 periods of lower corporate taxes did not at all result in higher capital investment, but, in fact, much lower investment.[42]

Japan in the 1970s and 1980s came to be admired throughout the world as a model of industrial efficiency and productive growth. Yet according to economist Robert Kuttner, Japan had the highest effective rates (that is the real rate paid, rather than the one called for by law) of corporate income tax—40 percent of corporate profits, compared to less than 15 percent in the United States.[43]

The issue of incentives is complicated. Different groups of people in the economy respond differently to money incentives, and individuals respond differently also. There are undoubtedly some kinds of economic activity that will go on or not go on depending on the rate of profit. But others will go on, so long as there is *some* profit, because it is not easy to move capital from one enterprise to another.

Certainly, some of the most useful things done in society are done by people whose monetary incentives are very low: teachers, social workers, nurses, musicians, actors, and writers. There are many people who do their work because they love it. And the quality of their work will not change because their salaries go up or down. However, they must be given some *minimum* economic reward, enough to allow them to survive, in order for them to continue their work.

How about doctors? If doctors who now make $100,000 a year would only make $60,000 a year, would the quality of their work get worse? Surely doctors do their best because of the motivation that brought them into medicine in the first place—the enormous satisfaction in making people well. Those who entered the profession to become rich are probably not the best in their field. The fortunes that some doctors make distort their interest—perhaps induce them to do unnecessary but expensive surgery or to spend their time doing cosmetic surgery for the rich instead of life-saving medicine for the general population.

What is usually missed in discussions on incentives is that there are all sorts of incentives other than *money* to bring out the best work in people: the respect and admiration of others and an increased self-respect and self-satisfaction. George Bernard Shaw pointed out that while there are menial and unsatisfying jobs that are not admired and don't give a feeling of great accomplishment, people will do them with the proper incentive of *freedom*—that is, people asked to do such necessary jobs should perhaps be given a twenty-hour week as incentive.

A good society will use incentives—money and time, for example—in all sorts of imaginative ways to bring out the best in people and get the most accomplished for society. But to reward the rich with the incentive of high profits, no matter what work they do or what contribution they make to society, and to punish the poor by withholding the necessities of life (a disincentive, to force them to work) is both unjust and inefficient.

It is true that the profit motive, in the history of capitalist development, has stimulated great industrial progress. Karl Marx, even while he looked forward to the disappearance of capitalism, acknowledged that it had brought the greatest increase in the productive forces of society, that it was a "progressive" stage in history. And it produced many useful, worthwhile things.

But it has also had the most terrible human consequences. In 1984 a company making an intrauterine birth control device—the Dalkon Shield—came before a court in Minneapolis to make a monetary settlement for complaints from women who had been badly injured internally by the device. According to the judge, addressing the company officials:

> Under your direction your company has in fact continued to allow women, tens of thousands of them, to wear this device—a deadly depth charge in their wombs, ready to explode at any time. . . . The only conceivable reasons you have not recalled this product are that it would hurt your balance sheet.[44]

The drive for profit is ruthless. Because its chief motive is making money, it may not matter *how* it makes this money, what it produces, and what happens to human beings in the process. The capitalist drive for profit is based on what has been called "rational self-interest," and the idea is that if everyone pursues their rational self-interest, the economy will grow and the world will be a better place.

However, as one economist said, "If . . . it pays to pollute a lake, then rational self-interested agents will pollute. If crime is a rational response to poverty wages, we can expect crime to rise."[45] Certainly, the most terrible violence, and the proliferation of dangerous drugs, including tobacco and alcohol, are provoked by the profit motive.

In the year 1989 the new American president, George Bush, announced a "war" on drugs. He was talking about the illegal traffic in cocaine and similar substances. He did not speak of tobacco, which brings immense profits to certain huge corporations and which is far more murderous than cocaine. Indeed, American tobacco companies, not content with poisoning people in this country, sought and received government help to push tobacco onto foreign markets, especially in Third World countries. All under the guise of "free trade."

The slogan *free trade* recalled the "open-door policies" of the eighteenth and nineteenth centuries, when Western countries forced China to accept their products and when England fought the Opium War to compel the Chinese to allow British companies to bring opium into China. In 1989 the United States was putting pressure on Thailand, which banned tobacco, to accept American tobacco exports. But there was a strong dissenting voice from Dr. C. Everett Koop, who was leaving his job as surgeon general of the United States, in obvious disagreement with many government policies. Koop told a public hearing in Washington:

> Years from now, I'm afraid that our nation will look back on this application of free trade policy and find it scandalous, as the rest of the world does now. . . . At a time when we are pleading with foreign governments to stop the export of cocaine, it is the height of hypocrisy for the United States to export tobacco.[46]

Koop ended his testimony with devastating statistics. "Last year, in the United States, 2000 people died from cocaine. In the same year, cigarettes killed 390,000 people."

The reckless drive for profit also produces toxic chemicals that cause cancer. They pollute our rivers, poison our air, and dirty our oceans.[47] Enormous numbers of fish are dying. Birds are dying. The rush of the automobile industry for profits (why do we need 100 million vehicles in a country of 250 million people?) has loaded the air with chlorofluorocarbons and produced the greenhouse effect, which may have catastrophic effects on the entire world in the next generation. Profits also stimulate the war industries, producing nuclear and other weapons.

Those who want government to stop interfering with "the market," with corporate profit, or with poverty should look back to the nineteenth and early twentieth centuries, to what happened before the labor laws were passed. I have described in *A People's History of the United States* the conditions in American cities at the end of the Civil War:

> The cities to which the soldiers returned were death traps of typhus, tuberculosis, hunger, and fire. In New York, 100,000 people lived in the cellars of the slums; 12,000 women worked in houses of prostitution to keep from starving; the garbage, lying two feet deep in the streets, was alive with rats. In Philadelphia, while the rich got fresh water from the Schuylkill River, everyone else drank from the Delaware, into which 13 million gallons of sewage were dumped every day. In the Great Chicago Fire in 1871, the tenements fell so fast, one after another, that people said it sounded like an earthquake.

Between the Civil War and World War I, the profit system produced remarkable results in the industrialization of the country. It also produced inhuman living conditions. The sweatshops of New York, where the new waves of immigrants from Eastern Europe went to work, were described by the poet Edwin Markham in 1907:

> In unaired rooms, mothers and fathers sew by day and by night. . . . And the children are called in from play to drive and drudge beside their elders. . . . Is it not a cruel civilization that allows little shoulders to strain under these grown-up responsibilities, while in the same city, a pet cur is jeweled and pampered and aired on a fine lady's velvet lap on the beautiful boulevards?[48]

With profit the motive, 500 garment factories sprang up in New York. A woman described the conditions of work in those days of laissez-faire, saying there were

> dangerously broken stairways . . . windows few and so dirty. . . . The wooden floors that were swept once a year. . . . Hardly any other light but the gas jets burning by day and by night . . . the filthy, malodorous lavatory in the dark hall. No fresh drinking water . . . mice and roaches. . . .
>
> During the winter months . . . how we suffered from the cold. In the summer we suffered from the heat. . . .
>
> In these disease-breeding holes we, the youngsters, together with the men and women, toiled from seventy and eighty hours a week! Saturdays and Sundays included! . . . A sign would go up on Saturday afternoon: "If you don't come in on Sunday, you need not come in on Monday." . . . We wept, for after all, we were only children.[49]

One wonders how those children would have responded to the bland claim that they lived in "capitalism and freedom," that they were, as Milton Friedman described the system, "protected from coercion by the employer because of other employers for whom he can work."

Was it not the profit motive that caused the employers of the Triangle Shirtwaist Company of New York City, in the year 1911, to keep the doors locked during working hours so that the company could keep track of its employees? On March 25 of that year a fire broke out, and with the doors locked, the workers on the eighth, ninth, and tenth floors were trapped. They burned to death or jumped from windows; 146 of them, mostly women, died.

In the year 1904 approximately 27,000 workers were killed on the job, in manufacturing, mining, railroads, and agriculture. The need to save money and to increase profits, led one employer to substitute lead powder for talcum to mark the designs in his embroidery factory. A report of the New York State Factory Commission in 1912 told of the effect of this on one employee:

> Sadie had been a very strong, healthy girl, good appetite and color; she began to be unable to eat. . . . Her hands and feet swelled, she lost the use of one hand, her teeth and gums were blue. When she finally had to stop work, after being treated for months for stomach trouble, her physician advised her to go to a hospital. There the examination revealed the fact that she had lead poisoning.

The conditions of that time produced bitter criticism of the profit system, of capitalism. The idea of socialism had not yet been corrupted by Soviet Russia. Socialism was the dream of many—Eugene Debs, Upton Sinclair, Jack London, Helen Keller, and more than 100,000 who joined the Socialist party. There were over 1,000 socialist officeholders in over 300 towns and cities. Perhaps a million people read the socialist newspapers.

Jack London turned from his popular adventure stories to write a political novel, *The Iron Heel.* Through his characters, he comments on the economic system: "Let us not destroy those wonderful machines that produce efficiently and cheaply. Let us control them. Let us profit by their efficiency and cheapness. Let us run them for ourselves. That, gentlemen, is socialism."

The great worldwide interest in socialism—which continues, despite the way the original dream has been distorted in a number of countries around the world—is due, I believe, to what people have seen happen in capitalism—that the profit motive has had some terrible human consequences. People turned to socialism because of the belief that human beings—once their essential needs are taken care of—can be motivated to work and create by considerations other than monetary profit: self-respect, the respect of others, compassion for others, and community spirit.

Moving Toward Justice

The American economic system is enormously productive, but shamefully wasteful and unjust. The contrasts between rich and poor, the flaunted luxury of the very wealthy alongside decaying cities, the pressure on everyone to make lots of money—there must be a connection between all that and the great number of violent crimes in this country, the frighteningly widespread use of drugs, the alcoholism, the mental illness, and the broken families.

The odds are stacked heavily against the poor—black and white. There was a study in the 1970s by the Carnegie Foundation, on the futures of American children. Looking at two children, both with average IQs but with different backgrounds, the researchers found that one of them, the son of a lawyer in the top tenth of the income structure, was four times as likely to enter college as the other, son of a custodian in the bottom tenth. He was twelve times as likely to complete college, and twenty-seven times as likely to end up in the top tenth of income at middle age.[50]

We need fresh thinking, new approaches. The old formulas for socialism have been discredited by the experience of "socialism" in the Soviet Union and Eastern Europe. But the standard praise of capitalism is not warranted by the human results of the American system. On the other hand, the mixed socialist and capitalist economies of Sweden, Norway, Denmark, and New Zealand have succeeded in achieving a certain degree of economic justice, a high standard of living available, without too much inequality, to the entire population.[51]

We need to start figuring out the arrangements, the principles, the practices, and the forms of production and distribution that will give our economic system both efficiency and justice, thinking boldly and bypassing the old ideologies. Economics is very complicated, even for economists. You can tell that by how often they are surprised by a sudden turn of events—the stock market collapses, the dollar plunges or rises, foreign trade diminishes or increases. And how, when they are interviewed on television to give the public their wisdom, they speak glibly but seem as mystified as everyone else.

It is up to the public to say to the technicians—the economists, the planners, and the managers—what the public wants done and what principles to follow. Let the experts figure out *how* to do it and have the public check up constantly on their suggestions. The people of the nation will need to reach some consensus (we will not get unanimity, because there are powerful interests opposed to change) on certain goals.

What might these goals be?

A real "war on poverty" is required (that was the phrase used by the government in the sixties, but it was a minor skirmish). The objective should be to make sure that every man, woman, and child in the United States has adequate food, decent housing, free medical care, a free college education if they want it and can't afford it. We need a real war on pollution: to clean up the air, the rivers, the lakes, and the beaches in a few years, in time for the next generation to enjoy the fresh beauty of nature.

There should be useful work guaranteed for everyone who wants to work. And every kind of work, however unskilled or however unwanted by "the market" (I am thinking of dishwashers, janitors, poets, painters, musicians, actors, and housewives among others) should be paid close to the average wage of working people in the country.

All these things can be done, because this country is brimming over with natural and human resources that have been either unemployed or badly employed. There are enormous parts of the national wealth—millions of people, hundreds of billions of dollars—used for absurd purposes, to produce stupid luxuries or vicious weapons.

Corporate profit, not social need, has determined what shall be produced. Huge amounts of steel, concrete, and human labor have gone into the building of skyscrapers in every city, which are used for banks, insurance companies, offices, or luxury apartments. Those ingredients could have gone into the building of homes in every city, for families in desperate need of a good place to live, except for the profit motive of builders.

That's where society comes in—through the federal government or local government or independent housing authorities—who will pay the builders and then, if necessary, subsidize the rents, so that we have no more homeless people or slum tenements in this country.

This will require an almost total turnaround in priorities and a measure of national and local planning. The money is there ($300 billion a year for useless or wrongly used weapons), but it needs to be used to subsidize the establishment of a decent standard of living for every person and the turning of our cities and countrysides into beautiful places.

Such subsidies are not something new in this country. We already do this with our military establishment. We subsidize everything in the military—the buildings, the weapons, the transport systems, and the personnel—and pay for it with public funds. We plan for what is needed and it all comes out of the national budget, paid for by taxes. We have a kind of socialism for military needs and capitalism for civilian needs.

Our nation experimented with a sort of "socialism" in the thirties when, desperately trying to escape economic disaster, the government planned and subsidized activities that the market, that is the profit seekers of the business world, would not pay for. The government paid young people to plant trees and build roads. It paid men to clean up parks and streets. It paid artists to paint murals on public buildings all over the country. It subsidized theater people, who put on exciting plays, and writers, who wrote beautiful guidebooks for the states.[52]

That kind of planning, the use of public funds for good purposes, did not diminish our liberties. Democracy was enhanced by bringing large numbers of people into useful service, by making the work of artists available to people who never could afford them.

There is no need to do away with private business or with profit or with competition. They can all play their part in an organized national economy that has a certain critical measure of planning and large areas of free enterprise.

At some point the planning would need to become global, because it is impossible to confine economic justice within national boundaries. The enormous disparity between the richest and the poorest countries cannot continue if we care about justice. It was estimated in the mid-1980s that in every year 15 million children around the world died of malnutrition or sickness.[53]

It will take a massive redistribution of resources to do away with this situation. The international organizations have so far been dominated by the national interests of the superpowers. The World Bank, for instance, has granted loans to Third World countries on condition that they use it to grow cash crops, to sell abroad and thus make money to pay off their foreign debts. The result has been less food grown for the consumption of their own people and mass starvation.[54]

In 1974 U.S. food aid was cut off to Bangladesh and other countries. By that summer Bangladesh could not pay for any more food because the price of wheat had tripled. It had contracted to buy 230,000 tons of wheat from the United States. The wheat was ready and the ships were ready to load it, but Bangladesh had run out of money and the wheat was not sent. A few months later there was famine and mass starvation in Bangladesh. The economist Emma Rothschild comments: "United States officials observed these commercial proceedings, but the Government chose not to intervene in the workings of free enterprise."[55]

It seems clear that, if justice is to be done, the rigid ideological insistence on "free enterprise," the fears of planning, of socialism, of interfering with the market, will have to be replaced by a willingness to plan, to experiment, and to take care of people's needs outside the money system.

President Reagan, early in his administration, was part of a "North-South Summit" of twenty-three nations meeting in Mexico to discuss the problems of poor nations. Mexican writer Carlos Fuentes related an exchange between Reagan and the leader of Tanzania, Julius Nyerere:

> Mr. Reagan . . . still insists that private enterprise do the job from scratch, which is not possible. When Reagan said that the problems of agriculture and food production could be solved only by private enterprise, Nyerere immediately shot back: "But Mr. President, you have the most heavily subsidized agriculture in the world. . . . It is an agriculture propped up by state interventionism, so what are you talking about?"[56]

The fear of the United States of socialist planning, the insistence that Third World nations depend on private enterprise, was reemphasized by President George Bush almost as soon as he took office in early 1989. Clearly, there is much resistance, among powerful interests devoted to making money, to the kinds of bold steps needed to bring about justice inside nations and in the world. Citizens of the various countries, rich and poor, will have to organize themselves as a force to turn the national and world priorities toward equality and economic democracy.

Reason, Representation, or Struggle?

In 1971 Harvard University philosopher John Rawls wrote *A Theory of Justice,* which led to years of discussion among political philosophers.[57] Rawls believes there is too much inequality, and he has worked out an elaborate philosophical argument for a just distribution of wealth.

He omits, however, one crucial problem: the real world of harsh conflict that surrounds every issue of economic justice. That real world is one of class difference and class conflict. A reasoned argument is not enough to persuade a billion-dollar corporation.

The establishment tries very hard to shut that out of public consciousness. During the 1988 presidential campaign candidate George Bush said, "I must say that I've been disturbed, as I've witnessed my opponent's campaign over the several past weeks, at the increasing appeals to class conflict. In my view, there is no place in American public life for philosophies that divide Americans one from another on class lines and that excite conflict among them."[58]

Confronting that real world of class conflict requires two things. First, we need to get a consensus of agreement among *most* people on the goal of basic equality. A minority of affluent, powerful opponents will oppose this. This is a class society and there will be class conflict. But if we can get a consensus among *most* people, they might organize themselves in such a way as to win that conflict.[59]

The consensus will be on the principle of equality. I'm not speaking of perfect equality; it's impractical and worrisome to many people to paint a picture of a perfect *leveling* of the situation. I mean equal access for every human being on earth to the fundamental necessities of existence: food, housing, medical care, education, civil liberties, useful work, and respect, with these things distributed according to need. And beyond that, a reasonable equality in income, using small differences as incentives when needed.[60]

Getting that consensus is not easy in a society where the dominant ideology is shaped by the people who have the wealth and the power to overwhelm the mass media and the educational system with their ideas. It will be necessary (this essay is such an attempt) to show the falseness of that ideology, with all its arguments against a radical reorganizing of society: the glorification of the present system ("the market . . . the profit motive . . . the money incentive . . . entitlement to wealth"), the putting down of the poor and less financially successful people ("they're lazy . . . they're not intelligent . . . they deserve what they get"), and the use of scare words ("socialism . . . communism").

In fact, we are not impossibly far from having such a consensus. During the 1988 presidential campaign, a *New York Times*/CBS News poll reported:

> Three-fourths of the public favors spending more for education and anti-drug programs. More than two-thirds favor more spending for the homeless, and half favor spending increases for daycare. But fewer than one-fifth of those surveyed want to spend more on military programs.[61]

One of the things said most often about the United States is that there is very little class consciousness. But there is strong evidence that this view is mistaken. Back in 1964 the Survey Research Center of the University of Michigan asked people, "Is the government run by a few big interests looking out for themselves?" About 26 percent of those polled answered yes. But by 1972, 53 percent answered yes. And long after the war was over, in 1984, the year that Ronald Reagan was reelected president, a poll by the Harris organization showed that 74 percent of the public believed "a small group of insiders run the country."[62]

Truth is, *class consciousness* is a slippery term, making it hard to decide whether American workers are class conscious. Most blue-collar and white-collar workers certainly know that there are employers and workers, rich and poor, powerful and powerless and that "a small group of insiders run the country." They have not translated this consciousness into the formation of a working-class party such as in England, France, Italy, Spain, etc. They have suffered many defeats at the hands of the employer class. But the fact that there is consciousness of their situation creates a basis for future action.

For his book *Working*, Studs Terkel spent three years interviewing hundreds of people: farmers, miners, receptionists, telephone operators, actors, truck drivers, garbage men, mechanics, janitors, policemen, welders, cabdrivers, hotel clerks, bank tellers, secretaries, supermarket workers, athletes, musicians, teachers, nurses, carpenters, and firemen. He found pride in work, but also "a scarcely concealed discontent" and, compared to his interviews of workers in the thirties, more people who said, "the system stinks."[63]

It seems that very many people understand the existence of injustice and the need for change. But they consider themselves helpless, and this is probably the greatest obstacle to social change.

History comes in handy in this situation. People can learn from the history of social struggle (a history that is largely omitted in the traditional learning that takes place in our schools and in the society) how seemingly powerless people were able to bring about changes in their own situation and changes in public policy. The history of the civil rights movement, the antiwar movement, the women's movement, and the labor movement can inspire people to create new movements for change.

History does show us how hard it is to challenge those in authority, those with great wealth and great power. It shows how many battles have been lost in class conflict in this country. But we also learn that at certain times in history, surprising, unpredicted victories became possible when ordinary people organized, risked, sacrificed, and persisted.

Those victories for social justice did not come through the normal workings of the political system. It is useful, even necessary, to work through the regular channels as far as they can take you. But they have never taken us very far. The very poor seem to understand that. In 1969 a Senate committee was investigating hunger in the South. A black woman was on the stand and Senator Ellender of Louisiana was questioning her. She suddenly said to him, "I want to ask *you* a few questions."

SENATOR: Don't get smart with me!
WOMAN: How come you can't do anything for us?
SENATOR: We just make the laws.
WOMAN: What good is the laws?
SENATOR: It is up to the Executive Branch to enforce them. There are three branches of government, Executive, Legislative, Judicial, and . . .
WOMAN (INTERRUPTING): You got all those branches of government to go through before something gets done. No wonder we starving!

However, the establishment of representative government, voting for Congress and for the president, created the possibility (although the political system itself would be controlled by money) of a legislative response to public pressure. And it was such pressure—coming out of social struggle—that brought about whatever economic reforms we see now in our economic system. Indeed, there is no country in the world that can match the United States in the number and intensity of labor struggles.

Take the eight-hour day. It was achieved for most workers, not through the legislative process, but through many years of bitter struggle. Hundreds of thousands of workers won the eight-hour day by going on strike long before it was enacted into law in the 1930s. In the year 1886 when the labor movement decided to make its big push to reduce the working day to eight hours, there were 1,400 strikes, involving 500,000 workers.

The period between 1877 and 1914 saw a series of bitter labor struggles take place throughout the country in rebellion against intolerable working conditions and starvation wages. In 1877 there was a wave of railroad strikes in the east, suppressed finally by state militia and federal troops at the cost of a hundred lives.

In 1894 the Pullman Strike tied up the nation's railroads until it was crushed by court injunctions and soldiers. In 1892 and again in 1902 there were strikes in the steel industry. In 1913–1914 there was the long, violent strike in the coal country of southern Colorado. The workers were almost always defeated by the power of the corporation, with the collaboration of the government.

In 1912, however, came a rare labor victory, in Lawrence, Massachusetts, in the strike of mostly immigrant textile workers, many of them women, against the powerful American Woolen Company. Strikers were beaten, jailed, and killed. Their children went hungry. But they organized mass picketing, chains of 7,000 and 10,000 men and women in endless picket lines. The company finally surrendered and agreed to give higher wages and overtime pay.

In the 1880s and 1890s farmers all over the United States, pushed to the wall by banks, railroads, and merchants, organized a vast network of alliances that became the Populist Movement, involving in one way or another several million farm families, north and south. They elected candidates to state legislatures and to Congress.

In response to these movements and out of the desire of the establishment to make its control more secure, certain reforms took place in the twentieth century. Congress passed laws to regulate the railroads. It set up a Federal Trade Commission supposedly to control the growth of monopolies. There was a constitutional amendment to enable the rich to be taxed more heavily through a graduate income tax. A number of states passed laws regulating wages and hours and providing for safety inspection of factories and compensation for injured workmen.

In the 1930s, in the midst of economic crisis, the country was in turmoil. There were general strikes tying up San Francisco and Minneapolis; riots of the unemployed; organizations of tenants all over the country stopping evictions; and massive strikes in the steel, rubber, and auto industries. There were more than 500 sit-down strikes in 1937 alone.[64]

The reform legislation of the New Deal—unemployment compensation, Social Security, work programs, minimum wages—was very much a reaction, not only to the economic situation, but to that wave of strikes and the threat of growing radicalism in the country. It is clear that the passage of the National Labor Relations Act, which for the first time gave trade unions legal standing and an official machinery for dealing with work grievances, was a response to the widespread disruption of labor struggles, an attempt to pacify the class conflict in the workplace.[65]

Frances Piven and Richard Cloward, in their book *Poor People's Movements,* look at the workers' movements of the thirties and explain the last-minute decision to support the National Labor Relations Act:

> Roosevelt and his advisors had originally thought of labor concessions primarily in terms of unemployment relief and insurance, old age pensions, and wages and hours protections. But rank-and-file agitation set new terms, and the terms would have to be met if labor was to be kept in line.[66]

Under the Social Security Act of 1935 the program Aid to Families with Dependent Children (AFDC) was created. But the states made it difficult for families to get this help. It was not until a disruptive, threatening welfare rights movement developed in the 1960s that AFDC began to give significant help to desperately poor families, most of them consisting of single mothers taking care of their children.[67]

Americans often point with pride to the high standard of living of the working class—the families that own their own homes, a car, and a television and can afford to go away on vacation. All of this—the eight-hour day, a fairly decent wage, and vacations with pay—did not come about through the natural workings of the market, or through the kindness of government. It came about through the direct action of workers themselves in their labor struggles or through the response of state and national governments to the threat of labor militancy.

None of this has been sufficient to bring about economic justice in this country of wealth and poverty, gigantic production and colossal waste, glittering luxury and miserable slums. If we are going to make the radical changes to produce a situation we can call economic justice, much more will be required. People will have to organize and struggle, to protest, to strike, to boycott, to engage in politics, to go outside of politics and engage in civil disobedience, to act out (as blacks did when they simply went into places where they were excluded) the equalization of wealth.

Only when wealth is equalized (at least roughly) will liberty be equalized. And only then will justice be possible in this country. Only then can we finally make real the promise of the Declaration of Independence, to give all men—and women and children—the equal right to "life, liberty, and the pursuit of happiness."

CHAPTER

EIGHT

Free Speech: Second Thoughts on the First Amendment

Growing up in the United States, we are taught that this is a country blessed with freedom of speech. We learn that this is so because our Constitution contains a Bill of Rights, which starts off with the First Amendment and its powerful words:

> Congress shall make no law respecting an establishment of religion, or prohibiting the free exercise thereof; or abridging the freedom of speech, or of the press; or the right of the people peaceably to assemble, and to petition the government for a redress of grievances.

The belief that the First Amendment guarantees our freedom of expression is part of the ideology of our society. Indeed, the faith in pledges written on paper and the blindness to political and economic realities seem strongly entrenched in that set of beliefs propagated by the makers of opinion in this country. We can see this in the almost religious fervor that accompanied the year of the Bicentennial, 200 years after the framing of the Constitution.

In 1987, from newspapers, television, radio, from the pulpits and the classrooms, from the halls of Congress, and in the statements issued by the White House, we heard praise of that document drawn up by the Founding Fathers. *Parade* magazine, read by several million people, printed a short essay by President Ronald Reagan. In it he said,

> I can't help but marvel at the genius of our Founders. . . . They created, with a sureness and originality so great and pure that I can't help but perceive the guiding hand of God, the first political system that insisted that power flows from the people to the state, not the other way around.

That same year, the newspapers carried large advertisements for "The Constitution Bowl," announced by the official Commission on the Bicentennial, to be made of "Lenox fine ivory China" showing the official flowers of the thirteen original states, and "bordered with pure 24 karat gold . . . a masterpiece worthy of the occasion." It was available for $95. A beautiful bowl indeed. And it was a perfect representation of the Constitution—elegant, but empty, capable of being filled with good or bad by whoever possessed the power and the resources to fill it.

So it has been with the First Amendment. The First Amendment was adopted in 1791, as part of the Bill of Rights, in response to criticism of the Constitution when it was before the public for ratification. Needing nine of the thirteen states to ratify it, The Constitution was approved by very small margins in three crucial states: Virginia, Massachusetts, and New York. Promises were made that when the first government took office, a Bill of Rights would be added, and so it was. Ever since then it has been hailed as the bedrock of our freedoms.

As I am about to argue, however, to depend on the simple existence of the First Amendment to guarantee our freedom of expression is a serious mistake, one that can cost us not only our liberties but, under certain circumstances, our lives.

"*No Prior Restraint*"

The language of the First Amendment looks absolute. "Congress shall make no law . . . abridging the freedom of speech." Yet in 1798, seven years after the First Amendment was adopted, Congress did exactly that; it passed laws abridging the freedom of speech—the Alien and Sedition Acts.

The Alien Act gave the president the power to deport "all such aliens as he shall judge dangerous to the peace and safety of the United States."

The Sedition Act provided that "if any person shall write, print, utter, or publish . . . any false, scandalous and malicious writing or writings against the government of the United States, or either house of the Congress of the U.S. or the President of the U.S., with intent to defame . . . or to bring either of them into contempt or disrepute" such persons could be fined $2,000 or jailed for two years.

The French Revolution had taken place nine years earlier, and the new American nation, now with its second president, the conservative John Adams, was not as friendly to revolutionary ideas as it had been in 1776. Revolutionaries once in power seem to lose their taste for revolutions.

French immigrants to the United States were suspected of being sympathizers of their revolution back home and of spreading revolutionary ideas here. The fear of them (although most of these French immigrants had fled the revolution) became hysterical. The newspaper *Gazette of the United States* insisted that French tutors were corrupting American children, "to make them imbibe, with their very milk, as it were, the poison of atheism and disaffection."[1]

The newspaper *Porcupine's Gazette* said the country was swarming with "French apostles of Sedition . . . enough to burn all our cities and cut the throats of all the inhabitants."

In Ireland revolutionaries were carrying on their long struggle against the English, and they had supporters in the United States. One might have thought that the Americans, so recently liberated from English rule themselves, would have been sympathetic to the Irish rebels. But instead, the Adams administration looked on the Irish as troublemakers, both in Europe and in the United States.

Politician Harrison Gray Otis said he "did not wish to invite hordes of wild Irishmen, nor the turbulent and disorderly of all parts of the world, to come here with a view to disturb our tranquility, after having succeeded in the overthrow of their own governments." He worried that new immigrants with political ideas "are hardly landed in the United States, before they begin to cavil against the Government, and to pant after a more perfect state of society."[2]

The Federalist party of John Adams was opposed by the Republican party of Thomas Jefferson. It was the beginning of the two-party system in the new nation. Their disagreements went back to the Constitution and the Bill of Rights, to battles in Congress over Hamilton's economic program. The tensions in the country were heightened at this time by an epidemic of yellow fever, with discontented citizens rioting in the streets.

Jefferson, a former ambassador to France, was friendly to the French Revolution, while Adams was hostile to it. President Adams, in the developing war between England and France, was clearly on the side of the English, and one historian has called the Sedition Act "an internal security measure adopted during America's Half War with France."[3]

Republican newspapers were delivering harsh criticism of the Adams administration. The newspaper *Aurora* in Philadelphia (edited by Benjamin Bache, the grandson of Benjamin Franklin) accused the president of appointing his relatives to office, of squandering public money, of wanting to create a monarchy, and of moving toward war. Even before the Sedition Act became law, Bache was arrested and charged on the basis of common law with libeling the president, exciting sedition, and provoking opposition to the laws.

The passage of the Sedition Act was accompanied by denunciations of the government's critics. One congressman told his colleagues, "Philosophers are the pioneers of revolution. They . . . prepare the way, by preaching infidelity, and weakening the respect of the people for ancient institutions. They talk of the perfectability of man, of the dignity of his nature, and entirely forgetting what he is, declaim perpetually about what he should be."[4] The statement about what man "is," could have been taken straight from Machiavelli.

The atmosphere in the House of Representatives in those days might be said to lack some dignity. A congressman from Vermont, Irishman Matthew Lyon, got into a fight with Congressman Griswold of Connecticut. Lyon spat in Griswold's face, Griswold attacked him with a cane, Lyon fought back with fire tongs, and the two grappled on the floor while the other members of the House first watched, then separated them. A Bostonian wrote angrily about Lyon: "I feel grieved that the saliva of an Irishman should be left upon the face of an American."[5]

Lyon had written an article saying that under Adams "every consideration of the public welfare was swallowed up in a continual grasp for power, in an unbounded thirst for ridiculous pomp, foolish adulation, and selfish avarice." Tried for violation of the Sedition Act, Lyon was found guilty and imprisoned for four months.

The number of people jailed under the Sedition Act was not large—ten—but it is in the nature of oppressive laws that it takes just a handful of prosecutions to create an atmosphere that makes potential critics of government fearful of speaking their full minds.

It would seem to an ordinarily intelligent person, reading the simple, straightforward words of the First Amendment—"Congress shall make no law . . . abridging the freedom of speech, or of the press."—that the Sedition Act was a direct violation of the Constitution. But here we get our first clue to the inadequacy of words on paper in ensuring the rights of citizens. Those words, however powerful they seem, are interpreted by lawyers and judges in a world of politics and power, where dissenters and rebels are not wanted. Exactly that happened early in our history, as the Sedition Act collided with the First Amendment, and the First Amendment turned out to be poor protection.[6]

The members of the Supreme Court, sitting as individual circuit judges (the new government didn't have the money to set up a lower level of appeals courts, as we have today) consistently found the defendants in the sedition cases guilty. They did it on the basis of English common law. Supreme Court Chief Justice Oliver Ellsworth, in a 1799 opinion, said, "The common law of this country remains the same as it was before the Revolution."[7]

That fact is enough to make us pause. English common law? Hadn't we fought and won a revolution against England? Were we still bound by English common law? The answer is yes. It seems there are limits to revolutions. They retain more of the past than is expected by their fervent followers. English common law on freedom of speech was set down in Blackstone's *Commentaries*, a four-volume compendium of English common law. As Blackstone put it:

> The liberty of the press is indeed essential to the nature of a free state, but this consists in laying no *previous* restraint upon publications, and not in freedom from censure for criminal matter when published. Every freeman has an undoubted right to lay what sentiments he pleases before the public; to forbid this is to destroy the freedom of the press; but if he publishes what is improper, mischievous, or illegal, he must take the consequences of his own temerity.[8]

This is the ingenious doctrine of "no prior restraint." You can say whatever you want, print whatever you want. The government cannot stop you in advance. But once you speak or write it, if the government decides to make certain statements "illegal," or to define them as "mischievous" or even just "improper," you can be put in prison.

An ordinary person, unsophisticated in the law, might respond, "You say you won't stop me from speaking my mind—no prior restraint. But if I know it will get me in trouble, and so remain silent, that is prior restraint." There's no point responding to common law with common sense.

That early interpretation of the First Amendment, limiting its scope to no prior restraint, has lasted to the present day. It was affirmed in 1971 when the Nixon administration tried to get the Supreme Court to stop the publication in the *New York Times* of the Pentagon Papers, the secret official history of the U.S. war in Vietnam.[9]

The Court refused to prevent publication. But one of the justices held up a warning finger. He said, we are making this decision on the basis of no prior restraint; if the *Times* goes ahead and prints the document, there is a chance of prosecution.

So, with the doctrine of no prior restraint, the protection of the First Amendment was limited from the start. The Founding Fathers, whether liberal or conservative, Federalist or Republican—from Washington and Hamilton to Jefferson and Madison—believed that seditious libel could not be tolerated, that all we can ask of freedom of speech is that it does not allow prior restraint.[10]

Well, at least we have that, a hopeful believer in the First Amendment might say: They can't stop free expression in advance. It turns out, however, that such optimism is not justified. Take the case of a book, *The C.I.A. and the Cult of Intelligence*, written by Victor Marchetti, a former CIA agent, and John Marks, a journalist. The book exposed a number of operations by the CIA that did not seem to be in the interests of democracy and that used methods an American might not be proud of. The CIA went to court asking that the publication of the book be stopped, or at least, that some 225 passages, affecting "national security" (or as Marchetti and Marks said, embarrassing the CIA) be omitted from the book.

Did the judge then invoke no prior restraint and say, We can't censor this book in advance; take action later if you like? No, the judge said, I won't order 225 deletions from the book; I'll only order 168 deletions.

Another bit of surgery on any citizen's innocent assumption that the First Amendment meant what it said. The book was published in 1972 with the court-ordered deletions. But the publisher left blank spaces, sometimes entire blank pages, where the deletions were made. It is, therefore, an interesting book to read, not only for what it tells about the CIA, but what it tells about the strength of the First Amendment.[11]

Or take the case of another CIA agent, Frank Snepp, who wrote a book called *Decent Interval*, a sharp critique of the actions of the U.S.

government and the CIA during the last-minute evacuation of American forces from Saigon in 1975. Snepp's book was not stopped from publication, but the CIA sued Snepp for violation of his contract, in which he had agreed to submit his writings for CIA approval before publication. Snepp argued the agreement only applied to material classified secret and he had not used any classified material in his book.

The Supreme Court ruled six to three (in an atmosphere of secrecy—no briefs were submitted, no oral argument took place) that even without an agreement the CIA had a right to stop publication because "the government has a compelling interest in protecting the secrecy of information important to our national security." Because the book was already published, the Court ruled that all its royalties must go to the U.S. government. Any citizen who reads *Decent Interval* can decide whether Snepp in any way hurt "national security" by what he wrote or if that scary phrase was once again being used to prevent a free flow of ideas.[12]

Free Speech and National Security

The powerful words of the First Amendment seem to fade with the sounds of war, or near war. The Sedition Act of 1798 expired, but in 1917 when the United States entered World War I, Congress passed another law in direct contradiction of the amendment's command that "Congress shall make no law . . . abridging the freedom of speech, or of the press." This was the Espionage Act of 1917.

Titles of laws can mislead. While the act did have sections on espionage, it also said that persons could be sent to prison for up to twenty years if, while the country was at war, they "shall wilfully cause or attempt to cause insubordination, disloyalty, mutiny, or refusal of duty in the military or naval forces of the United States, or shall wilfully obstruct the recruiting or enlistment service of the U.S."[13]

This was quickly interpreted by the government as a basis for prosecuting anyone who criticized, in speech or writing, the entrance of the nation into the European war, or who criticized the recently enacted conscription law. Two months after the Espionage Act was passed, a Socialist named Charles Schenck was arrested in Philadelphia for distributing 15,000 leaflets denouncing the draft and the war. Conscription, the leaflets said, was "a monstrous deed against humanity in the interests of the financiers of Wall Street. . . . Do not submit to intimidation."

Schenck was found guilty of violating the Espionage Act, and sentenced to six months in prison. He appealed, citing the First Amendment: "Congress shall make no law . . ." The Supreme Court's decision was unanimous and written by Oliver Wendell Holmes, whose reputation was that of an intellectual and a liberal. Holmes said the First Amendment did not protect Schenck:

> The most stringent protection of free speech would not protect a man in falsely shouting fire in a theatre and causing a panic. . . . The question in every case is whether the words used are used in such circumstances and are of such a nature as to create a clear and present danger that they will bring about the substantive evils that Congress has a right to prevent.[14]

It was a clever analogy. Who would think that the right of free speech extended to someone causing panic in a theater? Any reasonable person must concede that free speech is not the only important value. If one has to make a choice between someone's right to speak, and another person's right to *live,* that choice is certainly clear. No, there was no right to falsely shout fire in a theater and endanger human life.

A clever analogy, but a dishonest one. Is shouting fire in a crowded theater equivalent to distributing a leaflet criticizing a government policy? Is an antiwar leaflet a danger to life, or an attempt to save lives? Was Schenck shouting "Fire!" to cause a panic, or to alert his fellow citizens that an enormous conflagration was taking place across the ocean? And that they or their sons were in danger of being thrown into the funeral pyre that was raging there? To put it another way, who was creating a clear and present danger to the lives of Americans, Schenck, by protesting the war, or Wilson, by bringing the nation into it?

Also prosecuted under the Espionage Act was Socialist leader Eugene Debs, who had run against Wilson for the presidency in 1912 and 1916. Debs made a speech in Indiana in which he denounced capitalism, praised socialism, and criticized the war: "Wars throughout history have been waged for conquest and plunder. . . . And that is war in a nutshell. The master class has always declared the wars; the subject class has always fought the battles."[15]

Debs's indictment said that he "attempted to cause and incite insubordination, disloyalty, mutiny and refusal of duty in the military forces of the U.S. and with intent so to do delivered to an assembly of people a public speech." Debs spoke to the jury:

> I have been accused of obstructing the war. I admit it. Gentlemen, I abhor war. I would oppose war if I stood alone. . . . I have sympathy with the suffering, struggling people everywhere. It does not make any difference under what flag they were born, or where they live.

He was convicted and sentenced to ten years in prison, the judge denouncing those "who would strike the sword from the hand of this nation while she is engaged in defending herself against a foreign and brutal power."

When the case came to the Supreme Court on appeal, again Oliver Wendell Holmes spoke for a unanimous court, affirming that the First Amendment did not apply to Eugene Debs and his speech. Holmes said Debs made "the usual contrasts between capitalists and laboring men . . . with the implication running through it all that the working men are not concerned in the war." So, Holmes said, the "natural and intended effect" of Debs's speech would be to obstruct recruiting.[16]

Altogether, about 2,000 people were prosecuted and about 900 sent to prison, under the Espionage Act, not for espionage, but for speaking and writing against the war. Such was the value of the First Amendment in time of war.

Socialist leader Kate Richards O'Hare was sentenced to five years in prison because, the indictment claimed, she said in a speech that "the women of the United States were nothing more nor less than brood sows, to raise children to get into the army and be made into fertilizer."[17]

A filmmaker was arrested for making the movie *The Spirit of '76* about the American Revolution, in which he depicted British atrocities against the colonists. He was found guilty for violating the Espionage Act because, the judge said, the film tended "to question the good faith of our ally, Great Britain." He was sentenced to ten years in prison. The case was officially called *U.S. v. Spirit of '76.*[18]

The Espionage Act remains on the books, to apply in wartime and in "national emergencies." In 1963 the Kennedy administration proposed extending its provisions to statements made by Americans overseas. Secretary of State Rusk cabled Ambassador Henry Cabot Lodge in Vietnam, saying the government was concerned about American journalists writing "critical articles . . . on Diem and his government" that were "likely to impede the war effort."

Free speech is fine, but not in a time of crisis—so argue heads of state, whether the state is a dictatorship or is called a democracy. Has that not proved again and again to be an excuse for stifling opposition to government policy, clearing the way for brutal and unnecessary wars? Indeed, is not a time of war exactly when free speech is most needed, when the public is most in danger of being propagandized into sending their sons into slaughter? How ironic that freedom of speech should be allowed for small matters, but not for matters of life and death, war and peace.

On the eve of World War II, Congress passed still another law limiting freedom of expression. This was the Smith Act of 1940, which extended the provisions of the Espionage Act to peacetime and made it a crime to distribute written matter or to speak in such a way as to cause "insubordination or refusal of duty in the armed forces." The act also made it a crime to "teach or advocate" or to "conspire to teach or advocate" the overthrow of the government by force and violence.

Thus in the summer of 1941, before the United States was at war, the headquarters of the Socialist Workers party was raided, literature seized, and eighteen members of the party were arrested on charges of "conspiracy to advocate overthrow of the government of the United States by force and to advocate insubordination in the armed forces of the U.S." The evidence produced in court against them was not evidence of the use of violence or the planning of violence, but their writings and teachings in Marxist theory.

Their crime, it appeared was that they were all members of the Socialist Workers party, whose Declaration of Principles, said the judge who sentenced them to prison, was "an application of Marxist theories and doctrines to . . . social problems in America."[19] The judge noted that in the raid of their headquarters a "large number of communistic books were seized." The appeal of the party to the federal courts lost, and the Supreme Court refused to take the case.[20]

The Communist party, a bitter rival of the Socialist Workers party and a supporter of World War II, did not criticize its prosecution. After the war, it was itself prosecuted under the Smith Act, and its leaders sent to prison. Here, again, the evidence was a pile of seized literature, the works of Marx, Engels, Lenin, and Stalin.

The First Amendment, said the Supreme Court, did not apply in this case. The "clear and present danger" doctrine laid down by Holmes was still a principle of constitutional law, and now Chief Justice Vinson gave it a bizarre twist. He said that while the danger of violent overthrow was not "clear and present," the conspiracy to advocate that in the future was a *present* conspiracy, and so, the conviction of the Communist leaders must stand.[21]

The First Amendment was being subjected to what constitutional experts call "a balancing test," where the right of free expression was continually being weighed against the government's claims about national security. Most of the time, the government's claim prevailed. And why should we be surprised. Does the Executive Branch not appoint the federal judges and the prosecutors? Does it not control the whole judicial process?

It seems to me that the security of the American people, indeed of the world, cannot be trusted to the governments of the world, including our own. In crisis situations, the right of citizens to freely criticize foreign policy is absolutely essential, indeed a matter of life and death. National security is safer in the hands of a debating, challenging citizenry than with a secretive, untrustworthy government. Still, the courts have continued to limit free debate on foreign policy issues, claiming that national security overrides the First Amendment.

For instance, in the spring of 1986 a debate on problems in the Middle East was scheduled in Cambridge, Massachusetts, between Harvard Law School professor Alan Dershowitz and Zuhdi Terzi, a Palestine Liberation Organization (PLO) observer at the United Nations. The State Department went into court to prevent Terzi from traveling from New York to Boston to participate in the debate, claiming that Terzi's appearance would hurt the U.S. government's policy not to recognize the PLO. The federal district court in Boston refused to stop Terzi, but the U.S. Court of Appeals accepted the government's argument, ordered Terzi to stay away, and the debate did not take place.[22]

Various court decisions have upheld the right of the government to bar many artists and writers from entering the United States because of their political views and activities, for example, the Nobel Prize–winning novelist Gabriel García Márquez and the Italian playwright Dario Fo. Their books could be read, but their voices could not be heard.

A Latin-American journalist Patricia Lara, a citizen of Colombia, was kept from entering the United States in 1986 to attend a journalistic awards ceremony at Columbia University. What was revealed in the legal proceedings was that the Immigration and Naturalization Service had a "lookout book" containing the names of 40,000 people who were to be kept out of this country on grounds of national security.[23]

Poet Margaret Randall gave up her American citizenship to live for seventeen years in Mexico, Cuba, and Nicaragua, but then married an American citizen and wanted to regain her citizenship and return to the United States. The Immigration and Naturalization Service insisted she could not return. In court, it quoted from five of her books, saying, "Her writings go beyond mere dissent . . . to support of Communist dominated governments." In short, she was being kept out because of her ideas. (After a long battle in the courts, she won her case in 1989.)

Again for reasons having to do with national security, the First Amendment has been declared to have "a different application" for men in the military service. This was the language used by Supreme Court Justice William Rehnquist in the Court's decision in affirming the court-martial conviction of Howard Levy, an army doctor who served during the Vietnam War.[24]

Levy had been charged under the Uniform Code of Military Justice as guilty of conduct "unbecoming an officer and a gentleman" and of harming "good order and discipline" in the armed forces. As a physician stationed at Fort Jackson, South Carolina, Levy had supposedly said the following to enlisted men:

> The United States is wrong in being involved in the Vietnam war. I would refuse to go to Vietnam if ordered to do so. . . . If I were a colored soldier and were sent I would refuse to fight. Special Forces personnel are liars and thieves and killers of peasants and murderers of women and children.

Freedom of speech is supposed to protect even the strongest of words, but these words were too strong for Justice Rehnquist, who saw them as hurting the necessary discipline of the armed forces. He said, "The fundamental necessity of obedience . . . may render permissible within the military that which would be constitutionally impermissible outside it."

Earlier in the Vietnam War, an army lieutenant named John Dippel had tried to pin the Declaration of Independence to the wall of his barracks. This was not permitted by the commander of the base, and the army's legal office in Washington advised Dippel that he had no First Amendment right to do this.[25]

Another Supreme Court decision, in 1980, ruled that a base commander in the military had a right to approve any written material circulated or posted on the base, saying, "While members of the military services are entitled to the protections of the First Amendment, the rights of military men must yield somewhat to meet certain overriding demands of discipline and duty."[26]

As popular protest asserted itself powerfully during the Vietnam War and helped bring it to a close, in the higher reaches of government, democracy itself came to be looked on with suspicion.

In 1975 Samuel Huntington, a Harvard political scientist and adviser to presidents, wrote a report for the Trilateral Commission, a group of powerful men from government and business in the United States, Japan, and Western Europe. Huntington pointed to the protest movements of the sixties, saying, "The essence of the democratic surge of the 1960's was a general challenge to existing systems of authority, public and private." Huntington worried about the United States losing its dominant position in the world and wrote of "an excess of democracy." He said there might be "desirable limits to the extension of political democracy."[27]

Police Powers and the First Amendment

As we have seen, the national government can restrict freedom of speech in relation to foreign policy, through judicial reinterpretations of the First Amendment. But what about *state* laws restricting freedom of speech or press? For over a century, the First Amendment simply did not apply to the states, because it says, "*Congress* shall make no law." The states could make whatever laws they wanted.

And they did. In the years before the Civil War, as abolitionists began to print antislavery literature, the states of Georgia and Louisiana passed laws declaring the death penalty for anyone distributing literature "exciting to insurrection" or with "a tendency to produce discontent among the free population . . . or insubordination among the slaves."

When in 1833 the Supreme Court had to decide if the Bill of Rights applied to the states, Chief Justice Marshall said that the intent of the Founding Fathers was that it should not.[28] Indeed, James Madison had proposed an amendment forbidding the states from interfering with various rights including freedom of speech, and the Senate defeated it.

Madison's intent seemed finally to become part of the Constitution with the passage of the Fourteenth Amendment in 1868, which said that no state "shall deprive any person of life, liberty, or property, without due process of law." But in 1894, someone wanting to make a speech on the Boston Common was arrested because he had not gotten a permit from the mayor as required by city law. When he claimed that the Fourteenth Amendment now prevented any state from depriving persons of liberty, including freedom of speech, the Supreme Court ruled unanimously that the mayor could "absolutely or conditionally forbid public speaking in a highway or public park," that the Fourteenth Amendment did not affect the "police powers" of the state.[29]

This was a localized version of the national security argument for limiting freedom of speech, and it prevailed until 1925. In that year, 137 years after the ratification of the Constitution, the Supreme Court finally said that the states could not abridge freedom of speech, because of the Fourteenth Amendment.[30] However, this still left freedom of speech as something to be balanced against the "police powers" of the states. In the years that followed, the balance would sometimes go one way, sometimes another, leaving citizens bewildered about how much they could depend on the courts to uphold their rights of free expression.

For instance, in 1949, after Chicago police arrested Father Terminiello, an anti-Semitic preacher who had attracted an angry crowd around his meeting hall, the Supreme Court ruled that the Terminiello had a First Amendment right to speak his mind, and the fact that this excited opposition should not be used as an excuse to stop his speech. It said that one "function of free speech under our system of government is to invite dispute."[31]

Shortly after that, however, Irving Feiner, a college student in Syracuse, New York, was making a street corner speech from a small platform, denouncing the mayor, the police, the American Legion, and President Truman, when one of his listeners said to a policeman standing by, "You get that son-of-a-bitch off there before I do." The policeman arrested Feiner, and the Supreme Court upheld the arrest, saying this was not free speech but "incitement to riot," although the tumult and excitement around Terminiello's speech had been far greater than in Feiner's case.[32]

The uncertainty continues. In 1963 the Supreme Court overturned the arrest of 187 black students assembling peacefully on the grounds of the South Carolina state capitol to protest racial discrimination.[33] But three years later when a group of civil rights activists demonstrated peacefully on the grounds of a Tallahassee jail, the conviction was upheld. Justice Hugo Black said for the majority that people do not have a constitutional right to protest "whenever and however and wherever they please."[34]

The right to distribute leaflets on public streets has been affirmed by the Supreme Court on a number of occasions, even when the street was privately owned, as in 1946 when the Court upheld the right of Jehovah's Witnesses to distribute their literature in a company town.[35] It affirmed this conclusion (that when privately owned areas are open to public use, the First Amendment protections are not surrendered) in the 1968 case of union members distributing handbills about their labor dispute at a shopping mall.[36]

Four years later, however, when a group of people were arrested in a shopping mall for distributing leaflets against the Vietnam War, the Court said they were properly arrested. What was the difference between this case and the other? The union people, the Court said, were expressing themselves about an issue connected with the shopping center. But the Vietnam War had nothing to do with the shopping center, so those people had no First Amendment right to express themselves.[37]

For a long time, the public has been led to believe in the magic word *precedent.* The idea is that the courts follow precedents, that if a decision has been made in a case, it will not be overturned in similar cases. Lawyers and judges understand however, what laypeople often do not, that, in the rough-and-tumble reality of the courts, precedent has as much solidity as a Ping-Pong ball. All a court has to do is to find *some* difference between two cases and it has grounds for giving a different opinion.

In other words, judges can always find a way of making the decision they want to make, for reasons that have little to do with constitutional law and much to do with the ideological leanings of the judges. I would suspect that the decision against the Vietnam leafleters had much more to do with the justices' feelings about the war than with the fact that the shopping mall was not itself involved in the war.

What of the First Amendment rights of high-school students? Here again we find such conflicting decisions as to make us very dubious about the strength of the First Amendment. In the sixties, the Supreme Court said that school officials in Iowa could not prohibit students from wearing black arm bands to protest the Vietnam War. It said, "We do not confine . . . First Amendment rights to a telephone booth or the four corners of a pamphlet or to supervised and ordained discussion in a school classroom."[38]

We might have expected after this (if we had retained our innocence about the power of precedent) that the Court would not allow high-school officials to censor student publications. But in 1988, it ruled that a high-school principal in a suburb of St. Louis could cut out two pages of a student newspaper to eliminate stories on teenage pregnancy and on the effects of divorce on children.

The Court, straining to show the difference between this and the Iowa black arm band case, said, "The question whether the First

Amendment requires a school to tolerate particular student speech . . . is different from the question whether the First Amendment requires a school affirmatively to promote particular student speech."

As it had done in the case of soldiers speaking their minds, the Court found that students were not the same as ordinary citizens in their rights. "The public schools do not possess all of the attributes of streets, parks, and other traditional public forums." So the First Amendment, shaky enough for ordinary citizens, is even more feeble when the issue is the right of free speech of soldiers, foreigners, and high-school students.

To this list of groups exempt from the usual protections of the First Amendment we must add another: prisoners. In a decision that at first glance looked like a rejection of the right of prison authorities to read and censor the mail of prisoners, the Supreme Court said that the state of California could not do this . . . *except* when the prison officials decided it was necessary for reasons of security. In other words, it left the issue up to the same people who wanted the censorship in the first place.[39]

The point in all this recounting of cases is that citizens cannot *depend* on the First Amendment, as interpreted by the courts, to protect freedom of expression. One year the Court will declare, with inspiring words, the right of persons to speak or write as they wish. The next year they will take away that right.

A cloud of uncertainty hovers over how the Supreme Court will decide free speech cases. Nor is there any guarantee, if you decide to exercise your right of free expression by speaking in public or distributing literature, that the Supreme Court will even *hear* your case on appeal. It does not *have* to take appeals in free speech cases, and your chance of getting a hearing in the Supreme Court is about one out of eighty.

A young black man named Charles MacLaurin learned this by hard experience in the year 1963. That summer, he addressed a group of fifty black people in front of the courthouse in Greenville, Mississippi, protesting the arrest of several young black people who had been demonstrating against racial segregation. It was a peaceful meeting, in which MacLaurin criticized the conviction and urged that blacks register to vote to deal with such injustices. A police officer told McLaurin to move on. He said he had a right to speak and continued. He was arrested, charged with disturbing the peace and resisting arrest, found guilty by the local court, sentenced to six months in jail, and this was affirmed by the Mississippi Supreme Court.

When he appealed to the U.S. Supreme Court, he discovered the rule that most citizens (who grow up hearing again and again from some aggrieved person: "I'll take this to the Supreme Court!") don't know: Four of the nine justices must agree to take a case (in technical terms, to grant *certiorari*). Only three Supreme Court justices voted to take MacLaurin's case. By now, it was 1967, and so, four years after his conviction, he went to prison.

An even more serious problem with the First Amendment is that most situations involving freedom of expression never make it into the courts. How many people are willing or able to hire a lawyer, spend thousands of dollars, and wait several years to get a possible favorable decision in court. That means that the right of free speech is left largely in the hands of local police. What are policemen likely to be most respectful of—the Constitution, or their own "police powers"?

I was forced to think about this one day in 1961 when I was teaching at Spelman College and several black students showed up at my house to talk to me about their plan to go into downtown Atlanta to distribute leaflets protesting racial segregation in the city. They wanted to know from me, who taught a course in constitutional law, if they had a legal right to distribute leaflets downtown.

The law was plain. A series of Supreme Court decisions made the right to distribute leaflets on a public street absolute. It would be hard to find something in the Bill of Rights that was more clear cut than this.

I told my students this. But I knew immediately that I must tell them something else: that the law didn't much matter. If they began handing out leaflets on Peachtree Street and a white policeman (all police were white in Atlanta at that time) came along and said "Move!" what could they do? Cite the relevant Supreme Court cases to the policeman? "In *Lovell v. Griffin,* sir, as well as in *Hague v. C.I.O.* and *Largent v. Texas* . . ."

What was more likely at such a moment, that the policeman would fall prostrate before this recitation of Supreme Court decisions? Or that he would finger his club and repeat, "Move on!" At that moment the great hoax in the teaching of constitutional law, the enormous emphasis on the importance of Supreme Court decisions, would be revealed. What would decide the right of free expression of these black students in Atlanta in 1961, what would be more powerful—the words in the Constitution, or the policeman's club?

It wasn't until I began to teach constitutional law in the South, in the midst of the struggle against racial segregation, that I began to understand something so obvious that it takes just a bit of thought to see it, something so important that every young person growing up in America should be taught it: Our right to free expression is not determined by the words of the Constitution or the decisions of the Supreme Court, but by who has the *power* in the immediate situation where we want to exercise our rights.

One of those immediate situations is the street. Another is the workplace.

Free Speech on the Job

As we have seen, for more than a hundred years it was only Congress that was forbidden by the First Amendment to curtail freedom of speech and press. Then in 1925 the Supreme Court wrote freedom of speech into the Fourteenth Amendment and ruled that states could not violate that freedom. But nothing in the Constitution says that private employers may not limit the free speech of their employees.

Not many Americans distribute political pamphlets or speak on street corners, but most Americans work for employers, in situations where to speak their full minds might result in losing their jobs. And while political speakers might have recourse to the courts—weak as that protection is—speakers on the job have no constitutional support.[40]

In 1971 a man named Louis McIntire, who had worked for sixteen years as a chemical engineer with the DuPont Corporation in Texas, published a novel cowritten with his wife that satirized a chemical company. After the book came out, he was fired. He could sue for damages, but he had no constitutional right to his job.

David Ewing, an editor of the *Harvard Business Review,* discussing this case in *The Nation* wrote, "Corporate employees do not enjoy, and have never enjoyed, such basic guarantees of the Bill of Rights as free speech, free press and due process of law—at least, in activities that concern their employers."[37]

Staughton Lynd, a distinguished young historian and a professor at Yale University, visited North Vietnam shortly after the United States began its massive intervention there. He was a strong opponent of our government's actions. Shortly after he returned from his trip, he lost his job at Yale and, despite his impressive record as scholar and teacher, had such difficulty getting another teaching position anywhere that he left the profession and, in his forties, went to law school.

It was clear that his statements on the war, his opposition to American policy, his visit to North Vietnam, and his writings had resulted in his being, in effect, blacklisted in his chosen profession. We had been colleagues together at Spelman College, and when I was a professor at Boston University, I suggested to a member of the history department that they consider Staughton Lynd to fill a vacant faculty position. A senior member of the department said to me, "Oh, Lynd. I was on his doctoral committee at Columbia. A brilliant young man. But no, there's no point in our proposing him. He will never make it through the administration."

When Lynd had finished law school and went to work for a firm of labor lawyers, he wrote a little booklet addressed to working people to give them simple advice on labor law. The booklet started off with the suggestion: "You don't need a labor lawyer." When the book appeared, Lynd was dismissed from his job with the labor firm.

Chuck Atchison, a forty-year-old quality-control inspector for a construction company that built a nuclear energy plant in Texas, spoke out publicly in 1982 about numerous violations of safety regulations at the plant. He was fired. He lost his house, couldn't find a job in his field, and at one point walked the highway picking up beer cans to sell for scrap aluminum. The Bill of Rights could give him no protection.[42]

With no constitutional protection, employees sometimes look to the National Labor Relations Board (NLRB) to protect their rights. But this has proved a very flimsy defense against the power of employers. In the mid-1980s, a truck driver in Michigan was fired from his job because he insisted on inspecting his rig after it was involved in an accident where its brakes malfunctioned. He appealed to the NLRB, which is supposed to protect members of labor unions against "unfair labor practices." But the NLRB said the truck driver, because he was not a union member, could be legally dismissed from his job.[43]

Private colleges and universities do not fall within the scope of the First Amendment, and so their employees are without constitutional protection for their freedom of speech. Arlyn Boudreau, a nurse for twenty years, had worked for seven years with the Boston University Health Clinic when she and several other workers there began to protest the health and working conditions at the clinic. She and another worker were called in by the clinic director and told that they could either resign and get severance pay, or be fired without pay. They refused and were fired.

The official reason given for firing Mrs. Boudreau was "insubordination." *Webster's Dictionary* defines *subordinate* as "placed in a lower class or rank; inferior in order, nature, importance; submissive to authority." She refused to be submissive to authority, insisted on speaking her mind, and was without a job.

Standing in front of the clinic on a picket line protesting the situation, she said, "It's the first time I've been on a picket line in my life. I feel like such a radical I can't believe it. My three daughters, aged 17, 19, and 20, they all came out to picket too." It was 1975, and the nation was getting ready to celebrate the bicentennial of the Declaration of Independence. Mrs. Boudreau kept her independence, although she didn't keep her job.[44]

In reality, the difference between working for a private institution, with no constitutional protection, and working for a public institution, where the First Amendment is supposed to operate, is insignificant. In either case, the power of the employer in the immediate situation is the critical factor, and what legal redress there is must be exercised at the expense of thousands of dollars and years of time, and still remain uncertain in its outcome.

The writer Jonathan Kozol taught in a public school in a black district of Boston. Presumably, his freedom of speech was covered by the Fourteenth Amendment ("nor shall any state deprive any person of life, liberty, or property, without due process of law") because the Supreme Court in 1925 interpreted this clause to cover freedom of speech. But one day in the 1960s Kozol recited something to his class by the black poet Langston Hughes called "Ballad of the Landlord." The poem begins:

> Landlord, landlord,
> My roof has sprung a leak.
> Don't you 'member I told you about it
> Way last week?

The poem goes on to describe the man complaining about the steps, broken down too, whereupon the landlord threatens to cut off his heat, to evict him, to throw his furniture out in the street. The man threatens to hit the landlord. Then come the last stanzas.

Police! Police!
Come and get this man!
He's trying to ruin the government
and overturn the land!

Copper's whistle!
Patrol bell!
Arrest.

Precinct station.
Iron cell.
Headlines in press:

MAN THREATENS LANDLORD
TENANT HELD NO BAIL
JUDGE GIVES NEGRO 90 DAYS IN COUNTY JAIL

Jonathan Kozol was removed from his teaching job by the school committee, which said in its report: "It has been established as a fact that Mr. Kozol taught the poem 'Ballad of the Landlord' to his class and later distributed mimeographed copies of it to his pupils for home memorization."[45]

It is a special irony that in schools and colleges—supposed to be special places for the free dissemination of ideas—it can be dangerous to express yourself. The power of the high-school principal or the school board or the university president or the board of trustees is far more important than the words of the First Amendment.

At Boston University, under the dictatorial power of its president John Silber in the 1970s and 1980s, faculty who spoke their minds were in danger of losing their jobs or their pay raises.[46] Students who expressed themselves freely feared losing scholarships or being suspended. Administrators who differed with Silber might not hold their jobs long. Members of the board of trustees who wanted to stay on the board learned not to disagree with the president's policies.

Student newspapers were required to submit to censorship. Students and faculty who picketed peacefully in front of university buildings were photographed by the campus police. A news director of the university radio station was asked to resign because he refused to censor a broadcast of a speech that criticized President Silber. The Civil Liberties Union of Massachusetts reported that it had never received such a volume of complaints about any institution as about Boston University.

A remarkable student named Yosef Abramowitz, a Zionist who was also active against apartheid in South Africa and who was a member of a group asking that Boston University divest itself of its stock in corporations connected with South Africa, learned firsthand about free speech at the university. One day he hung a banner from his dormitory window, with one word scrawled large on it: "Divest." When he returned to his room at the end of the day, the banner was missing. This happened several times more and he kept putting it back. Then he got a letter from the University Housing Office, telling him he would be evicted from his room if he continued to hang the sign from his window.

Yosef asked the Civil Liberties Union for help, and they secured a lawyer for him. Although private universities are not covered by the First Amendment, a new civil rights law passed in Massachusetts in 1980 seemed to cover civil liberties at private institutions also.

In court, Boston University pretended that it was not at all interested in Abramowitz's message. All it cared about was the aesthetic effect of a banner hanging from his window. Abramowitz's lawyer brought a number of students to the stand who testified that they had the most ugly things hanging from their windows for months (an inflated yellow chicken, for one) and no one had said a word to them. The university's case began to look ludicrous. The judge found that Boston University had violated Yosef Abramowitz's freedom of speech and ordered it to stop harassing him about his banner.

What happened after that shows how the holders of power in any situation are very little perturbed by the law. Abramowitz proceeded to hang three Divest signs from his windows. The university then announced that the court decision only applied to him, and that any other student who hung a banner outside his or her window would face disciplinary action.

There is another institution where the restriction of free speech is especially ironic: the courtroom. The United States has long prided itself on "due process of law," which includes the right to have a lawyer and the right to a fair trial. But in the courtroom itself, the judge controls what can be said in the trial, by excluding any witnesses he chooses to exclude, or any testimony he considers irrelevant.

We thus have a peculiar situation in a country where people vote freely in elections, where a Bill of Rights exists, and something we call *democracy* operates in the society. In the everyday institutions of that democracy—in the schools, in the workplaces, on the military bases, in the courtrooms—freedom of speech is restricted by the power of the people who dominate those institutions.

Secret Police in a Democracy

In our country, so proud of its democratic institutions, a national secret police has operated for a long time, in a clandestine world where the Constitution can be ignored. I am referring to the Federal Bureau of Investigation and the Central Intelligence Agency. It was a CIA official Ray Cline who, when there was talk of the CIA's activities violating the First Amendment, told Congress, "It's only an amendment."[47]

We might comfort ourselves with the thought that the FBI and the CIA are not as fearsome as the KGB of the Soviet Union or the death squads that have operated in right-wing dictatorships supported by the United States—El Salvador, for instance. The *scale* of terror is not comparable. A radical critic of American foreign policy is not likely to be picked up in the middle of the night, immediately imprisoned, or taken out and shot. (Although it is sobering to recall that the FBI conspired with Chicago police in 1969 to murder the black leader Fred Hampton in his bed.)

But should citizens who cherish democracy use the standards of totalitarian states to measure their freedom? We want something better than to be able to say we're not as bad as *those* countries.

The actual apprehension of dissidents is on a much smaller scale in our country compared to theirs. But the mere existence of organizations secretly collecting information on citizens must have a chilling effect on the free speech of everyone. The FBI, according to a Senate report of 1976, has files on 500,000 Americans.

However, the FBI goes far beyond the collection of information. We learned this from a mysterious raid in 1971 on FBI offices in the town of Media, Pennsylvania (its perpetrators have not yet been found). The FBI files were ransacked and then leaked to a small radical magazine that published them. Many of the documents were headed with the word *COINTELPRO*, and only later was it discovered what that stood for: Counter Intelligence Program. The Senate committee investigating the FBI in the mid-seventies wrote in its report:

> COINTELPRO is the FBI acronym for a series of covert action programs directed against domestic groups. In these programs, the Bureau went

> beyond the collection of intelligence to secret action designed to "disrupt" and "neutralize" [the FBI's words] target groups and individuals. The techniques . . . ranged from the trivial (mailing reprints of *Readers Digest* articles to college administrators) to the degrading (sending anonymous poison-pen letters intended to break up marriages) and the dangerous (encouraging gang warfare and falsely labeling members of a violent group as police informers.)[48]

The program began in 1956, according to the Senate committee, ending in 1971 because of the threat of public exposure. (The raid on the Media office took place on March 8, 1971; the FBI decided to terminate COINTELPRO April 27, 1971.) The Senate report said,

> In the intervening 15 years the Bureau conducted a sophisticated vigilante operation aimed squarely at preventing the exercise of First Amendment rights of speech and association, on the theory that preventing the growth of dangerous groups and the propagation of dangerous ideas would protect the national security and deter violence.

Again, the excuse of national security. James Madison, back in 1798, had warned about this in a letter to Thomas Jefferson: "Perhaps it is a universal truth that the loss of liberty at home is to be charged to provisions against danger real or pretended from abroad."[49]

In a totalitarian state, we assume that the head of state is aware of the operations of his secret police. In a country like the United States, however, the higher officials may claim that they don't know what is going on. Former Attorney General Katzenbach said he didn't know, but couldn't have stopped it anyway. Officially, the attorney general is higher in rank than the director of the FBI, but the FBI has a power that attorneys general, and even presidents, have been afraid to touch.

It should not be thought that the president or the attorney-general strongly disapproved of these activities, illegal as many of them were. J. Edgar Hoover's successor Director Clarence Kelley told the Senate committee "the FBI employees . . . did what they felt was expected of them by the President, the Attorney General, the Congress, and the people of the United States." How the FBI knew what "the people" wanted is not clear. But the bureau did have a fairly good idea of what the president wanted and what would get support in Congress.

There is a long record, at least from 1953 to 1973, of illegal opening of citizens' mail by the FBI. There is also a long record of illegal break-ins, "black bag jobs," sometimes called "surreptitious entry." The report of the Senate committee concluded:

> We cannot dismiss what we have found as isolated acts which were limited in time and confined to a few willful men. The failures to obey the law and, in the words of the oath of office [of the president], to "preserve, protect and defend" the Constitution have occurred repeatedly throughout administrations of both political parties going back four decades.[50]

In 1973 staff assistant in the White House Tom Huston drew up a plan, approved by Nixon, that included wiretapping, mail coverage, and "surreptitious entry." He said, "Use of this technique is clearly illegal. It amounts to burglary. It is also highly risky and could result in great embarrassment if exposed. However, it is also the most fruitful tool and can produce the kind of intelligence which cannot be obtained in any other fashion."[51]

One wonders about the files on those 500,000 people (or is it 1 million or 2 million—how can we tell, because the FBI operates in secret). We know from the records of the loyalty investigations of the 1950s that the FBI filed reports on government employees who had been seen entertaining black people, or who had been seen at a concert where Paul Robeson sang, and so on.

One employee was told, "We have a confidential informant who says he visited your house and listened in your apartment to a recorded opera entitled *The Cradle Will Rock,* and that the opera followed along the lines of a downtrodden laboring man and the evils of the capitalist system."[52]

The FBI also maintained files on a number of famous American writers. (This was disclosed when journalist Herbert Mitgang managed to get documents under the Freedom of Information Act.) There was a file on Ernest Hemingway, whom the FBI labeled a drunk and a Communist. The novelists John Steinbeck *(The Grapes of Wrath)* and Pearl Buck *(The Good Earth)* were in the FBI records as people who promoted the civil rights of blacks. John Dos Passos *(U.S.A.)*, William Faulkner *(The Sound and the Fury)*, and Tennessee Williams (*A Streetcar Named Desire*) were all on the list. About Sinclair Lewis, on whom there was a dossier of 150 pages, the FBI said his novel *Kingsblood Royal* was "propaganda for the white man's acceptance of the Negro as a social equal."[53]

There were more serious FBI files than these—the ones kept on members of radical groups, whose names were put on a "Security Index," which at one time listed 15,000. The people on this list were to be picked up and detained without trial in case of a "national emergency." In 1950 Congress passed an Emergency Detention Act, which provided for a set of detention centers (perhaps more accurately, concentration camps) for those people on the Security Index. And although the Act was repealed in 1971, the FBI continued its index.[54]

When the former head of the FBI's Racial Intelligence Section was asked if during the fifteen years of COINTELPRO anyone in the FBI questioned the legality of what was being done, he replied, "No, we never gave it a thought."[55]

As part of the COINTELPRO the FBI in 1970 tried to discredit Jean Seberg, an actress (famous for her part in the French film *Breathless*) who was a sympathizer of the Black Panther party. The FBI suggested she be "neutralized." Seberg was in her seventh month of pregnancy when she read in a newspaper that she had become pregnant by a member of the Black Panther party. It was a false story planted by the FBI. The shock of the story led to premature labor, and the child was born dead. She tried to commit suicide, according to her husband, every year on the anniversary of her baby's death, and in 1979 she did kill herself.[56]

It is hard to tell how many people lost their lives as a result of COINTELPRO, but documents from FBI files, obtained under the Freedom of Information Act, indicated that in the late 1960s and early 1970s, when the FBI was trying to break up the Black Panther organization, nineteen Black Panthers across the nation were killed by law-enforcement officials or by one another in internal feuds (some of which were provoked by the FBI).[57]

One of those deaths was of Fred Hampton, the Chicago Black Panther leader. It turned out that his personal bodyguard, William O'Neal, was an FBI infiltrator, who gave his FBI contact, Roy Mitchell, a detailed floor plan of the apartment occupied by Fred Hampton and others. In a predawn raid, Chicago police fired hundreds of bullets into the apartment, and Hampton, asleep in his bed, was killed. There was an FBI memorandum from the Chicago field office on December 8, 1969 (a few days after the raid):

> Prior to the raid, a detailed inventory of the weapons and also a detailed floor plan of the apartment were furnished to local authorities. . . . The raid was based on the information furnished by the informant.[58]

Although the COINTELPRO was declared suspended in 1971, the FBI continued to keep tabs on organizations that were carrying on nonviolent political activities, but in opposition to official government policy. In 1988 it was revealed (in documents given to the Center for Constitutional Rights under the Freedom of Information Act) that the FBI had infiltrated and kept records on hundreds of organizations in the United States that were opposed to President Reagan's policies in Central America. There were 3,500 pages of files. The excuse was that the FBI was concerned about terrorism. But the records showed a concern for what these organizations were saying, in speech and in writing. A dispatch from the FBI field office in New Orleans in 1983 said:

> It is imperative at this time to formulate some plan of action against CISPES [Committee in Support of the People of El Salvador] and specifically against individuals who defiantly display their contempt for the U.S. government by making speeches and propagandizing their cause.[59]

What happens to members of the secret police who engage in illegal acts? Hardly anything. If there is a particularly flagrant set of actions that are exposed to the public, there may be a token prosecution of one or two minor figures. But they certainly will not be sent to prison, as would ordinary people who intercepted mail or broke into people's homes.

We have as evidence the case of Mark Felt and Edward Miller, two FBI agents, who were the only FBI men prosecuted despite the evidence of thousands of illegal acts brought out by the Senate committee investigating the FBI. Felt and Miller were convicted of authorizing nine illegal break-ins at homes of friends and relatives of members of the Weather Underground (a radical offshoot of the sixties student movement).

There was no evidence that any of these friends and relatives had broken the law, but FBI agents broke in, photographed personal papers, including diaries, statements of political philosophy, and love letters. None of this turned up evidence that helped them find members of the Underground. Felt and Miller could have received maximum prison sentences of ten years. They were not sent to prison, but were fined $5,000 and $3,500, respectively.[60]

The most striking evidence that the FBI was not acting against terrorism or for national security, but was in fact interfering with the First Amendment rights, was its harassment of Martin Luther King, Jr.

In 1961–1962 there were mass demonstrations by black people in Albany, Georgia, against racial segregation there. The Southern Regional Council, an Atlanta research group, sent me (I was teaching at Spelman College in Atlanta at that time) to Albany to report on the situation there. My report was critical of the federal government in general and the FBI in particular. It said, "With all the clear violations by local police of constitutional rights, with undisputed evidence of beatings by sheriffs and deputy sheriffs, the FBI has not made a single arrest on behalf of Negro citizens."[61]

The 1976 Senate committee report on the FBI referred to my report on the Albany situation:

> Before even receiving the full report, Bureau officials were describing it as "slanted and biased" and were searching their files for information about the report's author.
>
> Shortly after the Report was issued, newspapers quoted Dr. King saying that he agreed with the Report's conclusions that the FBI had not vigorously investigated civil rights violations in Albany. FBI headquarters was immediately notified of Dr. King's remarks.

It was not long after this that the FBI began its serious surveillance of King. According to the Senate report: "From December, 1963 until his death in 1968, Martin Luther King Jr. was the target of an intensive campaign by the Federal Bureau of Investigation to 'neutralize' him as an effective civil rights leader."[62] William Sullivan, the FBI man in charge of this operation, told the committee, "No holds were barred. We have used similar techniques against Soviet agents. . . . We did not differentiate. This is a rough, tough business."

Of course, King was not a Soviet agent but an American citizen exercising his constitutional right to speak, write, and organize. The FBI tried its best to stop him from doing this effectively. It tried to discredit him in various ways, tried to get universities to withhold honorary degrees from him, to prevent the publication of articles favorable to him, to find "friendly" journalists to print unfavorable articles. It put microphones in his hotel rooms and offered to play the recordings to reporters. In May of 1962 the FBI included King on its "Reserve Index" as a person to be rounded up and detained in event of a national emergency.[63]

It should be noted that all this cannot be attributed simply to the racial bias of FBI Director J. Edgar Hoover. The Kennedy administration and the Johnson administration collaborated with the FBI. David Garrow, biographer of King, referred to Attorney General Robert Kennedy's "unquestioning acceptance of the Bureau's reports on King." And Burke Marshall, head of the Civil Rights division of the Department of Justice, defended Robert Kennedy's action in authorizing the FBI to eavesdrop on King's conversations.[64]

What all of this indicates is that, despite the Constitution, despite the First Amendment and its guarantees of free speech, American citizens must fear to speak their minds, knowing that their speech, their writings, their attendance of meetings, their signing of petitions, and their support of even the most nonviolent of organizations may result in their being listed in the files of the FBI, with consequences no one can surely know. It was Mark Twain who said, "In our country we have those three unspeakably precious things: freedom of speech, freedom of conscience, and the prudence never to practice either."

The Control of Information

We have not yet come to perhaps the most serious issue of all in regard to freedom of speech and press in the United States. Suppose all of the restrictions on freedom of speech were suddenly removed—the Supreme Court's limitations on the absolute words of the First Amendment, the power of the local police over people wanting to express themselves, the fear of losing one's job by speaking freely, and the chill on free speech caused by the secret surveillance of citizens by the FBI. Suppose we could say anything we want, without fear. Two problems would still remain. They are both enormous ones.

The first is Okay, suppose we can say what we want—how many people can we reach with our message? A few hundred people, or 10 million people? The answer is clear: It depends on how much money we have.

Let's say no one can stop us from getting up on a soapbox and speaking our mind. We might reach a hundred people that way. But if we were the Procter and Gamble Company, which made the soapbox, we could buy prime time for commercials on television, buy full-page ads in newspapers, and reach several million people.

In other words, freedom of speech is not simply a yes or no question. It is also a "how much" question. And *how much* freedom we have depends on how much money we have, what power we have, and what resources we have for reaching large numbers of people. A poor person, however smart, however eloquent, truly has very limited freedom of speech. A rich corporation has a great deal of it.

The writer A. J. Liebling, who wrote about freedom of the press, put it this way, "The person who has freedom of the press is the person who owns one."[65] Owning a press gives you a lot more freedom of speech than having to write a letter to your local newspaper, hoping the editor publishes it. It takes more and more money to own a newspaper, and even if you owned one, it is harder and harder to prevent it being taken over by some giant corporation. At the end of World War II, more than 80 percent of the daily newspapers in the United States were independently owned. Forty years later only 28 percent were independent, the rest owned by outside corporations. And fifteen huge corporations controlled half of the nation's newspaper business.[66]

Three television networks (CBS, ABC, and NBC) control about three-fourths of the prime time on television. With 90 million households owning TV sets, that gives those networks enormous influence on the American mind. Ten publishing companies have half of the $10 billion in book sales. Four giants dominate the movie business.

Mergers and consolidations have created huge media empires, in which ordinary business corporations have bought out publishers, television stations, and newspapers. For instance, International Telephone and Telegraph (IT&T) merged with ABC television in the mid-sixties. Time, Inc. and Warner Communications, Inc. joined in the 1980s to form the world's largest media firm, worth $18 billion. Ben Bagdikian, dean of the Graduate School of Journalism at the University of California, Berkeley, and author of *The Media Monopoly,* summarized the situation: "When 50 men and women, chiefs of their corporations, control more than half the information and ideas that reach 220 million Americans, it is time for Americans to examine the institutions from which they receive their daily picture of the world."[67]

Not only is the usefulness of the First Amendment dependent on wealth, but when occasionally a state legislature tries to remedy the situation slightly, the corporations plead the First Amendment. This is what happened in 1977 when the Massachusetts legislature said corporations could not spend money to influence a public referendum. The idea behind the law was that corporations could so dominate the debate around a public issue as to make freedom of speech on that issue meaningless for people without money.

The corporation lawyer, arguing before the Supreme Court, said,

"Money is speech." (He might have added, "And we have lots of money, so we should have lots of speech.") The Supreme Court decided heroically that the First National Bank of Boston should not be deprived of its First Amendment rights by limiting its use of money to influence a referendum.[68]

The Supreme Court is clearly reluctant to put meaning in the First Amendment by recognizing the great inequality of resources and trying to do something about that. Back in 1969 it unanimously upheld the Federal Communications Commission's "fairness doctrine," which said people attacked on the air had a right to respond.[69] But since then the Court has refused to interfere with the moneyed powers in broadcasting and their ability to keep off the air views they don't like.

In 1973 the Supreme Court decided that CBS had a right to refuse an ad placed by a group of business executives who opposed the war in Vietnam. Even the liberal Justice William O. Douglas went along with the majority, arguing that the government should not interfere with the right of CBS to sell time to whomever it wanted. In saying that, of course, it was approving the right of CBS to interfere with the access of concerned citizens to television time.[70]

Douglas argued that "TV and radio . . . are entitled to live under the laissez faire regime which the First Amendment sanctions." He was succumbing to the basic flaw in all of laissez-faire theory: It pretends to leave people free by keeping government out of a situation and ignores the fact that they are then left to the mercy of the rich in society.

The fairness doctrine itself, which is at least a step toward insisting that the broadcast media give time for opposing views, was considerably weakened by Congress in 1959, when it exempted news conferences and debates. This means that the president or any of his staff can hold news conferences, say whatever they want to a huge television audience, with no opportunity for rebuttal by political critics of the president. It also means that in the campaign for president, the debates between contenders can be limited to the Republican and Democratic parties, excluding minor parties. The Democratic party challenged the provision on news conferences, but the Supreme Court would not hear its appeal. The Socialist Workers party also went to court, claiming its presidential candidate had a right to be heard by the public. The Court refused to take the case.[71]

The second enormous problem for free speech is this: Suppose no one—not government, not the police, not our employer—stops us from speaking our mind, *but we have nothing to say.* In other words what if we do not have sufficient information about what is happening in the country or in the world and do not know what our government is doing at home and abroad? Without such information, having the freedom to express ourselves does not mean much.

It is very difficult for the ordinary citizen to learn very much about what is going on, here or in other countries. There is so much to know. Things are so complicated. But what if, in addition to these natural limitations, there is a deliberate effort to keep us from knowing? In fact, that is the case, through government influence on the media, through self-censorship of the media (being prudent, as Mark Twain said), and through the government's lies and deceptions.

There is no democratic conscience at work when the government decides that it must manipulate the press on behalf of its foreign policy objectives. An editor of *Strategic Review* (A. G. B. Metcalf, also chairman of the board of trustees of Boston University), a right-wing publication dealing with military strategy, delivered a stern warning to the media in 1983:

> In a free democracy where every act, every appointment, every policy is subject to public questioning and public pressure, the mass media have a special responsibility for not impairing, in the name of free speech, the credibility of its duly elected leadership upon whose success in a dangerous world the maintenance of that freedom depends. . . . This is a matter which—in the name of the First Amendment—has gotten completely out of hand.[72]

It's the old argument of national security. It goes like this: We are in a dangerous conflict with a ruthless foe; our leaders are taking care of us in this conflict, so don't criticize them too much. Sure, we have a free press, but it must behave responsibly. Trust our leaders.

Metcalf is a private citizen, but undoubtedly he reflected some of the thinking in the highest circles of the government. Rather than trust the press to be responsible on its own, our government, for a long time, has tried to use the press as an adjunct to official policy. Sometimes it fails. Sometimes it succeeds. Here are a few examples of how it was done.

In 1954 the U.S. government was secretly planning to overthrow the democratically elected government of Guatemala, which had decided to take back land from the United Fruit Company. A *New York Times* correspondent there, Sidney Gruson, thought it was the job of the press to report what it saw. His reports became troublesome. CIA Director

Allen Dulles contacted his old Princeton classmate, Julius Ochs Adler, business manager of the *Times,* and Gruson was transferred to Mexico City.[73]

In late 1960 the editor of *The Nation* magazine, Carey McWilliams, was informed by a Latin American specialist at Stanford University, just returned from Guatemala, that Cuban exiles were being trained in that country by the United States for an invasion of Cuba. McWilliams wrote an editorial on this and sent copies to all the major news media, including the Associated Press (AP) and United Press International (UPI). Neither the AP nor the UPI used the story. Nine days later, the *New York Times* reported that the president of Guatemala denied rumors of any pending invasion.[74]

The press went on playing the role of adjunct to the government, even though the evidence of a U.S. sponsored invasion began to grow. *Time* magazine (which later confirmed that it was a CIA operation) at first talked of Castro's "continued tawdry little melodrama of invasion." This was right in line with the statement by the U.S. ambassador to the United Nations James J. Wadsworth, who said the Cuban charge of a planned invasion was "empty, groundless, false and fraudulent."

The White House asked the magazine *New Republic* not to print a planned story about the invasion preparations, and it complied. Arthur Schlesinger, Jr., later referred to this as "a patriotic act which left me slightly uncomfortable."[75]

Four days before the invasion began, Kennedy told a press conference, "There will not be under any conditions an intervention in Cuba by the U.S. armed forces." Kennedy knew that the CIA was using Latin Americans for the invasion. But he also knew that American pilots were flying some of the planes in the invasion. Four of those pilots were killed, but the circumstances of their deaths were withheld from their families. By the time of that press conference, the evidence of U.S. complicity in the invasion was clear, yet the press did not challenge Kennedy.

When the *Times* Latin American correspondent Tad Szulc prepared a story that the CIA was behind the invasion plans, and that the invasion itself was imminent, the big guns of the *Times*—publisher Orvil Dryfoos, editor Turner Catledge, and columnist James Reston—got together to edit Szulc's story to eliminate references to the CIA and to the imminence of the invasion. Instead of a headline running over four columns, it was given a one-column headline.

In their 1963 essay on the press and the Bay of Pigs, Victor Bernstein and Jesse Gordon wrote,

> The press had a right to be angry. It had been lied to, again and again, by President Kennedy, Allen W. Dulles, Dean Rusk, and everyone else. . . . But it also had the duty to be ashamed. No law required it to swallow uncritically everything that officialdom said. On the very day the American-planned, American-equipped expedition was landing at the Bay of Pigs, Secretary Rusk told a group of newsmen: "The American people are entitled to know whether we are intervening in Cuba or intend to do so in the future. The answer to that question is no." Where was the editorial explosion that should have greeted this egregious lie?

The general manager of the Associated Press, retiring in 1963, said, "When the President of the United States calls you in and says this is a matter of vital security, you accept the injunction."[76]

The slavishness of the major media (with a few heroic exceptions) to the power and the bullying of government goes a long way toward nullifying that right declared in the First Amendment, "the freedom of the press." More instances of government influence on the media include the following.

1. When CBS correspondent Daniel Schorr managed to get a copy of the House of Representatives report on the CIA in 1976 (a report suppressed and withheld from the public), he was investigated by the Justice Department and then fired by CBS.

2. At one time the CIA secretly owned hundreds of media outlets and also used the services of at least fifty individuals who worked for news organizations in this country and abroad, including *Newsweek*, *Time*, the *New York Times*, United Press International, CBS News, and various English-language newspapers all over the world.[77]

3. After Ray Bonner, Central American correspondent for the *New York Times*, wrote a series of articles critical of U.S. policy in El Salvador in 1982, he was removed from his post.[78]

4. In 1981 a new one-hour series titled *Today's FBI* began on national television. The program got official approval and support from William Webster, the director of the FBI, who was given veto power over all the scripts.[79]

5. A CBS television show on the Vietnam War called *Tour of Duty* was given free use by the Pentagon of all sorts of military facilities,

including helicopters, planes, and personnel. In return, the Pentagon was allowed to review and veto the scripts. The producer of the show, Ron Schwary, said, "The outlines are sent to Washington, and if they approve them, they're written and then the final approval is made through the project officer here."[80]

6. In the 1980s a number of documentary films were labeled as propaganda by the U.S. Information Agency (USIA) and denied the certificates that would enable them to be sent abroad. One of them was about children and drug problems. It had won an Emmy award and a prize at the American Film Festival but the USIA said it "distorts the real picture of youth in the U.S." A film on the historical roots of the Nicaraguan revolution was also refused certification because, the USIA said, it gave "an inaccurate impression of U.S. policy toward Nicaragua today."[81]

7. President Jimmy Carter tried to discourage the *Washington Post* from printing a story about CIA payments to King Hussein of Jordan.[82]

8. Also in the Carter era, a dispatch in the *New York Times* related, "The White House made several calls to officials of CBS News late last week to try to delete a long segment from the '60 Minutes' news program about American relations with the Shah of Iran and on the activities of Savak, the deposed Shah's secret police force." (The CIA had helped train the Savak, which was notorious for its use of torture and general brutality.)[83]

9. In the spring of 1988 it was disclosed that the FBI was asking librarians to report suspicious behavior by library users. The American Library Association listed eighteen libraries that in the last two years were approached by the FBI. For instance, at the University of Maryland, FBI agents asked for information on the reading habits of people with foreign-sounding names.[84]

10. During Reagan's administration, CBS News management kept toning down White House correspondent Lesley Stahl's coverage of the president. Her scripts were changed a number of times to make her stories less critical of Reagan.

11. A documentary film made by Japanese scientists who rushed to Hiroshima just after the bombing to record the effects of the bombing on the city's residents was confiscated by the American army and then finished. But the film was not allowed to be shown until 1967. It was nicknamed in Japan "the film of illusion," because it was not supposed to exist.[85]

12. When in 1981 the U.S. government leaked documents designed to

prove that the Cubans, with the aid of the Soviet Union, were suddenly sending large amounts of arms to El Salvador—a claim that turned out to be a great deception—CBS correspondent Diane Sawyer and others reported it without a critical examination. It was an attempt to portray the rebellion in El Salvador as a foreign operation rather than arising from the terrible conditions in that country. *National Wirewatch*, a newsletter for editors of wire-service dispatches, criticized the wire services for "heeding in lock-step fashion" the "party line from Washington on Communist infiltration."[86]

In general, according to *Washington Post* writer Mark Hertsgaard, during Reagan's presidency the press, although claiming objectivity, "was far from politically neutral—largely because of the overwhelming reliance on official sources of information."[87] Hertsgaard said the press and television were "reduced . . . to virtual accessories of the White House propaganda apparatus." The role of a critical press was especially important at that time, because the supposed opposition party, the Democrats, "were a pathetic excuse for an opposition party—timid, divided, utterly lacking in passion, principle, and vision."

All this is not just a recent phenomenon. During World War II, the U.S. government put all sorts of pressure on the black press to support the war. Attorney General Francis Biddle pointed to news stories in the black press about racial clashes between white and black soldiers and said this hurt the war effort; he threatened to close down the black newspapers.[88]

The evidence is powerful that the government has tried, often successfully, to manipulate the press. But, as Noam Chomsky has said, "It is difficult to make a convincing case for manipulation of the press when the victims proved so eager for the experience."[89]

In short the First Amendment without information is not of much use. And if the media, which are the main source of information for most Americans, are distorting or hiding the truth due to government influence or the influence of the corporations that control them, then the First Amendment has been effectively nullified.

Nevertheless, it would be wrong to say that in the United States we have no freedom of speech, no freedom of the press. There are totalitarian countries all over the world in which one can say that. In the Soviet Union, before Gorbachev's *glasnost* policies opened things up, such a flat statement would have been accurate. Here the situation is too complicated for that.

Perhaps the difference between totalitarian control of the press and democratic control of the press can be summed up by the observation of Edward Herman and Noam Chomsky in their book *Manufacturing Consent:* In Guatemala dissident journalists were murdered; in the United States they were fired or transferred.[90]

By reading the mainstream press carefully (the inner pages, the lower paragraphs, the quick one-day mention) it is possible to learn important things. Occasionally, there is a burst of boldness, as when the *New York Times,* the *Washington Post,* and the *Boston Globe* printed, in defiance of the government, the Pentagon Papers, revealing embarrassing facts about the Vietnam War. From time to time, honest, courageous pieces of reporting appear in the big newspapers.

A dissident media exists in the United States. Its editors and writers are not jailed. But they are starved for resources, their circulations limited. On the air, there is a glimmer of independence in cable television, which, of course, has only a small corner of the viewing population. There are small local radio stations (for example, WBAI in New York and Radio Pacifica on the West Coast) that run programs not heard on national radio.

Public radio and television teeters between constant caution and occasional courage. The *MacNeil-Lehrer NewsHour,* the leading news program of national public television, concentrates on caution. It loads its programs with establishment spokesmen and cannot discuss any major issue without bringing in government officials and members of Congress. It is open to ultraconservatives, but not to radicals. For instance, it has never put on the air the leading intellectual critic of American foreign policy, a man who is a world-renowned scholar, Noam Chomsky. It would be as if, throughout the post–World War II period, Jean-Paul Sartre had been blacklisted in France and could not be heard by any mass audience. Courage was shown by Bill Moyers, who interviewed Noam Chomsky in two extraordinary sessions on public broadcasting.

We mislead ourselves if we think that "public television," because it has no commercial advertising, is therefore *free.* It depends on government funding, and it worries about corporate donations. Here is an Associated Press dispatch that appeared in the *New York Times* under the headline "Public Broadcasting Head Eyes Donors."

> William Lee Hanley Jr., the new chairman of the Corporation for Public Broadcasting, wants to make educational radio and television programs such

a good investment for American businesses that they will readily donate more money.[91]

The problem with free speech in the United States is not with the *fact* of access, but with the degree of it. There is *some* access to dissident views, but these are pushed into a corner. And there is *some* departure in the mainstream press from government policy, but it is limited and cautious. Some topics are given big play, others put in the back pages or ignored altogether. Subtle use of language, emphasis, and tone make a big difference in how the reading public will perceive an event.

Herman and Chomsky in *Manufacturing Consent* document this with devastating detail. They point out how the American press paid much attention to the genocide in Cambodia (which deserved attention, of course), but ignored the mass killings in East Timor, carried on by Indonesia with U.S. military equipment. They note the very large attention given to Arab terrorism and the small attention given to Israeli terrorism. They comment on the sensational coverage of the break-in of Democratic party headquarters (Watergate) and the very tiny coverage of the much more extensive series of break-ins by the FBI of the headquarters of the Socialist Workers party.

There is difference of opinion in the American mainstream press, but it is kept within bounds, just as there is difference between Republican and Democratic parties, but also within bounds. It is a puny pluralism that gives us a choice between Democrats and Republicans, *Time* and *Newsweek*, CBS, ABC, and NBC, MacNeil-Lehrer and William Buckley.

On a very small scale, I got a taste of American freedom of the press—its positive side and its limits—back in the mid-1970s. The *Boston Globe*, in the more open atmosphere created by Vietnam and Watergate and the increased skepticism of government, invited me and young Boston radical Eric Mann (he had spent time in prison for trashing the offices of Harvard's Center for International Affairs) to alternate in writing a weekly column. We were to be the left counterpart of George Will and William Buckley, conservatives whose columns appeared regularly on the *Globe*'s Op-ed page.

And indeed, our columns appeared, uncensored, for more than a year. Probably no big-city newspaper in the country went as far as the *Globe* in opening its pages to radical views. But then two things happened. A column by Eric Mann critical of Israel was not run. When we went to the *Globe* building to protest, the person who regularly received our column explained to us sadly that the *Globe* had to think about its Jewish advertisers.

Not long after that, on Memorial Day 1976, I submitted my column as usual. It was not a traditional Memorial Day statement, celebrating military heroism and past wars, but a passionate (I would like to think) statement against war. It certainly did not fit in neatly with the usual Memorial Day pictures of veterans with caps and flags and the tributes to patriotism. The column didn't get printed. When I inquired, I was told that, in fact, no column of mine would appear again. There was a new editor of the op-ed page, who explained that the page needed less political material and more family columns. Buckley and Will, I noted, continued to appear. They seemed to constitute a family.

Lies, Deception, Secrecy

When the government acts in secrecy, free speech is thwarted, and democracy undermined. With World War II over, the two victorious nations, the United States and the Soviet Union, immediately became rivals in a race for world power. The cold war was on. In such an atmosphere, the openness of a democratic society was bound to suffer.

The National Security Council was created in 1947 to consult with the president on foreign policy. Established with it, presumably to feed it information and advise it, was the Central Intelligence Agency. *National Security Council Report #68,* prepared in early 1950 under the direction of Secretary of State Dean Acheson, called for a larger military establishment. It also said that people had to "distinguish between the necessity for tolerance and the necessity for just suppression." It worried about the "excess of tolerance degenerating into indulgence of conspiracy."[93]

The mood of the government became the mood of vigilantism, which might be expressed this way: We are good. Our enemy is evil. We mustn't tie our hands with the law, the Constitution, democratic procedures, or the ordinary rules of decency. In 1954 Lieutenant General James Doolittle, appointed by President Eisenhower to head a commission to advise him on foreign policy matters, reported back that what was needed was

> an aggressive covert psychological, political and paramilitary organization more effective, more unique and, if necessary, more ruthless than that em-

ployed by the enemy. No one should be permitted to stand in the way of the prompt, efficient, and secure accomplishment of this mission. . . . There are no rules in such a game. Hitherto acceptable norms of human conduct do not apply.[94]

The commission was just putting into frank language what the United States, like other imperial powers in the world, had been doing throughout its history, long before there was a "Communist threat." But there was something different now in the language of the Doolittle Commission—the word *covert.* It is always a tribute to the citizenry when a government must do its dirty deeds in secrecy. The phrase *covert operations* was defined in National Security Council memorandum #5412 of March 15, 1954, as "all activities . . . which are so planned and executed that any U.S. Government responsibility for them is not evident to unauthorized persons and that if uncovered the U.S. Government can plausibly disclaim any responsibility for them."[95]

When the Doolittle Commission made its report, covert actions had already begun. The CIA had already tried to influence elections in Italy (that *had* to be secret; wasn't this country always talking about "free elections"?). In 1953 the CIA successfully engineered a coup in Iran to overthrow the nationalist leader Mossadegh, because he was too unfriendly to our oil corporations. And in the very year of the report, the United States was preparing to overthrow the government of Guatemala.

The excuse for covert action is that telling the truth will endanger the country, while secrecy will save lives. But secrecy may result in the *taking* of people's lives, behind the backs of the public, which if it knew what was happening, might stop it. People were killed in the coup that put the shah back on the throne of Iran; many more were killed by the shah's police afterward. The secret operation in Guatemala resulted in a police state that later killed tens of thousands of Guatemalans. In the invasion of Cuba, thousands died. Secrecy did not save lives.

Nor did it save lives in Vietnam. The secret undermining of the elections that were supposed to take place in 1956 to unite Vietnam led to a hard division between North and South, and ultimately to a war that cost over a million lives. What if the American public had been told what the government recorded secretly in the Pentagon Papers—that the South Vietnamese government whose independence we were supposedly defending was "essentially the creation of the United States"? And that "only the Viet Cong had any real support and influence on a broad base in the countryside"? Perhaps the movement to stop the war would have started sooner and saved countless lives.

The covert actions in Chile that overthrew the democratically elected government of Salvador Allende in 1973 was, in part, a conspiracy between the CIA and IT&T, according to a 1975 Senate report.[96] It led to a murderous regime whose death squads killed thousands of Chileans and engaged in torture and mutilation. Suppose the American people had known that our government was interfering in an honest election and putting a military dictatorship in place? Might there not have been a public protest, and perhaps a change in policy?

Is not that one of the purposes of the First Amendment, to enable the free flow of information, so that policies in the interests of the citizenry can be pursued, so that a few people at the head of government cannot secretly, with no accountability to the public, do things that later make the citizenry ashamed of its own government?

It was the World War II experience that led influential American journalist Walter Lippmann to distrust public opinion, and, therefore, to support government secrecy: "The unhappy truth is that the prevailing public opinion has been destructively wrong at the critical junctures. The people have imposed a veto upon the judgments of informed and responsible officials."[97]

Years later, when the United States began military action in Vietnam, Lippmann knew it was wrong. His old words must have haunted him. Because here was a case when public opinion, once it learned what was happening in Vietnam, was right in wanting out, and the "informed and responsible officials" were continuing an unspeakably brutal war.

A huge mythology has been built up in the public mind about secrecy. Perhaps it is the fascination of spy stories or the childhood delight in secrets. But most of the secrets nations make a big fuss about are either not secret at all (the secret of the atomic bomb could not be secret for long) or, if disclosed, would hardly make any difference in the world situation.[98]

The cold war atmosphere after World War II has produced a kind of hysteria about secrecy. It led to the execution of the Rosenbergs for allegedly passing atomic information to the Soviets when such information could not have made any significant difference to the Soviet making of an atomic bomb.

Similarly, the press went wild over the "pumpkin papers"—documents supposedly stolen by Alger Hiss and given to Whittaker Chambers—but there was nothing of value, no important secrets, in those famous pumpkins, although they contributed to Hiss spending four years in prison.[99]

The arms race, the fascination with nuclear weapons, has led to secrecy that is dangerous to the public. From the *New York Times:*

> The Department of Energy said today that it was responsible, along with its predecessor, the Atomic Energy Commission, for keeping secret from the public a number of serious reactor accidents that occurred over a 28-year period at the Savannah River Plant in South Carolina.
>
> The Energy Department said the failure to disclose the problems illustrated a deeply rooted institutional practice, dating from the days of the Manhattan Project in 1942, which regarded outside disclosure of any incident at a nuclear weapons production plant as harmful to national security.[100]

The Iran-Contra Affair

Covert action and "plausible denial" once again became prominent news stories during the second Reagan administration. A dispatch in the foreign press led to disclosures that were enormously embarrassing to the White House. It is not a tribute to the American press that aside from a few isolated stories here and there, it did not do the kind of investigative work that would have exposed the "Iran-Contra" affair earlier.

The root of the situation was the Nicaraguan Revolution of 1979, in which the rebel Sandinistas overthrew the Somoza regime, a family dictatorship that was long the darling of the U.S. government. The revolutionaries were named after the Nicaraguan rebel Sandino, who in the 1920s and 1930s had led a guerrilla force against the dictatorship and against the occupation of Nicaragua by the U.S. Marines. Sandino signed a truce, then was lured to a spot where he was executed by the National Guard headed by Colonel Somoza, who established the Somoza dynasty in Nicaragua.

The Sandinistas, a coalition of Marxists, left-wing priests, and assorted nationalists, set about to give more land to the peasants and to spread education and health care among the very poor and long-oppressed people of Nicaragua. Almost immediately, the Reagan administration began to wage a secret war against them, hoping to get rid of a government that would not play ball as submissively as the Somozas did.

The covert war against Nicaragua consisted of organizing and training a counterrevolutionary force, the contras, many of whose leaders were former National Guard officers under Somoza. The contras seemed to have no popular support inside Nicaragua and so were based in Honduras, a very poor country dominated by the United States and dependent on U.S. economic and military aid. From Honduras, they moved across the border into Nicaragua, raiding farms and villages; killing men, women, and children; and committing many atrocities.

When one of the contras' public relations people, Colonel Edgar Chamorro, learned what they were doing—essentially acts of terrorism against poor Nicaraguan farmers—and saw that the CIA was behind the whole operation, he resigned, telling his story to the newspapers. He also testified before the World Court:

> We were told that the only way to defeat the Sandinistas was to use the tactics the Agency [the CIA] attributed to Communist insurgencies elsewhere: kill, kidnap, rob and torture. . . . Many civilians were killed in cold blood. Many others were tortured, mutilated, raped, robbed, or otherwise abused. . . .
>
> When I agreed to join . . . in 1981, I had hoped that it would be an organization of Nicaraguans, controlled by Nicaraguans. . . . [It] turned out to be an instrument of the U.S. government, and specifically of the CIA.[101]

One of the reasons for the secrecy of Reagan's operations in Nicaragua was that public opinion surveys showed that the American people were not in favor of U.S. military operations in Central America. He decided he could do certain things openly, like strangling the Nicaraguan economy with an embargo, which the law permitted him to do if he declared the situation a national emergency.

But other actions were to be taken secretly. In 1984 the CIA, using Latin American agents, put mines in the harbors of Nicaragua to blow up ships. Secretary of Defense Caspar Weinberger told ABC news, "The United States is not mining the harbors of Nicaragua." The deceptions multiplied after Congress, responding perhaps to common sense, public opinion, and the memory of our embroilment in Vietnam, passed the Boland Amendment in October 1984, making it illegal for the United States to support "directly or indirectly, military or paramilitary operations in Nicaragua."

The Reagan administration decided to ignore this law and to find ways to fund the contras secretly, by looking for "third-party support." Reagan himself solicited funds from Saudi Arabia, at least $32 million. The friendly government of Guatemala was used to get arms surreptitiously to the contras. Honduras was used, as always, for the final passage to the contra army on its soil. Israel, so dependent on the United States and, therefore, so dependable, was also used.[102]

All of this was illegal, but the only ones prosecuted were several of Reagan's aides. Reagan himself was kept out of it. It was a perfect example of plausible denial, where an operation is conducted by underlings, so that the president can simply deny he was involved and no one can prove it.

At Reagan's news conference November 19, 1986, when asked about the disclosure that weapons had been sent to Iran (supposedly a bitter enemy of the United States) and profits from this given to the contras, he told four lies: that the shipment to Iran consisted of a few token antitank missiles (it turned out to be 2,000), that the United States didn't condone shipments by third parties, that weapons had not been traded for hostages, and that the purpose of the operation was to promote a dialogue with Iranian moderates (the purpose was to help the contras).

In October 1986 when a transport plane that had carried arms to the contras was downed by Nicaraguan gunfire and the American pilot captured, the lies multiplied. Assistant Secretary of State Elliot Abrams lied. Secretary of State Schultz lied ("no connection with the U.S. government at all"). There was so much nonsense being told the public that even the patient *New York Times* became irritated and wrote in an editorial, "It may cross the reader's mind that Americans are learning more of the truth from Managua than Washington."[103]

The whole Iran-Contra affair is a perfect example of the double line of defense of the American establishment. The first defense is to lie. If exposed, the second defense is to investigate, but not too much; the press will publicize, but they will not get to the heart of the matter.

Neither the House-Senate committee that investigated the scandal (once the scandal was out in the open) nor the press nor the trial of Colonel Oliver North, who oversaw the contra aid operation, got to the critical questions: What is U.S. foreign policy all about? How are the president and his staff permitted to support a terrorist group in Central America to overthrow a government that, whatever its faults, is a great improvement over the terrible governments the United States has supported there over the years? What does the scandal tell us about democracy, about freedom of expression, about an open society?

Out of the much-publicized scandal came no powerful critique of secrecy in government or of the erosion of democracy by actions taken in secret by a small group of men safe from the scrutiny of public opinion.[104] The media, in a country with a First Amendment, kept the public informed only on the most superficial level.

There are scholarly pundits who shake their heads sadly at the idea that the public should be told the truth about foreign policy operations. In the midst of the Iran-Contra affair, Harvard professor James Q. Wilson came forward to warn that too much was being exposed. Wilson, a member of Reagan's Foreign Intelligence Advisory Board, wrote in the *New York Times*, "We may disagree over foreign policy, but hardly any American interests are served by extensive leaks about every sensitive operation we may wish to undertake."[105] Wilson did not like the Democratic party acting like an opposition party, as if it were a true two-party system. He had little to fear. The limits of Democratic opposition were revealed by a leading Democrat, Sam Nunn of Georgia who, as the investigation was getting under way, said, "We must, all of us, help the President restore his credibility in foreign affairs."

But Wilson seemed to deplore the fact that *some* Democrats were *somewhat* critical. He looked back nostalgically to a "bipartisan consensus" (the equivalent of the one-party system in a totalitarian state). What he worried about most was "a lack of national resolve to act like a great power."

Machiavelli would have agreed.

Taking Our Liberties

If the government deceives us and the press more or less collaborates with it—to keep us from knowing what is going on in the most important matters of politics: life and death, war and peace—then the existence of the First Amendment will not help us. Unless, of course, we begin to act as citizens, to put life into the amendment's promise of freedom of expression by what we do ourselves. British novelist Aldous Huxley *(Brave New World)* once said, "Liberties are not given; they are taken."

We, as citizens, want freedom of expression for two reasons. First, because in itself it is fundamental to human dignity, to being a person, to independence, to self-respect, to being an important part of the world, and to being alive. Second, because we badly need it to help change the world and to bring about peace and justice.

We should know by now that we cannot count on the courts, the Congress, or the presidency, to assure us the freedom to speak, to write, to assemble, and to petition. We cannot count on the government or the mainstream press to give us the information necessary to be active, critical citizens. And we cannot count on those who own the media to give us the opportunity to reach large numbers of people.

Therefore, it seems Huxley is right; we will have to *take* our liberties. Historically, that has always been the case. Despite the Sedition Act after the American Revolution, in which some people were jailed for criticizing the government, hundreds of other pamphleteers and writers insisted, at the risk of prison, on writing as they pleased. They *took* their liberty.[106]

We need to remind ourselves of individuals who have insisted on their freedom to speak their minds. Emma Goldman was a feminist and anarchist of the early twentieth century whose views on patriotism, (agreeing with Samuel Johnson, "the last refuge of a scoundrel"), on preparedness for war ("violence begets violence"), on marriage ("it has nothing to do with love; it is an insurance contract"), on free love ("what is love if it is not free?") and on birth control ("a woman should decide for herself whether or not she wants a baby") outraged many people and certainly the authorities.

She lectured all over the United States, and wherever she went, the police were there to stop her. In one month, May 1909, police broke up eleven meetings at which she spoke. She was arrested again and again. But she kept coming back.

In San Francisco, she spoke to 5,000 people on patriotism; the crowd stood between her and the police, and the police retreated. When she came back to San Francisco the following year, the police broke up the meeting, using their clubs on members of the audience.

In East Orange, New Jersey, police blocked the entrance to the lecture hall. She spoke to her audience on the lawn. In San Diego, a mob kidnapped her lover and manager and tarred and feathered him. She insisted on coming back to San Diego to speak the next year.

When she lectured on birth control and the use of contraceptives, she was repeatedly arrested. But she refused to stop.

She opposed U.S. entrance into World War I, as most Socialists and anarchists did. She knew she was in danger for encouraging young men to resist the draft, but she continued to speak. She was tried and imprisoned for two years, and when she came out of prison she was deported from this country. But she continued to speak her mind on American events—the Tom Mooney case and the case of Sacco and Vanzetti—flinging her thoughts across the ocean, during her long exile in Europe.[107]

In the decade before World War I, the Industrial Workers of the World (IWW), a radical trade union, was organizing all workers—skilled and unskilled, men and women, native born and foreign—into "One Big Union." IWW organizers, going to speak in cities in the far west to miners and lumberjacks and mill workers, were arrested again and again. They refused to stop. They engaged in what they called "Free Speech Fights": when one of them was put in jail, hundreds of others would come into that town and speak and be arrested until the jails could not hold them and they were released. But they refused to be silent.

This is always the price of liberty—taking the risk of going to jail, of being beaten and perhaps being killed.

There is another risk for people speaking and organizing in the workplace: loss of one's job. Historically, the only way workers, subject to the power of a foreman or an employer, could have freedom of expression, was to join with other workers and form a union so that they could collectively defend themselves against the power of the employer.

Freedom of the press depends on the energy and persistence of people in developing their own newspapers, magazines, and pamphlets, to say things that will not appear in the mainstream press. Throughout American history, these little publications, pressed for money, have managed to form a kind of underground press.

The Populist Movement of the late nineteenth century spread literature throughout the farm country, north and south. The Socialist press of the early twentieth century was read by 2 million people. Black people, taking a cue from the first abolitionist newspaper printed by a black man in 1829, developed their own newspapers, because they knew they could not depend on the orthodox press to tell the truth about the race situation in the United States.

When in the 1950s journalist I. F. Stone decided he could not count on having an outlet in the regular press, he published his own little four-page newspaper. *I. F. Stone's Weekly* contained information unavailable elsewhere, which Stone, in Washington, D.C., put together by reading obscure government documents and the *Congressional Record;* it soon became a famous source of reliable facts. The first rule of journalism, Stone declared, is that "governments lie," and so alternate sources of information are desperately needed if we are to have a democracy.

The movements of the sixties—the black movement, the antiwar movement, the women's movement, and the prisoners' rights movement—produced an enormous underground press. There were 500 underground high-school newspapers alone.

Soldiers against the Vietnam War put out their newspapers on military bases around the country. By 1970 there were fifty of them: *About Face* in Los Angeles; *Fed Up* in Tacoma, Washington; *Short Times* at Fort Jackson, South Carolina; *Last Harass* at Fort Gordon, Georgia; *Helping Hand* at Mountain Home Air Base, Idaho.

Underground newspapers sprang up during the war in cities all over the country. In early 1969 J. Edgar Hoover instructed his field offices to target these publications. FBI agents raided and ransacked the offices of newspapers in San Diego, Philadelphia, Phoenix, Jackson, and other places. Advertisers were persuaded to withdraw. One landlord after another agreed to evict newspapers from their offices. The Underground Press Syndicate and Liberation News Service became targets of FBI infiltrators.[108]

By 1972 these attacks badly crippled the underground press. But slowly it made its way back and today around the country community newspapers continue to print material not found in the regular media.

In the past few years, a new form of free speech has become important: "whistle-blowing." A whistle-blower is a person who risks his or her job with the government or with a large corporation to expose truths that have been kept under wraps.

For instance, Pentagon employee A. Ernest Fitzgerald embarrassed his employer in 1969 by telling Congress that a transport plane ordered by the air force would cost $2 billion more than it expected to pay. Fitzgerald was dismissed from the Pentagon, then reinstated but given lesser assignments.

Dr. Jacqueline Verrett, of the Bureau of Foods of the Food and Drug Administration, granted an interview with a television reporter. She was told never to speak to the press again. She was warned (in her words), "not to answer my phone but to get someone else to answer it and say I wasn't there."

Nevertheless, Fitzgerald and Verrett continued to speak their minds.[109] So did others. A safety engineer with the Ford Motor Corporation exposed the fact that Ford, to save money, had chosen a gas tank that was prone to rupture under stress. Peter Faulkner, an engineer, exposed faults in a nuclear device made by General Electric. He was called in to discover why he had such "deep-seated hostility." Then he was fired. But he published a book about his experience.[110]

It takes courage to divulge information embarrassing to the government, especially when there are laws that can be used to imprison you for doing that. Daniel Ellsberg faced 130 years under the Espionage Act for photocopying the 7,000 pages of the Pentagon Papers and sending them to the newspapers, to expose the truth about the war in Vietnam. But he went ahead.

It is impossible to judge the impact of those papers on the public, but it is reasonable to assume that the several million people who read the *Times*, the *Washington Post*, and the *Boston Globe* learned things about the war they had not known before. This, along with all the other disclosures about the war going on at the time, helped turn public opinion against the war. But Ellsberg, and codefendant Tony Russo, had to risk prison to make the First Amendment come alive.

During the Vietnam War, with the government lying and with the press slow in getting past official propaganda, a whole network of techniques was developed to spread information about the war. There were teach-ins on college campuses, alternative newspapers, rallies, picket lines, demonstrations, petitions, ads in newspapers, and graffiti on walls.

In Southeast Asia an alternative news organization was created—*Dispatch News Service*—which sent out news items revealing what the government was keeping secret, like the story of the My Lai massacre.

The thousands of acts of civil disobedience during the war were acts of communication, small works of art, appealing to the deepest feelings of people. Art plays a critical role in any social movement, because it intensifies the movement's messages. It tries to make up for the lack of money and resources by passion and wit. It communicates through music, drama, speech, demonstrative action, drawings, posters, songs, surprise, sacrifice, and risk.

During the Vietnam War, a very successful commercial artist (Seymour Chwast) turned his talents to the antiwar movement, and produced a poster with a simple design and eight large words printed on it: WAR IS GOOD FOR BUSINESS. INVEST YOUR SON.

It was chilling and powerful. It was just part of the work of hundreds of thousands of people all over the country, speaking to millions of people in many different ways, bringing life to the First Amendment and an end to a war.

CHAPTER

NINE

Representative Government: The Black Experience

Amid the enthusiastic celebrations in 1987 surrounding the Bicentennial of the Constitution, novelist James Michener wrote,

> The writing of the Constitution of the United States is an act of such genius that philosophers still wonder at its accomplishment and envy its results. Fifty five typical American citizens . . . fashioned a nearly perfect instrument of government. . . . Their decision to divide the power of the government into three parts—Legislative, Executive, Judicial—was a master stroke.[1]

In the abolitionist movement of the early nineteenth century, there was no such enthusiasm. William Lloyd Garrison, editor of *The Liberator,* held up a copy of the Constitution before several thousand people at a picnic of the New England Anti-Slavery Society and burned it, calling it "a covenant with death and an agreement with hell," and the crowd shouted "Amen!"

Ex-slave Frederick Douglass, invited to deliver a Fourth of July speech in 1852, told his white audience,

The rich inheritance of justice, liberty, prosperity and independence, bequeathed by your fathers, is shared by you, not by me. The sunlight that brought light and healing to you, has brought stripes and death to me. This Fourth of July is yours, not mine. You may rejoice, I must mourn.

During our 1987 celebrations, former Chief Justice Warren Burger, chairman of the Bicentennial Commission, delivered the usual superlatives to the Founding Fathers and the Constitution. But the sole black Supreme Court Justice Thurgood Marshall spoke this way:

In this bicentennial year, we may not all participate in the festivities with flag-waving fervor. Some may more quietly commemorate the suffering, struggle, and sacrifice that has triumphed over much of what was wrong with the original document, and observe the anniversary with hopes not realized and promises not fulfilled.[2]

Historian Leon Litwack has written:

It had been the genius of the Founding Fathers to sanction, protect, and reinforce the enslavement of black men and women. . . . It had been the genius of the founders to build safeguards for slavery into the Constitution without even mentioning slavery by name. The legitimization of slavery was the price of the new federal union, and the Founding Fathers shared . . . the assumption that blacks were culturally and genetically unsuited for democracy.[3]

Today, Americans still celebrate the Constitution; they learn in school about checks and balances and what Michener called "the master stroke" of dividing the government into Executive, Legislative, and Judicial branches. We hold elections, vote for president and representatives in Congress, and think *that* is democracy. Yet for black people in this country, none of those institutions—not the Constitution, not the three branches of government, not voting for representatives—has been the source of whatever progress has been made toward racial equality.

Before we rush to conclude that representative government has worked for white people in this country, but not for blacks, we should consider it is the special gift of oppressed groups to reveal universal truths. French writer Fourier said that you could tell the state of progress in any society by looking at the condition of women, and George

Bernard Shaw said you could measure the condition of society by the treatment of its prisoners.

The history of blacks in the United States exposes dramatically the American political system. What that history makes clear is that our traditional, much-praised democratic institutions—representative government, voting, and constitutional law—have never proved adequate for solving critical problems of human rights.

Theories of representative government became prominent in the seventeenth and eighteenth centuries, when monarchies and feudal arrangements were being challenged by rising classes of merchants and manufacturers. People were moving into cities and the new middle classes wanted more power in government.

The new way of thinking was expressed by John Locke. He was an adviser to the Whig party, which wanted to diminish the power of the king and increase that of Parliament. Locke wrote about the advantages of representative government. His name is associated with the idea of the "social contract", under which the community—wanting more order, less trouble, and more safeguards for life, liberty, and property—agrees to choose representatives who would accomplish these purposes.

Locke said that in ancient times in "the state of nature" people got along quite well, but this was disrupted by money, commerce, and greed. Monarchy didn't help, because kings acted as if they were in a state of nature, not responsible to the community. Now, Locke said, you needed settled law, judges, and a stable society based on the will of the majority represented by the legislature.[4] The legislature would be the supreme power, Locke proposed, but it had to abide by the terms of the contract, to promote peace, safety, and the public good. If the government ever seriously violated the contract, rebellion might be justified, Locke said.

Therefore, although written in the 1680s, Locke's statements almost have the idealistic ring of the Declaration of Independence. But there is something suspect about his theory. It pretends that there is some nice unified community that agrees to set up this constitutional government. In reality, there was no such unity, neither in England nor in the American colonies. There were rich and poor, and the poor are never in a position to sign a contract on equal terms with the rich. Indeed, they are not usually consulted when a contract is drawn up, after which they are told: "*We* agreed on this." So while it may sound good that property and liberty will be protected by representative government, in reality it is the property and liberty of the wealthy and powerful that is most likely to be protected.[5]

We get a clue to the reality behind Locke's liberal theory when we look at his activities. In the 1660s, he was given the job of writing a constitution for the Carolinas (not yet North and South Carolina). His constitution set up a feudal-type system, in which eight barons would own 40 percent of the land; one of these barons would be governor. Locke's constitution also contained this clause: "every freeman of Carolina shall have absolute power and authority over his negro slaves, of what opinion or religion soever." This last part was to take care of the claim that Christianized slaves might be freed.

The American revolutionists had probably not read John Locke.[6] They didn't have to. They were moved by similar circumstances: the necessity to overthrow monarchical rule, to put forth a rhetoric that would win popular support, and then to set up a government that would be more democratic than a monarchy. It would be a representative government (a revolutionary idea at that time), but one that would represent the interests of the wealthy classes most of all. And so, the Declaration of Independence, a masterpiece of rhetorical idealism, was followed by the Constitution, a masterpiece of ambiguous practicality.

That combination of rhetoric and ambiguity appeared in the Bill of Rights itself, in the Fifth Amendment, which says no person shall be deprived of "life, liberty, or property" without due process of law. The white person might be thankful that "liberty" was safe, but the black slave, knowing he or she was "property," might well be unimpressed. Indeed, when the Supreme Court in 1857 had to decide between Dred Scott's liberty and his former master's property, it decided for property and declared Dred Scott a nonperson, to be returned to slavery.

Those were not "fifty-five typical American citizens" (James Michener's phrase) who drew up the Constitution. At that convention, there was no representation of black people, who at that time numbered about one-fifth of the population of the states. There was no representation of women, who were about half the population, and certainly no representation of Indians, whose land all of the colonists were occupying.

The Indians, like blacks, were not looked on as human beings by those who were fighting a revolution in the name of freedom. Six months after the battles of Lexington and Concord, the Massachusetts legislature proclaimed monetary rewards for dead Indians: "For every scalp of a male Indian brought in . . . forty pounds. For every scalp of such female Indian or Male Indian under the age of twelve years that shall be killed . . . twenty pounds."

The Constitution was blatant in its representation of the interests of the slaveholders. It included the provision (Article IV, Section 2) that escaped slaves must be delivered back to their masters. Roger Sherman pointed out to the Convention that the return of runaway horses was not demanded with such specific concern, but he was ignored.

In eighty-five newspaper articles *(The Federalist Papers)*, arguing for the ratification of the Constitution among New York State voters (blacks, women, Indians, and whites without property were excluded), James Madison, Alexander Hamilton, and John Jay were quite frank. Madison wrote (as we noted in the chapter "Economic Justice") that representative government was a good way of calming the demand of people "for an equal division of property, or for any other improper or wicked object." It would accomplish this by creating too big a nation for a revolt to spread easily and by filtering the anger of rebels through their more reasonable representatives.

The authors of *The Federalist Papers* explained, more candidly than any other political leaders of the nation have done since, what the institution of *representative government* is really for. As they put it (it is not clear whether Madison or Hamilton wrote this), speaking of the usefulness of the Senate:

> I shall not scruple to add that such an institution may be sometimes necessary as a defence to the people against their own temporary errors and delusions. . . . There are particular moments in public affairs when the people, stimulated by some irregular passion, or some illicit advantage, may call for measures which they themselves will afterwards be the most ready to lament and condemn. In these critical moments, how salutary will be the interference of some temperate and respectable body of citizens in order to check the misguided career, and to suspend the blow meditated by the people against themselves, until reason, justice, and truth can regain their authority over the public mind?[8]

That passage suggests that whites as well as blacks, men as well as women, might look with suspicion on the claims of modern representative government—that while it indeed is an improvement over monarchy, and may be used to bring about some reforms, it is chiefly used by those holding power in society as a democratic facade for a controlled society and a barrier against demands that threaten their interests.

The experience of black people reveals this most clearly, but there is instruction in it for every citizen. The Constitution did not do away with slavery; it legalized it. Congress and the president (including later the antislavery but politically cautious Abraham Lincoln) had other priorities that came ahead of abolishing slavery. Billions of dollars were invested in southern slaves, and northern political leaders, wanting to keep what power they had, did not want to rock the national boat.

It became clear to those who wanted to abolish slavery that they could not depend on the regular structures of government. So they began to agitate public opinion. This was dangerous not just in the South, where blacks were enslaved, but in the North, where they were segregated and denied the right to vote, their children excluded from public schools, and they were treated as inferiors in every way.[9]

A free black man in Boston, David Walker, wrote the pamphlet *Walker's Appeal,* a stirring call for resistance, in 1829:

> Let our enemies go on with their butcheries. . . . Never make an attempt to gain our freedom . . . until you see your way clear—when that hour arrives and you move, be not afraid or dismayed. . . . They have no more right to hold us in slavery than we have to hold them. . . . Our sufferings will come to an end, in spite of all the Americans this side of eternity. . . . 'Every dog must have its day,' the American's is coming to an end.

Georgia offered $1,000 to anyone who would kill David Walker. One summer day in 1830, David Walker was found dead near the doorway of the shop where he sold old clothes. The cause of death was not clear.

From the 1830s to the Civil War, antislavery people built a movement. It took ferocious dedication and courage. White abolitionist William Lloyd Garrison, writing in *The Liberator,* breathed fire: "I accuse the land of my nativity of insulting the majesty of Heaven with the greatest mockery that was ever exhibited to man." A white mob dragged him through the streets of Boston in chains, and he barely escaped with his life.

The Liberator started with twenty-five subscribers, most of them black. By the 1850s, it was read by more than 100,000. The movement had become a force.

Black abolitionists were central to the antislavery movement. Even before Garrison published *The Liberator,* a black periodical, *Freedom's Journal,* had appeared. Later, Frederick Douglass, ex-slave and abolitionist orator, started his own newspaper, *North Star.* A conference of blacks in 1854 declared "it is emphatically our battle; no one else can fight it for us."

The Underground Railroad brought tens of thousands of slaves to freedom in the United States and Canada. Harriet Tubman, born into slavery, had escaped alone as a young woman. She then made nineteen dangerous trips back into the South, bringing over 300 slaves to freedom. She carried a pistol and told the fugitives, "You'll be free or die."

When the Fugitive Slave Act was passed by Congress in 1850, blacks, joined by white friends, took the lead in defying the law, in harboring escaped slaves, in rescuing captured slaves from courtrooms and police stations. After the act was passed, Reverend J. W. Loguen, who had escaped from slavery on his master's horse, had gone to college, and had become a minister in Syracuse, New York, spoke to a meeting in that city:

> The time has come to change the tones of submission into tones of defiance—and to tell Mr. Fillmore (President Millard Fillmore, who signed the law) and Mr. Webster (Senator Daniel Webster of Massachusetts, who supported the law), if they propose to execute this measure upon us, to send on their blood-hounds. . . . I received my freedom from Heaven, and with it came the command to defend my title to it. . . . I don't respect this law—I don't fear it—I won't obey it! It outlaws me, and I outlaw it. . . . I will not live a slave, and if force is employed to re-enslave me, I shall make preparations to meet the crisis as becomes a man.[10]

No more shameful record of the moral failure of representative government exists than the fact that Congress passed the Fugitive Slave Act, the president signed it, and the Supreme Court approved it.[11]

The act forced captured blacks to prove they were not someone's slave; an owner claiming him or her needed only an affidavit from friendly whites. For instance, a black man in southern Indiana was taken by federal agents from his wife and children and returned to an owner who claimed he had run away nineteen years ago. Under the act more than 300 people were returned to slavery in the 1850s.

The response to it was civil disobedience. "Vigilance committees" sprang up in various cities to protect blacks endangered by the law. In 1851 a black waiter named Shadrach, who had escaped from Virginia, was serving coffee to federal agents in a Boston coffeehouse. They seized him and rushed him to the federal courthouse. A group of black men broke into the courtroom, took Shadrach from the federal marshals, and saw to it that he escaped to Canada. Senator Webster denounced the rescue as treason, and the president ordered prosecution of those who had helped Shadrach escape. Four blacks and four whites were indicted and put on trial, but juries refused to convict them.[12]

Federal agents were sent to Boston right after the passage of the Fugitive Slave Law to apprehend William and Ellen Craft, who were famous escapees from slavery. They had disguised themselves as master and servant (she was light skinned and dressed as a man) and had taken the railroad north. Boston was full of defiance. White abolitionist minister Theodore Parker hid Ellen Craft in his house and kept a loaded revolver on his desk. A black abolitionist concealed William Craft. He stacked two kegs of gunpowder on his front porch. The local vigilance committee warned the federal marshals it was not safe to remain in Boston, and they left town.

In Christiana, Pennsylvania, in September 1851, a slaveowner arrived from Maryland with federal agents, to capture two of his slaves. There was a shoot-out with two dozen armed black men determined to protect the fugitives, and the slaveowner was shot dead. President Fillmore called out the marines and assembled federal marshals to make arrests. Thirty-six blacks and five whites were put on trial. A jury acquitted the first defendant, a white Quaker, and the government decided to drop the charges against the others.

Rescues took place and juries refused to convict. In Oberlin, Ohio, a group of students and one of their professors organized the rescue of an escaped slave; they were not prosecuted.

A white man in Springfield, Massachusetts, had organized blacks into a defense group in 1850. His name was John Brown. In 1858, John Brown and his band of white and black men made a wild, daring effort to capture the federal arsenal at Harper's Ferry, Virginia, and set off a slave revolt throughout the South. Brown and his men were hanged by the collaboration of the state of Virginia and the national government. He became a symbol of moral outrage against slavery. The great writer Ralph Waldo Emerson, not an activist himself, said of John Brown's execution: "He will make the gallows holy as the cross."

What Garrison had said was necessary—"a most tremendous excitement" was shaking the country. The abolitionist movement, once a despised few, began to be listened to by millions of Americans, indignant over the enslavement of 4 million men, women, and children.

Nevertheless when the Civil War began, Congress made its position clear, in a resolution passed with only a few dissenting votes: "This war is not waged . . . for any purpose of . . . overthrowing or interfering with the rights of established institutions of those states, but . . . to preserve the Union."

As for President Lincoln, his caution, his politicking around the issue of slavery (despite his personal indignation at its cruelty) had been made clear when he campaigned for the Senate in 1858. At that time he told voters in Chicago: "Let us discard all this quibbling about . . . this race and that race and the other race being inferior, and therefore they must be placed in an inferior position."

But two months later, in southern Illinois, he assured his listeners: "I will say, then, that I am not, nor ever have been, in favor of bringing about in any way the social and political equality of the white and black races. . . . I as much as any other man am in favor of having the superior position assigned to the white race."[13]

The abolitionists went to work. To their acts of civil disobedience and of armed resistance, they added more orthodox methods of agitation and education. Petitions for emancipation poured into Congress in 1861 and 1862. Congress, responding, passed a Confiscation Act, providing for the freeing of slaves of anyone who fought with the Confederacy. But it was not enforced.

When the Emancipation Proclamation was issued at the start of 1863, it had little practical effect. It only declared slaves free in states still rebelling against the Union. Lincoln used it as a threat to Confederate states: if you keep fighting, I will declare your slaves free; if you stop fighting, your slaves will remain. So, slavery in the border states, on the Union side, were left untouched by the proclamation. The *London Spectator* remarked drily, "The principle is not that a human being cannot justly own another, but that he cannot own him unless he is loyal to the United States."[14] Still, the moral impact of the proclamation was strong. It came from Lincoln's military needs, but also from the pressures of the antislavery movement.

By the summer of 1864 approximately 400,000 signatures asking legislation to end slavery had been gathered and sent to Congress. The First Amendment's right "to petition the government for a redress of grievances" had never been used so powerfully. In January 1865 the House of Representatives, following the lead of the Senate, passed the Thirteenth Amendment, declaring slavery unconstitutional.

The representative system of government, the constitutional structure of the modern democratic state, unresponsive for eighty years to the moral issue of mass enslavement, had now finally responded. It had taken thirty years of antislavery agitation and four years of bloody war. It had required a long struggle—in the streets, in the countryside, and on the battlefield. Frederick Douglass made the point in a speech in 1857:

> Let me give you a word of the philosophy of reforms. The whole history of the progress of human liberty shows that all concessions yet made to her august claims have been born of struggle. . . . If there is no struggle there is no progress. Those who profess to favor freedom and yet deprecate agitation, are men who want crops without plowing up the ground. They want rain without thunder and lightning. They want the ocean without the awful roar of its many waters. The struggle may be a moral one; or it may be a physical one; or it may be both moral and physical, but it must be a struggle. Power concedes nothing without a demand. It never did and it never will.

A hundred years after the Civil War, Frederick Douglass's statement was still true. Blacks were being beaten, murdered, abused, humiliated, and segregated from the cradle to the grave and the regular organs of democratic representative government were silent collaborators.

The Fourteenth Amendment, born in 1868 of the Civil War struggles, declared "equal protection of the laws." But this was soon dead—interpreted into nothingness by the Supreme Court, unenforced by presidents for a century.

Even the most liberal of presidents, Franklin D. Roosevelt, would not ask Congress to pass a law making lynching a crime. Roosevelt, through World War II, maintained racial segregation in the armed forces and was only induced to set up a commission on fair employment for blacks when black union leader A. Philip Randolph threatened a march on Washington. President Harry Truman ended segregation in the armed forces only after he was faced with the prospect—again it was by the determined A. Philip Randolph—of black resistance to the draft.

The Fifteenth Amendment, granting the right to vote, was nullified by the southern states, using discriminatory literacy tests, economic intimidation, and violence to keep blacks from even registering to vote. From the time it was passed in 1870 until 1965, no president, no Congress, and no Supreme Court did anything serious to enforce the Fifteenth Amendment, although the Constitution says that the president "shall take care that the laws be faithfully executed" and also that the Constitution "shall be the Supreme Law of the land."

If racial segregation was going to come to an end, if the century of humiliation that followed two centuries of slavery was going to come to an end, black people would have to do it themselves, in the face of the silence of the federal government. And so they did, in that great campaign called the civil rights movement, which can roughly be dated from the Montgomery Bus Boycott of 1955 to the riot in Watts, Los Angeles, in 1965, but its roots go back to the turn of the century and it has branches extending forward to the great urban riots of 1967 and 1968.

I speak of roots and branches, because the movement did not suddenly come out of nowhere in the 1950s and 1960s. It was prepared by many decades of action, risk, and sacrifice; by many defeats; and by a few victories. The roots go back at least to the turn of the century, to the protests of William Monroe Trotter; to the writings of W. E. B. Du Bois; to the founding of the National Association for the Advancement of Colored People (NAACP); to the streetcar boycotts before World War I; to the seeds sown in black churches, in black colleges, and in the Highlander Folk School of Tennessee; and to the pioneering work of radicals, pacifists, and labor leaders.[15]

It is a comfort to the liberal system of representative government to say the civil rights movement started with the Supreme Court decision of 1954 in *Brown v. Board of Education of Topeka.* That was when the Supreme Court finally concluded that the Fourteenth Amendment provision of "equal protection of the laws" meant that public schools had to admit anyone, regardless of color. But to see the origins of the movement in that decision gives the Supreme Court too much credit, as if it suddenly had a moral insight or a spiritual conversion and then read the Fourteenth Amendment afresh.

The amendment was no different in 1954 than it had been in 1896, when the Court made racial segregation legal. There was just a new context now, a new world. And there were new pressures. The Supreme Court did not by itself reintroduce the question of segregation in the public schools. The question came before it because black people in the South went through years of struggle, risking their lives to bring the issue into the courts.

Local chapters in the South of the NAACP had much to do with the suits for school desegregation. The NAACP itself can be traced back to an angry protest in Boston in 1904 of the black journalist William Monroe Trotter against Booker T. Washington. Washington, a black educator, founder of Tuskegee Institute, favored peaceful accommodation to segregation. Trotter's arrest and his sentence of thirty days in prison aroused that extraordinary black intellectual W. E. B. DuBois, who wrote later, "when Trotter went to jail, my indignation overflowed. . . . I sent out from Atlanta . . . a call to a few selected persons for organized determination and aggressive action on the part of men who believe in Negro freedom and growth."[16] That "call to a few persons" started the Niagara Movement—a meeting in Niagara, New York, in 1905 that led to the founding of the NAACP in 1911.

Many years later, with the legal help of the NAACP, the Reverend Joseph DeLaine rallied the black community in Clarendon County, South Carolina, to bring suit in the *Brown* case. Because of this, Reverend DeLaine was fired from his teaching job. So were his wife, two of his sisters, and a niece. He was denied credit from any bank. His home was set ablaze, while the fire department stood by and watched. When gunmen fired at his house in the night, he fired back, and then, charged with felonious assault, he had to flee the state. His church was burned to the ground, and he was considered a fugitive from justice.[17]

It seems a common occurrence that a hostile system is made to give ground by a combination of popular struggle and practicality. It had happened with emancipation in the Civil War. In the case of school desegregation, the persistence of blacks and the risks they took became joined to a practical need of the government. The *Brown* decision was made at the height of the cold war, when the United States was vying with the Soviet Union for influence and control in the Third World, which was mostly nonwhite.

Attorney General Herbert Brownell, arguing before the Supreme Court, asked that the "separate but equal" doctrine, which allowed segregation in the public schools, "be stricken down," because "it furnishes grist for the communist propaganda mills, and it raises doubt, even among friendly nations, as to the intensity of our devotion to the democratic faith."[18]

In outlawing school segregation, the Supreme Court declared that integration should proceed "with all deliberate speed," and indeed, the executive branch was very *deliberate* in enforcing the decision. Eleven years later, by 1965, over three-fourths of the school districts in the South remained segregated. It was not until the urban riots of 1965, 1967, and 1968 that the Supreme Court finally said the "all deliberate speed" injunction was no longer "constitutionally permissible" and then desegregation of schools in the Deep South began to speed up.[19]

By the provision of the Fourteenth Amendment for equal protection, there should have been no segregation of the buses in Montgomery, Alabama, in 1955. If the amendment had meaning, Rosa Parks should not have been ordered out of her seat to give it to a white person; she should not have been arrested when she refused. But the federal government was not enforcing the Constitution. The checks and balances were check-mated and out of equilibrium, and the black population of Montgomery had to get rid of bus segregation by their own efforts.

They organized a citywide boycott of the buses. Black people, old and young, men and women, walked miles to work. One of those people, an elderly lady who walked several miles to and from her job, was asked if she was tired. She replied, "My feets is tired, but my soul is rested."[20]

In Montgomery the struggle for rights meant not only mass meetings, car pools, and wearying walks. It meant going to jail. A hundred leaders of the boycott were indicted by the city. It meant facing daily violence and the threat of violence. Bombs exploded in four black churches. A shotgun blast was fired through the front door of the home of twenty-seven-year-old Martin Luther King, Jr., a minister and a leader of the Montgomery boycott. Later King's home was bombed.

Finally, the government responded. In November 1956, a year after the boycott began, the Supreme Court outlawed segregation on local bus lines.[21] A glimmer of what went on that year is conveyed in a journalist's account of one of the mass meetings in Montgomery during the boycott:

> More than two thousand Negroes filled the church from basement to balcony and overflowed into the street. They chanted and sang; they shouted and prayed; they collapsed in the aisles and they sweltered in an eighty-five degree heat. They pledged themselves again and again to "passive resistance". Under this banner they have carried on . . . a stubborn boycott of the city's buses.

Why did four black college students have to sit at a "whites only" lunch counter in Greensboro, North Carolina, on February 1, 1960, and be arrested? Why did there have to be a "sit-in movement" to end discrimination in restaurants, hotels, and other public places throughout the South?[22] Was it not the intent of the Thirteenth Amendment, as Justice John Harlan said back in 1883, to remove not only slavery but the "badges" of slavery? Was it not the intent of the Fourteenth Amendment to make all blacks citizens, and did not the Constitution (Article 4, Section 2) say that "the citizens of each State shall be entitled to all privileges and immunities of citizens in the several States"?[23]

But Harlan was alone in his opinion. He was overruled by the other members of the Supreme Court, who said that the Fourteenth Amendment was directed at the *states* ("no state shall") and, therefore, *private* persons, businesses, and corporations could discriminate as they like. (But did not the *state* enforce discrimination by arresting people who protested it? The court was silent on that.)

So it would take a struggle to relieve black parents of the problem of telling their little children that they could not sit at *this* lunch counter, use *this* water fountain, enter *this* building, or go to *this* movie theater. It would take sit-ins in city after southern city. There would be beatings and arrests. There would be in the year 1960 sit-ins and demonstrations in a hundred cities involving more than 50,000 people, and over 3,600 demonstrators would spend time in jail.

Columnist for the *Atlanta Constitution* Ralph McGill was at first wary of the sit-in movement as too provocative. Later, however, he wrote in his book *The South and the Southerner*, "No argument in a court of law could have dramatized the immorality and irrationality of such a custom as did the sit-ins."[24]

There was an electric effect of all this on black people around the country. Bob Moses, who would later become an organizer of the movement in Mississippi, told how, sitting in his Harlem apartment, he saw on television the pictures of the Greensboro sit-in:

> The students in that picture had a certain look on their faces, sort of sullen, angry, determined. Before, the Negro in the South had always looked on the defensive, cringing. This time they were taking the initiative. They were kids my age, and I knew this had something to do with my own life.[25]

The young black veterans of the sit-ins from the Deep South, along with some blacks from the North and a few whites, formed a new organization, the Student Nonviolent Coordinating Committee (SNCC). They became the "point" people (to use a military term: those who go ahead into enemy territory) for the civil rights movement in the Deep South.

In the spring of 1961 the Congress of Racial Equality (CORE) organized the "Freedom Rides": whites and blacks rode together on buses throughout the South to try to break the segregation pattern in interstate travel. The two buses that left Washington, D.C., on May 4, 1961, headed for New Orleans, never got there. In South Carolina, riders were beaten. In Alabama, a bus was set afire. Freedom Riders were attacked with fists and iron bars. The southern police did not interfere with any of this violence, nor did the federal government. FBI agents watched, took notes, did nothing.

CORE decided to call off the rides. SNCC, younger, more daring (more rash, some thought) decided to continue them. Before they started out, they called the Department of Justice in Washington to ask for protection. A SNCC staff member, Ruby Doris Smith (one of my students at Spelman College) told me about the phone call: "The Justice Department said no, they couldn't protect anyone, but if something happened, they would investigate. You know how they do."

The Kennedy administration, in touch with Governor John Patterson of Alabama, accepted his promise that he would protect the Riders. He broke his promise; his police did not show up until the Riders were beaten bloody. A black student from Nashville was knocked unconscious by a group using baseball bats. A young white man was pounded with fists and sticks, was left bleeding, and was given no medical attention for two hours. Ruby Doris Smith recalled that SNCC organizer John Lewis (who was to go through this many times, and years later would be elected to Congress) was beaten, blood coming from his mouth.

The response of the federal government was less than adequate. The Riders, despite all that, were going on to Jackson, Mississippi. Attorney General Robert Kennedy made a deal with the governor of Mississippi: Don't let the Freedom Riders get beaten; just arrest them. So they were arrested, busload after busload arriving in Jackson; 300 were arrested by the end of the summer and sent to Parchman State Penitentiary, some of them beaten, all of them abused. It was a feeble act by the most powerful government on earth, refusing to enforce its own laws, allowing mobs to do violence to citizens peacefully riding buses, allowing local police to neglect their function of protecting people against assault.

The law was clear. Presumably, representative government had done its work by enacting the Fourteenth Amendment, which called for equal protection of the law. In 1887 Congress had enacted the Interstate Commerce Act, which barred discrimination in interstate travel, and the courts had reinforced this in the 1940s and 1950s. But it took the Freedom Rides and the embarrassing publicity surrounding them that went around the world to get the federal government to do something. In November 1961, through the Interstate Commerce Commission, it issued specific regulations, asking that posters be put on all interstate terminals and establishing the right of travel without segregation.

Even that was not seriously enforced. Two years later, in Winona, Mississippi, a group of blacks who used the white waiting room were arrested and brutally beaten. Constitutional government did not exist for them.

In the small southwest Georgia town of Albany, in the winter of 1961 and again in 1962, mass demonstrations took place against racial segregation. Of the 22,000 blacks in Albany, 1,000 of them went to jail. The Southern Regional Council, a research group in Atlanta, sent me down to Albany to do a report on the events there.

I found that blacks were doing no more than exercising their constitutional rights—marching, assembling, and speaking. Yet they were jailed and beaten—a pregnant black woman was kicked and lost her baby, a white civil rights worker had his jaw broken, a black lawyer was clubbed bloody by the local sheriff—and the U.S. government did nothing to interfere.[26]

I knew what the Constitution said, and that was enough to make me sure President John Kennedy and his brother Attorney General Robert Kennedy were not abiding by their oaths of office. I looked up the statutes. There was a law passed after the Civil War, now in the books as Title 18, Section 242 of the U.S. Code, which made it a crime for any official to willfully deprive any persons of their constitutional rights. That law was not being used to protect blacks in the South.

I had another opportunity to see if the federal government would enforce its own laws in November 1963 when I traveled to Selma, Alabama, to participate in Freedom Day. It was a day when black people in Dallas County were being organized to come to Selma, the county seat, and register to vote. It was a dangerous thing for a black person to do in Dallas County, and so a mass meeting was held the evening before in a black church, with speeches designed to build people's courage for the next day. Novelist James Baldwin came and so did comedian Dick Gregory, who tried to diminish fear with laughter. And there were the thrilling voices of the Selma Freedom Singers.

The next day, black men and women, elderly people, and mothers carrying babies lined up in front of the county courthouse where the voting registrar had his office. The street was lined with police cars. Colonel Al Lingo's state troopers were out in force, carrying guns, clubs, gas masks, and electrified cattle prods. Sheriff Jim Clark had deputized a large group of the county's white citizens, who were there, also armed. It looked like a war.

The federal building in Selma was across the street from the county courthouse. When two SNCC workers climbed up on the steps of the building and held up signs facing the courthouse that read Register to Vote, Sheriff Jim Clark and his deputies mounted the steps and dragged them off into police cars.

That federal building also housed the local FBI. Two FBI agents were out on the street taking notes. Two representatives of the Justice Department's Civil Rights Division were also there. We were all watching the arrest of two men for standing on federal property urging people to register to vote, I turned to one of the Justice Department lawyers. "Don't you think federal law has just been violated?" I asked.

The Justice man said, "Yes, I suppose so."

"Are you going to do something about that?"

"Washington is not interested."

A few moments later, two SNCC fellows—Chico Neblett and Avery Williams—tried to bring food and drink to people on the line who had been waiting a long time in the hot sun. They were intercepted by state troopers, knocked to the ground, jabbed with electric prods (I watched their bodies jump with each jolt), and taken off to jail. (Months later Avery Williams showed me the burns, which were still there on his leg.)

James Baldwin and I walked into the FBI office, whose windows looked out on the street and could take in the whole scene. We asked the FBI man in charge if he was going to arrest any of the state troopers or deputies for violating federal law. He shrugged and said he had no power to make an arrest.

It was a lie. The FBI has the power to make arrests when federal law is violated before its eyes. Would its agents let a bank robber do his work and just watch and take notes? They would apprehend a bank robber, but not a local southern policeman violating a black man's constitutional rights. When I wrote an article for the *New Republic* on what happened in Selma, pointing to the failure of the U.S. government to enforce its own laws, Burke Marshall of the Justice Department replied. He defended the federal government's inaction, speaking mystically of "federalism," which refers to the division of power between states and federal government. But the Fourteenth Amendment had made a clear statement about that division of power and gave the federal government the right to forbid the states from doing certain things to its citizens. And a number of laws were on the books to buttress the Fourteenth Amendment.

Indeed there was a powerful statute, going way back to the Revolutionary period and then reinforced after the Civil War. I found this when I was puzzling over the inaction of the federal government. This was Title 10, Section 333 of the U.S. Code, which says,

> The President, by using the militia or the armed forces, or both *or by any other means* shall take such measures as he considers necessary to suppress, in a State, any . . . domestic violence, unlawful combination, or conspiracy, if it . . . opposes or obstructs the execution of the laws of the United States (emphasis added).

There was no real need for more laws. The ones already on the books might well be enough, if there were a president determined to enforce them. But the passage of new laws is always an opportunity for a politician to make a statement to the public that says, Look, I'm doing something.

Kennedy had not planned to introduce new civil rights legislation. But in the late spring of 1963 he put his force behind a new, sweeping civil rights law, designed to outlaw segregation in public accommodations, eliminate segregation in state and local facilities, provide for fair employment regardless of race, and also put a bit more teeth into the federal government's actions against discrimination in schools and in voting.

What had changed Kennedy's mind was the mass demonstrations in Birmingham, Alabama, in the spring of 1963. These were organized by Martin Luther King and the Southern Christian Leadership Conference, along with local black leaders like Fred Shuttlesworth. Thousands of children marched in the streets, against firehoses and billy clubs and police dogs. The photos of police brutality, of children being smashed against the wall by high-power hoses, of a boy being attacked by a police dog, went around the world.

The demonstrations spread beyond Birmingham. In the ten weeks following the children's march, over 3,000 people were arrested in 758 demonstrations in 75 southern cities. By the end of 1963, protests had taken place in 800 cities across the country. Congress debated furiously the provisions of the new civil rights law, which it finally passed, after a year of debate and filibuster—the longest debate on any bill in history. That became the Civil Rights Act of 1964.[28]

The same summer that the new law was being debated, events in Mississippi revealed the limits of the federal government's commitment to racial equality, how little meaning there was to that end product of representative government, the federal statute.

The civil rights groups working together in Mississippi—SNCC, CORE, and SCLC—decided that they needed help and that they should call on young people from all over the country to come to Mississippi in the summer of 1964. The plan was to engage in an all-out effort to end segregation, to register black Mississippians to vote, to encourage local black people by showing how much national support they had.

Everyone connected with the plan knew it would be dangerous. Black people in Mississippi faced that danger every day, all their lives. In early June 1964, on the eve of what was to be called "The Mississippi Summer," movement organizers rented a theater near the White House, and a busload of black Mississippians traveled to Washington to tell a "jury" of prominent citizens (writers Paul Goodman, Joseph Heller, Murray Kempton, Robert Coles, Sarah Lawrence President Harold Taylor, black community activist Noel Day, and others) about the violence they had experienced in Mississippi.

Constitutional lawyers testified at that hearing. They pointed out that the federal government had sufficient legal authority—in the Constitution and in the statutes—to protect the civil rights workers and black Mississippians, from harm.

The transcript of the hearing was sent to President Lyndon Johnson and to Attorney General Robert Kennedy, along with a request to send federal marshals to Mississippi to protect all those working there for civil rights. There was no response from the White House. The busload of Mississippians went back home.

Twelve days after the hearing, three young people with the summer project—a black Mississippian named James Chaney and two whites from the North, Michael Schwerner and Andrew Goodman—disappeared while on a trip to Neshoba County, Mississippi, to investigate the burning of a black church. Chaney and Schwerner were staff members of CORE. Goodman was a summer volunteer and had just arrived in Mississippi hours before.

Two days later their burned station wagon was found, but no trace of the three men. On August 4, forty-four days after their disappearance, their bodies were found buried on a farm. James Chaney had been brutally beaten, so badly that a pathologist examining him said he had "never witnessed bones so severely shattered, except in tremendously high speed accidents such as aeroplane crashes." All three had been shot to death.[29]

In 1988 a film called *Mississippi Burning* was seen throughout the country. It was the story of the FBI search for the murderers of Chaney,

Goodman, and Schwerner. It portrayed the FBI as the heroes of the investigation that led to the discovery of the bodies and the prosecution of a number of Neshoba County men. One of those prosecuted was Deputy Sheriff Cecil Price, who had arrested them for speeding and then released them from jail in a prearranged plan to have them murdered. Price and several others were found guilty, spent a few years in prison.[30]

Those of us who were involved in the Mississippi Summer were angered by the movie. We knew how the FBI, again and again, had failed to do its duty to enforce federal law where the rights of black people in the South were at stake, how many times they had watched bloody beatings and done nothing, how the law had been violated before their eyes and they made no move. And we knew how outrageously they had behaved, along with the entire federal government, when the three young men disappeared.

Mary King worked in the SNCC office in Atlanta in those tumultuous times. In her book *Freedom Song,* she gives a detailed account, almost hour by hour, of the communications between the Atlanta office of SNCC and the FBI, the Department of Justice, and the White House. Mary was in charge of the Communications Office in Atlanta that Sunday, June 21, 1964, when the telephone call came in that the three men were hours overdue on their return from Philadelphia.

Her chronology makes clear how maddeningly cold and unresponsive were the FBI and the Department of Justice in what was a matter of life and death for three men in the hands of murderous racists. On hearing that Goodman, Chaney, and Schwerner were hours overdue, she phoned every jail and detention center in the counties surrounding Philadelphia, phoned the FBI and the Justice Department. She worked at this until 2:30 A.M. Monday, then was relieved by another SNCC worker, Ron Carver, who kept phoning through the early hours, until Mary arrived again at 8:00 A.M. She says;

> An attorney in the Civil Rights Division of the Justice Department . . . said quite bluntly that he wasn't sure there was any federal violation and therefore wouldn't investigate the matter. . . . A call to John Doar (of the Justice Department's Civil Rights Division) . . . resulted in this response: 'I have invested the FBI with the power to look into this.' But FBI agent H.F. Helgesen in Jackson denied there had been any word from John Doar. . . . He could do nothing until contacted by the New Orleans FBI office. . . . An hour later . . . he said he had called New Orleans but had received

> no orders to investigate. . . . We got in touch with an FBI agent named Mayner in New Orleans, who said he had received no orders from Washington. . . . Between 1:45 P.M. and 2:45 P.M. on Monday, I attempted to contact John Doar and Burke Marshall (Assistant Attorney General in charge of the Civil Rights Division), but to no avail. . . . The FBI agent in Meridian still insisted that he had no orders to initiate an investigation. . . . At 5:20 P.M. John Doar finally called me back. . . . But he did not specifically address the question of whether the FBI was investigating.

Mary King continues, "Finally, at 10:00 P.M. on Monday, June 22, thirty-seven hours after they were last seen alive by a member of our staff, UPI carried a story that . . . the FBI was being ordered into the case."[31]

It may well be that there was no way of saving the lives of the three young men after their disappearance. But there had certainly been a way of *preventing* what happened, if the government had only met the movement's request that it station federal marshals in Mississippi, to be on the spot, to accompany people into dangerous situations like Neshoba County. Don't they send police to guard the payrolls of banks?

Most of all, the behavior of the FBI and the Justice Department in that situation tells something about the moral and emotional remoteness of liberal constitutional government from the deepest grievances of its citizens. It tells us how important is Frederick Douglass's admonition that those who want the rain of freedom must themselves supply the thunder and lightning.

Later that same summer the Democratic party refused to seat a black delegation from Mississippi that claimed 40 percent of the seats (the percentage of blacks in the state). Instead the Credentials Committee voted to give 100 percent of the Mississippi seats to the official white delegates. It was representative government for whites, exclusion for blacks.

By 1965 it was clear that despite the Fifteenth Amendment, which said that citizens could not be denied the right to vote on grounds of race, and despite the civil rights acts of 1957, 1960, and 1964, all concerned with voting in some way, blacks in the Deep South were still not being allowed to vote.

A little-noticed clause of the Fourteenth Amendment, Section 2, says that if citizens are unfairly denied the right to vote, the representation in Congress of that state can be reduced. This would be the job of the president, who officially gets the census and decides on the number of representatives from each state. But no president, liberal or conservative, Republican or Democrat, had ever invoked this part of the Constitution—although it would have been a powerful weapon against racial discrimination in voting.

In the spring of 1965 the Southern Christian Leadership Conference began a campaign for voting rights in Selma, Alabama, around the same time that President Lyndon Johnson was discussing with Congress a new voting rights bill. Martin Luther King, Jr., went to Selma to join the action.

On March 7, later called "Bloody Sunday," a column of civil rights activists, beginning the long walk from Selma to the state capital in Montgomery, was confronted by state troopers demanding they turn back. They continued to walk, and the troopers set on them with clubs, beating them viciously, until they were dispersed and the bridge was splattered with blood.

During that campaign in Selma, Jimmie Lee Jackson, a black man, was shot in the stomach by a state trooper and died hours later. James Reeb, a white minister from the North, was clubbed to death by angry whites as he walked down the street.

News of the violence occurring in Selma was carried across the nation and around the world. One of the incidents, described by a reporter for United Press International, conveys the atmosphere. The fact that it appeared in the *Washington Post* suggests that the government could not be oblivious to what was going on:

> Club-swinging state troopers waded into Negro demonstrators tonight when they marched out of a church to protest voter registration practices. At least 10 Negroes were beaten bloody. Troopers stood by while bystanders beat up cameramen.[32]

The Voting Rights Act of 1965, for the first time, took the registration of blacks out of the hands of racist registrars in areas with a record of discrimination and put the force of the federal government behind the right to vote. David Garrow, in his book *Protest at Selma*, calls the new law "a legislative enactment that was to stimulate as great a change in American politics as any one law ever has."[33] It resulted in a dramatic increase in black voters and the election of black officials all through the Deep South.

What is clear from Garrow's careful study is how the protest movement in Selma was crucial in bringing about the Voting Rights Act. He gives some credit to the federal courts, but he says, "black southerners were unable to experience truly substantial gains in voting rights until, through their own actions, they were able to activate the federal executive and Congress." Furthermore, "the national consensus in favor of that bill . . . was primarily the result of the very skillful actions of the SCLC in Selma."[34]

Voting brought some black Americans into political office. It gave many more the feeling that they now had political rights equal to that of whites. They were now *represented* in local government and in Congress, at least more than before.

But there were limits to what such representation could bring. It could not change the facts of black poverty or destroy the black ghetto. After all, black people in Harlem or the South Side of Chicago had the right to vote long ago; they still lived in Harlem or the South Side, in broken-down tenements, amid rats and garbage. Thirty to 40 percent of young blacks were unemployed. Crime and drugs are inevitable in that atmosphere.

So it is not surprising that almost exactly at the time the Voting Rights Act was being enacted in 1965, the black ghetto of Watts, Los Angeles, erupted in a great riot. Or that in 1967 there were disorders, outbreaks, and uprisings in over a hundred cities, leaving eighty-three people dead, almost all blacks. And in 1968, after Martin Luther King was assassinated, there were more outbreaks in cities all over the country, with thirty-nine people killed, thirty-five of them black.[35]

But riots are not the same as revolution. The *New York Times* reported in early 1978: "The places that experienced urban riots in the 1960s have, with a few exceptions, changed little, and the conditions of poverty have spread in most cities."

The constitutional system set up by the Founding Fathers, a system of representation and checks and balances, was a defense in depth of the existing distribution of wealth and power. By arduous struggle and sacrifice, blacks might compel it to take down its "whites only" signs here and there. But poverty remained as the most powerful barrier to equality.

That is the barrier Madison spoke of when he said the system being set up in the new United States of America would prevent "an equal division of property or any other improper or wicked object." It is the fact of *class,* however disguised it is by the procedures of modern liberal societies.

Representative government does not solve the problem of race. It does not solve the problem of class. The very *principle* of representation is flawed, as Jean Jacques Rousseau, living in prerevolutionary France in the mid-eighteenth century, pointed out. His book *The Social Contract* was a confusing, contradictory, difficult search for a more direct democracy, in which a majority could not vote a minority into slavery or poverty. "How have a hundred men who wish for a master the right to vote on behalf of ten who do not?"

Rousseau was typical of many political philosophers, in that no one could be sure what he meant. As one commentator notes: "Robespierre . . . the great revolutionary, always expressed love and admiration for the philosopher. . . . Marie Antoinette, whom Robespierre guillotined, loved and admired Rousseau. The Thermidorians, who guillotined Robespierre, loved and admired Rousseau."[36] Despite the confusion, Rousseau provokes us to think critically about the whole idea of representation. It is an idea that we grew up to accept without question because it was an advance over monarchy and is today much preferable to dictatorship.

But it has serious problems. No representative can adequately represent another's needs; the representative tends to become a member of a special elite; he has privileges that weaken his sense of concern over his constituents' grievances. The anger of the aggrieved loses force as it is filtered through the representative system (something Madison saw as an advantage in *Federalist #10*). The elected official develops an expertise that tends towards its own perpetuation. Representatives spend more time with one another then with their constituents, become an exclusive club, and develop what Robert Michels called "a mutual insurance contract" against the rest of society.[37]

We can see the difficulties in the United States, which has one of the most praised systems of representative government in the world. People have the right to vote, but the choices before them are so limited, they see so little difference between the candidates, they so despair of their vote having any meaning, or they are so alienated from society in general because of their own misery that roughly 50 percent of those eligible to vote do not vote in presidential elections and over 60 percent do not vote in local elections.[38]

Money dominates the election process. The candidate for national office either has to have millions of dollars or have access to millions of dollars. (In 1982 a senator from Minnesota spent $7 million on his campaign.) Money buys advertising, prime time on television, a public image. The candidates then have a certain obligation to those with money who supported them. They must *look* good to the people who voted for them, but *be* good to those who financed them.[39]

Voting is most certainly overrated as a guarantee of democracy. The anarchist thinkers always understood this. As with Rousseau, we might not be sure of their solutions, but their critique is to the point. Emma Goldman, talking to women about their campaign for women's suffrage, was not *opposed* to the vote for women, but did want to warn against excessive expectations:

> Our modern fetish is universal suffrage. . . . I see neither physical, psychological, nor mental reasons why woman should not have the equal right to vote with man. But that can not possibly blind me to the absurd notion that woman will accomplish that wherein man has failed. . . . The history of the political activities of men proves that they have given him absolutely nothing that he could not have achieved in a more direct, less costly, and more lasting manner. As a matter of fact, every inch of ground he has gained has been through a constant fight, a ceaseless struggle for self-assertion, and not through suffrage.[40]

Helen Keller, who achieved fame for overcoming her blindness and deafness and displaying extraordinary talents, was also a socialist, and wrote the following in a letter to a woman suffragist in England:

> Are not the dominant parties managed by the ruling classes, that is, the propertied classes, solely for the profit and privilege of the few? They use us millions to help them into power. They tell us, like so many children, that our safety lies in voting for them. They toss us crumbs of concession to make us believe that they are working in our interest. Then they exploit the resources of the nation not for us, but for the interests which they represent and uphold. . . . We vote? What does that mean? It means that we choose between two bodies of real, though not avowed, autocrats. We choose between Tweedledum and Tweedledee.[41]

In Chapter 6 I noted how the vote for president means so little in matters of foreign policy; after the president is elected he does as he likes. We should also note that voting for members of Congress is meaningless for the most important issues of life and death. That is not just because it is impossible to tell at election time how your representative will vote in a future foreign policy crisis. It is also because Congress is a feeble, often nonexistent factor in decisions on war and peace, usually following helplessly along with whatever the president decides. That fact makes a shambles of "representative" government.

One of the more creative political philosophers of this century, Hannah Arendt (*The Origins of Totalitarianism, Eichmann in Jerusalem*), aware of the flaws in representative government, argued in her book *On Revolution* for something called the "council system" of government. Its basis was not voting, but neighborhood councils all over, with anyone who wanted to join the discussion invited to do so, and then these councils would form a kind of federation to make regional and national decisions.[42]

At various points in history there have been brief experiences with this direct democracy. In ancient Athens (except that women, slaves, and foreigners were excluded) all citizens had a chance to participate in decision making. For three months in Paris in 1871, while the elected Commune of Paris met constantly, there were also continuous meetings of people all over the city to discuss issues and then register their views with the Commune. On the eve of the Russian revolution, there were Soviets (councils) of workers, peasants, and soldiers—whoever wanted to join the discussion—but the Soviets were replaced by the rule of the Communist party and the new Soviet Constitution provided for elections to a legislative body.

Direct democracy is possible in small groups, and a wonderful idea for town meetings and neighborhood meetings. There could be discussions in offices and factories, a workplace democracy that neither the commissars of the Soviet Union nor the corporate executives of the United States and often not even the trade union leaders in these countries allow today.

To make national decisions directly is not workable, but it is conceivable that a network of direct democracy groups could register their opinions in a way that would result in some national consensus. Lively participation and discussion of the issues by the citizenry would be a better, more democratic, more reliable way of representing the population than the present stiff, controlled system of electoral politics.

There is already experience with special democratic procedures. Many states have provisions for initiatives and referenda. Citizens, by petition, can initiate legislation, call for general referenda, change the laws and the Constitution. That leads to a lively discussion among the public and something close to a real democratic decision. Except that so long as there are wealthy corporations dominating the media with their money, they can virtually buy a referendum the way they now buy elections.

There is also the idea of proportional representation, so that instead of the two-party system of Democrats and Republicans monopolizing power (after all, a two-party system is only one party more than a one-party system), Socialists and Prohibitionists and Environmentalists and Anarchists and Libertarians and others would have seats in proportion to their following. National television debates would show six points of view instead of two.

The people who control wealth and power today do not want any real changes in the system. For instance, when proportional representation was tried in New York City after World War II and one or two Communists were elected to the City Council the system was ended.) Also, when one radical congressman, Vito Marcantonio, kept voting against military budgets at the start of the cold war era, but kept getting elected by his district time after time, the rules were changed so that his opponent could run on three different tickets and finally beat him.

Someone once put a sign on a bridge over the Charles River in Boston: If Voting Could Change Things, It Would Be Illegal. That suggests a reality. Tinkering with voting procedures—proportional representation, initiatives, etc.—may be a bit helpful. But still, in a society so unequal in wealth, the rich will dominate any procedure. It will take fundamental changes in the economic system and in the distribution of wealth to create an atmosphere in which councils of people in workplaces and neighborhoods can meet and talk and make something approximating democratic decisions.[43]

No changes in procedures, in structures, can make a society democratic. This is a hard thing for us to accept, because we grow up in a technological culture where we think: If we can only find the right mechanism, everything will be okay, then we can relax. But we can't relax. The experience of black people in America (also Indians, women, Hispanics, and the poor) instructs us all. No Constitution, no Bill of Rights, no voting procedures, no piece of legislation can assure us of peace or justice or equality. *That* requires a constant struggle, a continuous discussion among citizens, an endless series of organizations and movements, creating a pressure on whatever procedures there are.

The black movement, like the labor movement, the women's movement, and the antiwar movement, has taught us a simple truth: The official channels, the formal procedures of representative government have been sometimes useful, but never sufficient, and have often been obstacles, to the achievement of crucial human rights. What has worked in history has been *direct action* by people engaged together, sacrificing, risking together, in a worthwhile cause.

Those who have had the experience know that, unlike the puny act of voting, being with others in a great movement for social justice not only makes democracy come alive—it makes the people engaged in it come alive. It is satisfying, it is pleasurable. Change is difficult, but if it comes, that will most likely be the way.

CHAPTER
TEN

Communism and Anti-communism

In 1948 a series of pamphlets was distributed by the House Committee on Un-American Activities titled: *One Hundred Things You Should Know about Communism.* There were 100 questions and answers.

QUESTION 1: "What is Communism?"
ANSWER: "A system by which one small group seeks to rule the world."

When I came across this in my files (the committee probably had files on me, so it seemed to me I should have files on them), I thought these men had taken an advanced course in political theory, also in expository writing, to be able to sum up such a complicated theory in so few words.

Skipping a number of questions, we come to:

QUESTION 76: "Where can a Communist be found in everyday life?" (This question interested me because there had been times when I was in need of a Communist, and didn't know where to find one.)
ANSWER: "Look for him in your school, your labor union, your church, or your civic club (Really, everywhere.)"
QUESTION 86: "Is the YMCA a Communist target?"
ANSWER: "Yes, so is the YWCA."

Anti-communism is part of the dominant American ideology. I am not speaking of a rational critique of communism or of countries that are called Communist. I mean by *anticommunism* a hysterical fear that has led the United States to spy on its own citizens, to invade other countries, to tax the hard-earned salaries of Americans to pay for trillions of dollars of monstrous weapons.

That hysteria is not just historical fact, going back to the 1950s and what is called "McCarthyism." It continues. In 1987 Robert McFarlane, national security adviser to President Reagan, said that he was opposed to sending arms illegally to the contras, but "where I was wrong was not having the guts to stand up and tell the President that. . . . Because if I'd done that, Bill Casey (CIA director), Jeane Kirkpatrick (ambassador to the United Nations), and Cap Weinberger (secretary of defense) would have said I was some kind of commie, you know."[1]

The national security adviser to President Reagan "some kind of commie"? A bizarre idea. But perhaps McFarlane knew the extent of his boss's paranoia. Reagan, campaigning for the presidency in 1980, summed up the world situation: "Let us not delude ourselves. The Soviet Union underlies all the unrest that is going on. If they weren't engaged in this game of dominoes, there wouldn't be any hot spots in the world."

Twenty years earlier, in 1960, ex-President Harry Truman reacted to the lunch counter sit-ins of black students in the South by telling an audience at Cornell University that they were inspired by Communists. When he was asked for proof of this, Truman said he had none. "But I know that usually when trouble hits the country the Kremlin is behind it."

Anti-communism goes back at least to the Bolshevik revolution of 1917. But it became very intense after World War II, when another huge country, China, had a Communist revolution, and when the cold war with the Soviet Union was taking the form of a reckless buildup of weapons on both sides.

In that time, we came to expect bizarre things. For instance, Congressman Harold Velde of Illinois, a former FBI man and later chair of the House Un-American Activities Committee, spoke in the House in March 1950 opposing mobile library service in rural areas because, he said; "Educating Americans through the means of the library service could bring about a change of their political attitude quicker than any other method. The basis of Communism and socialistic influence is education of the people."

It was not just the junior senator from Wisconsin, Joseph McCarthy, who was spreading wild fears about communism. The young congressman from Massachusetts, John F. Kennedy, reacted to the Communist victory in China by saying; "The House must now assume the responsibility of preventing the onrushing tide of Communism from engulfing all of Asia."[2]

Talk of spies and traitors filled the air in the 1950s. Julius and Ethel Rosenberg were executed, found guilty of passing atomic secrets to Russian agents, although it is clear that even if they did the data were of minor value and the death sentence viciously cruel. There is on the record an extraordinary statement made after their deaths by General Leslie Groves, head of the Manhattan Project, to a secret meeting of the Atomic Energy Commission. Groves said; "I think that the data that went out in the case of the Rosenbergs was of minor value. I would never say that publicly. . . . I should think it should be kept very quiet, because . . . the Rosenbergs deserved to hang."[3]

It also appears, on the basis of FBI documents subpoenaed in the 1970s, that the death sentence was prepared for them in advance by collusion between the judge and the prosecution, and that the chief justice of the Supreme Court assured the attorney general he would call a full court session to override any single justice's stay of execution (which is what happened, after Justice William O. Douglas granted a last-minute stay).

The atmosphere of anti-communism spawned all sorts of odd incidents. A navy ensign was refused a commission in the naval reserve because he continued "closely to associate" with a former Communist—his mother. A young music teacher in Washington, D.C., was refused a license to sell secondhand pianos because he had pleaded the Fifth Amendment before the House Un-American Activities Committee.[4]

In 1947 an art exhibition, "Advancing American Art," which opened in Europe to rave reviews, was canceled by the State Department on the grounds that it was "un-American" and "radical." The artists Georgia O'Keeffe, Ben Shahn, and Robert Motherwell were among those whose work was in the exhibit. Michigan Congressman George Dondero said, "Modern art is Communistic because it is distorted and ugly, because it does not glorify our beautiful country. . . . [It] breeds dissatisfaction . . . and those who create and promote it are our enemies."

The textbook commissioner of Indianapolis said the story of Robin Hood (who stole from the rich and gave to the poor) should be removed from schools because, as she put it: "There is a Communist directive now to stress the story of Robin Hood."[5]

Hollywood actors were threatened with blacklisting if they did not give the names of people they believed were Communists. Actor, sailor, and adventurer Sterling Hayden, who played tough roles on the screen, was bullied into informing on fellow leftists by the committee and later he was angry at himself and at his interrogators. His autobiography *Wanderer* was published in 1963, dedicated to Rockwell Kent and Warwick M. Tomkins, "Sailormen, Artists, Radicals." In his book, Hayden addresses his former psychoanalyst, who apparently had advised him to cooperate with the committee; "I'll say this too, that if it hadn't been for you I wouldn't have turned into a stoolie for J. Edgar Hoover. I don't think you have the foggiest notion of the contempt I have had for myself since the day I did that thing. . . . Fuck it! And fuck you too."[6]

The famous "Hollywood Ten," including some of the most important directors and writers in the motion picture industry, refused to give names or to discuss their political affiliations, citing the First Amendment. They were sent to prison.

Joseph Papp, producer of Shakespeare-in-the-Park in New York City, was called before the committee, and there was this exchange:

> ARENS [STAFF DIRECTOR FOR THE COMMITTEE]: Do you have the opportunity to inject into your plays . . . any propaganda . . . which would influence others to be sympathetic with the communist philosophy?
>
> PAPP: Sir, the plays we do are Shakespeare's plays. Shakespeare said, "To thine own self, be true.
>
> ARENS: There is no suggestion here by this chairman or anyone else that Shakespeare was a Communist. That is ludicrous and absurd. That is the Commie line.[7]

Playwright Arthur Miller defied the committee. Questioned about whether "a Communist who is a poet" should have the right to advocate revolutionary ideas, he said; "I tell you frankly, sir, I think, if you are talking about a poem, I would say that a man should have the right to write a poem just about anything." Miller refused to give names, was cited for contempt, convicted, but won on appeal.

Some of the chairmen of the House Un-American Activities Committee ended up in prison for fraudulent activities of various kinds. The screenwriter Ring Lardner, Jr., one of the Hollywood Ten, recalled this encounter of his prison days:

> The blue prison fatigues hung loosely on the weary, perspiring man whose path across the quadrangle was about to meet mine. . . . He was custodian of the chicken yard at the Federal Correctional Institution, Danbury, Connecticut, and his name was J. Parnell Thomas, formerly chairman of the Committee on Un-American Activities of the House of Representatives.

Thomas had been convicted of defrauding the government by padding his payroll.[8]

Novelist Howard Fast was a member of the Communist party, who left it in anger at its support of the Soviet Union's invasion of Hungary. In his memoir, *The Naked God,* he recalled what had happened to him in the United States because of his support of communism:

> During this period I found my own destruction as a writer who had full and normal access to the American public. Bit by bit, that access was pared away; reviewers began to read Communist propaganda into things I had written; bookstores were reluctant to order my books; "public-spirited" individuals undertook movements to have my books banned; and *Citizen Tom Paine,* of all things, was thrown out of the New York City school system on the excuse of "purple passages."[9]

Fast was indignant at his own treatment, but contrasted it with the fate of dissident writers in the Soviet Union and elsewhere (silenced, tortured, put to death):

> In the United States, I was crippled in my function as writer. At great cost and financial loss, I had to publish my own books. From comparative wealth and success, I was reduced to a struggle for literary existence; and gradually my continuing work became less and less known.
>
> But beyond deprivation, these facts are important:
>
> 1. I continued to write.
> 2. I continued to live.
> 3. I continued to fight for my inalienable privilege of writing as I pleased.[10]

No doubt the punishment of radical intellectuals in the United States was mild compared to what happened to them in the Soviet Union. But it was not mild in human terms for the people involved, and not tolerable at all in a country claiming to be the home of liberty. The nation was deprived for years of the talents of some of its most extraordinary artists.

Paul Robeson, for instance, the black singer and actor, became a nonperson because of his sympathy with communism. His banishment reached the point where the 1950 edition of *College Football*, in listing the All-American football team for 1918, omitted his name. Robeson had been All-American that year, and so the book had to list a ten-man football team.[11]

The scientist Albert Einstein received a letter in 1953 from a New York school teacher (teachers were being dismissed for radical or Communist activities) who asked him for advice on dealing with congressional investigations. Einstein replied,

> What ought the minority of intellectuals to do against this evil? Frankly, I can only see the revolutionary way of noncooperation in the sense of Gandhi's. Every intellectual who is called before one of the committees ought to refuse to testify, i.e., he must be prepared for jail and economic ruin, in short, for the sacrifice of his personal welfare in the interest of the cultural welfare of his country.

Perhaps an example of this "noncooperation" was given by the German playwright Bertolt Brecht, who came to work in Hollywood for a while and was called before the House Committee on Un-American Activities. Brecht kept the committee off-guard and confused by his replies. To questions about the plays he had written, in which the committee saw sinister Communist ideas, Brecht asked his interrogator if he had read it in the original German. The committee, perhaps embarrassed, certainly baffled, let Brecht go. Someone later described the questioning of Brecht this way: "It was as if a zoologist were being cross-examined by apes."[12]

Despite all the absurdities, the congressional appropriations for the committee grew in the postwar period from $50,000 in 1945, to $800,000 in 1970. By then it had 754,000 names on three-by-five cards in its files.

No president, liberal or conservative, Republican or Democrat, ever called for the abolition of the committee. And the Supreme Court, even at its most liberal (the Warren Court of the 1960s) never used its opportunities to declare the committee unconstitutional on the obvious ground that its very purpose (to investigate "un-American propaganda activities") violated the First Amendment. The committee, therefore, was not simply a creature of the paranoid right in America; it was sustained and supported by liberals and conservatives of our two-party system, in all three branches of government.

In other words, anti-communism was as American as apple pie. It was a bipartisan policy. Democrat Harry Truman himself issued Executive Order #9835, requiring the Department of Justice to prepare a list of organizations it determined to be "fascist, communist, or subversive" or "seeking to alter the form of government of the United States by unconstitutional means." Not only membership in, but "sympathetic association" with any organization on this list was to be considered in determining disloyalty for government employees.[13]

That list of "subversive" groups grew longer and longer. By 1954, it counted in the hundreds, including, besides the Communist party and the Ku Klux Klan, the Chopin Cultural Center, the Committee for the Negro in the Arts, the Committee for the Protection of the Bill of Rights, the League of American Writers, and the Nature Friends of America.

Until the 1960s, the committee towered frighteningly over its witnesses. The strategy of the left in refusing to answer questions, counting on the First or the Fifth Amendment, may well have been wrong. It led many people to think they had something to hide. The movements of the sixties brought a new kind of witness, the brazen radical willing to answer all questions, seizing the initiative, and using the committee room as a forum for political declamations.

Dagmar Wilson, of "Women Strike for Peace" gave the committee a taste of the new defiance in 1962. Six years later, in December 1968, Tom Hayden and Rennie Davis of the Students for a Democratic Society, and Dave Dellinger, long-time radical pacifist, appeared before the committee. They were all involved in the movement to stop the war in Vietnam. They answered all the questions, and then took the offensive. As in this encounter between the Committee and Tom Hayden:

> MR. CONLEY (A STAFF MEMBER OF THE COMMITTEE): Mr. Hayden, is it your present aim to seek the destruction of the present American democratic system?
>
> MR. HAYDEN: Well, I don't believe the present American democratic system exists. That is why we can't get together to straighten things out. You have destroyed the American democratic system by the existence of a committee of this kind.[14]

By 1970 the committee was losing its standing. It was being overwhelmed by the events of the sixties: the civil rights movement, the antiwar movement. It changed its name to the House Internal Security Committee and was not heard from very much.

Behind the absurdities, something serious had been taking place and was still going on—the attempt to shape the American mind so that people would react with automatic anger when they heard the word *communism,* so they would accept the huge military budgets (which doubled from 1950 to 1970 and then tripled from 1970 to 1980, going from $40 billion in 1950 to over $250 billion in 1980) and so they would accept wars and covert actions overseas if they were aimed at "communism."

In the history of the human race, we have often seen certain words used to stop thinking, to end rational discourse, to arouse hatred, words which are murderous. The words *Jew* and *nigger* have led to mass murder, lynchings, and enslavement. The words *Catholic, Protestant,* and *Moslem* have been used to inflame religious wars.

The word *Communist* in this country has been such a word. It has been used to justify the support by the United States of military dictatorships in Chile, the Philippines, Iran, El Salvador, and other places. When Ferdinand Marcos and his wife Imelda, who ran the Philippines ruthlessly and accumulated a huge fortune, were finally overthrown in 1988 and fled the islands with enormous sums of money, their American friend, Doris Duke, the tobacco heiress, praised them as "my country's vital outpost in combating Communism."[15]

In Guatemala, when the newly elected President Arbenz decided to confiscate large amounts of land owned by the United Fruit Company, the CIA and United Fruit both began preparing for the overthrow of the Arbenz government. Tom McCann, who was a Public Relations officer for United Fruit, wrote later about his orders: "Get out the word, a Communist beachhead has been established in the Western Hemisphere."[16] United Fruit created a book, *Report on Guatemala,* which said the new government was a Moscow-directed conspiracy. The book was sent to every congressman. And in 1954 a CIA-backed invasion overthrew Arbenz and set up a right-wing dictatorship that murdered tens of thousands of people.

Chile was the victim of a military coup in 1973, overthrowing its elected president, Salvador Allende. The United States had been working secretly against Allende, a moderate Socialist, since 1964. A staff member of the Senate Select Committee investigating the CIA, Karl Inderfurth, testified in 1975 that the CIA had set up a special group to carry on a massive propaganda campaign against Allende. It was, he said, "a scare campaign . . . it relied heavily on images of Soviet tanks and Cuban firing squads."

Another staff member of the Senate committee, Gregory Treverton, testified that, despite the public propaganda, the United States government knew through the National Intelligence Estimate (NIE) that there was no significant threat of a Soviet military presence in Chile. But the campaign went forward.[17]

The need to "stop communism" was used to justify the invasion of Vietnam and to carry on there a full-scale war in which over a million people died. It was used to justify the bombing of peasant villages, the chemical poisoning of crops, the "search and destroy missions," the laying waste of an entire country. GI Charles Hutto, who participated in the massacre of Vietnamese peasants at My Lai, told army investigators: "I remember the unit's combat assault on My Lai 4. The night before the mission we had a briefing by Captain Medina. He said everything in the village was communist. So we shot men, women, and children."

The Vietnam War may have been a turning point, when more and more Americans, seeing where anticommunism had led us, began to be suspicious of government propaganda. Charles Hutto, speaking years after the war, by then married with two children, said, "I was 19 years old and I was always told to do what the Government . . . told me to do. But now I'll tell my son, if the Government calls, use your own judgment. Now I don't think there should even be a thing called war, because it messes up a person's mind."[18]

In the sixties, leaders of the civil rights movement—Martin Luther King and the young blacks in SNCC—refused to be bullied by accusations of Communist influence that appeared in the press. They knew it was fraudulent, designed to weaken the movement. Charles Sherrod, a SNCC activist in Albany, Georgia, reacted to journalists who spoke vaguely, ignorantly, of "Communist infiltration" of the civil rights movement. Sherrod said, "I don't care who the heck it is—if he's willing to come down in the front lines and bring his body along with me to die—then he's welcome."

Communism: A Rational Critique

We have been dealing with an irrational, hysterical anti-communism, which has had terrible consequences for human rights in this country and abroad. There is, however, a rational critique of communism that requires thoughtful discussion.

I would start such a critique with two statements that I have come to believe over the years.

1. The *ideal* of communism—a classless society of equal abundance for all, based on highly developed technology, a very short workday, and, therefore, the possibility of real freedom for individuals to develop their aesthetic and personal interests as they like; a society free of the coercive apparatus of the state, organized by associated collectives, based on workplaces and neighborhoods, repudiating racial or sexual supremacy; a genuine participatory democracy, with full opportunity for free expression of all ideas, devoid of national hatreds, national boundaries, and war—this remains a wonderful goal. Karl Marx, I believe, envisioned such a society. Millions of people in the world have been inspired by that ideal and have been willing to sacrifice and risk all for it.

2. The Soviet Union, transforming Marx's transitional dictatorship of the proletariat into a deeply entrenched dictatorship of party bureaucrats, has repeatedly betrayed the communist ideal. While it has achieved a certain amount of economic progress and instituted social programs—child care, universal health care, free education, retirement benefits, full employment—it has been brutal in its treatment of its own citizens, murdering peasants in large numbers during the process of collectivization; imprisoning, torturing, and executing those it considered dissidents, whether ordinary people, intellectuals, artists, or distinguished leaders of the 1917 Revolution. The term *police state* fits it very well, and this is intolerable to anyone who believes in democratic socialism. It has imitated the imperialist powers in invading other countries—Hungary, Czechoslovakia, and Afghanistan, killing thousands of people.

When I was a young shipyard worker, I read the *Communist Manifesto* and understood why this essay, written in 1848 by Karl Marx, age thirty, and Friedrich Engels, age twenty-eight, had excited millions of people all over the world for a hundred years. It analyzed society in a way immediately verifiable by our own experience. "The history of all hitherto existing society is the history of class struggle."

Even knowing only a little history of the United States, who could deny the truth of that? Wasn't money, yes, *class*, behind all the political conflicts in the country, however hidden, however "class" was glossed over in the United States? Didn't we have a long history of strikes, struggles between capital and labor, and even if workers didn't see this as "class struggle," wasn't it just that, and might they not one day understand it and try to defeat not just one employer, but the entire capitalist class?

It made great sense, the way Marx and Engels traced the history of human society. How first, with simple tools, people had to live cooperatively, in a kind of primitive communism. How this gave way to the first class divisions, as agriculture developed, and it became possible for people to produce a surplus beyond their needs, and, therefore, to be exploited by lords who took the land as their private domain. And how this feudalism, with the further development of agriculture, of trade, of money as medium of exchange, of commerce, of cities, and of manufactures, had to make way for a new form of social organization—capitalism—with strong national governments to subsidize the capitalist corporations and promote investments, and maintain *law and order*, so that the system could move ahead, despite the misery of the working people, without rebellion.

Then came the inspiring message of the *Manifesto*'s sweeping treatment of history: like all previous societies, capitalism was becoming outmoded. It had done a superb job in building up the forces of production, in developing technology and workers' skills, and in producing huge amounts of goods. But it had created an economy it could not control, that went into periodic crises, and the competition among capitalist nations led repeatedly to wars. The economy was now international, complex, and *social;* it was irrational to have it owned and controlled by individual capitalists or corporations. The ownership should now be social, to match the reality. The system should be internationalized to match the reality that national boundaries were only hampering progress and provoking wars.

Workers, drawn together by their common grievances, more and more exploited, would see all this. They would see that what they thought was eternal, the rules of capitalism, the ideas, the political forms, were temporary, had once come into existence, would one day come to an end. They would see that the family had been distorted by capitalism and by money and that women had been treated as instruments to produce children. They would organize to change the system, would renounce national hatred and build a new international, cooperative order. "Workers of the world, unite! You have nothing to lose but your chains."

The working class, the proletariat, would one day overthrow the capitalist system, and organize a socialist society, which, after a transition period of "dictatorship of the proletariat," would do away with the instruments of *the state*, the police, and the army. There would be no class conflict requiring this force. Production would be organized rationally, its products distributed according to people's needs. There would be no need to work more than a few hours a day, and people would be free to live their lives as human beings should. It would be the real beginning of human history. Everything that came before was prelude.

This was the message of the *Manifesto.*

Years after I started reading the words of Marx, I came across the *Economic and Philosophic Manuscripts.* He had written this five years before the *Manifesto,* when he was twenty-five and living in Paris and first developing his radical ideas. These manuscripts were never published during his lifetime. When they finally appeared, in the 1930s, the Soviet Union dismissed them as Marx's "immature" writings. In fact, the *Manuscripts,* as I read them, seemed to me among the most profound of Marx's writings.

In them, Marx discusses how labor, under capitalism, is "alienated," "estranged" from the laborer himself. That is, "the worker's activity is not his spontaneous activity. It belongs to another; it is the loss of his self." Indeed, how many people work at what they really want to do? Some scientists, artists, some skilled workers, some professionals. But most people are not free to choose their line of work. Economic necessity forces them into it. So it is "alienated labor."

Furthermore, most workers produce things or commodities for sale that they did not choose to produce and that belong to someone else—the capitalist who employed them. Workers are thus alienated from the product of their labor.

As they work under modern industrial conditions, the atmosphere is not of cooperation, but competition, not association, but isolation. Workers are, therefore, alienated from one another. And, brought together in cities and factories, they are alienated from *nature.*

Marx saw a communist society as changing all that. He wrote in the *Manuscripts* that communism is "the complete return of man to himself as a social being, a human being. . . . This communism . . . equals humanism. . . . It is the genuine resolution of the conflict between man and nature and between man and man."

(Marx's ideas on labor and capital are not outrageously *radical.* They are shared by many people who are far from Marxism in many ways. For instance, in 1981, Pope John Paul II issued a papal encyclical *On Human Work,* in which he said the Church believed in ownership and private property, but that "Christian tradition has never upheld this right as absolute and untouchable. . . . The right to private property is subordinated to the right to common use, to the fact that goods are meant for everyone.)"[19]

The ideas of Marx and Engels are profound in their analysis of modern society, inspiring in their vision of a future, truly human way to live. The Soviet Union, the first revolution to claim a Marxist heritage, has violated that vision, has given socialism a bad name, which it does not deserve. Marxism became there an *ideology,* an orthodoxy of the left, a fixed package of ideas. It was not subject to criticism. It would be interpreted by the experts of the Party, as religious fundamentalists saw the Bible as interpreted by the experts of the Church.

As I read Marx, I was struck by his essential humanism. He was misrepresented in the press and in much of American culture as a theoretician of violence and dictatorship. But it seemed clear to me that this was not true to Marx's writings. In one of his articles for the *New York Daily Tribune* Marx criticized capital punishment:

> Is there not a necessity for deeply reflecting upon an alteration of the system that breeds these crimes, instead of glorifying the hangman who executes a lot of criminals to make room only for the supply of new ones?[20]

The Soviet Union, China, and other nations calling themselves "Marxist" and yet practicing capital punishment should be embarrassed by this statement.

If I can put in one word what has always infuriated me in any person, any group, any movement, or any nation, it is: *bullying.* This was the essence of fascism, the stormtroopers smashing the windows of Jewish storekeepers; in this country, police using their clubs on people walking picket lines or peacefully demonstrating; the bullying by white racists of black children going to school; abroad, the bullying by powerful nations of weaker ones, whether Italy invading Ethiopia, the Nazis invading Czechoslovakia, the United States destroying Vietnam, or the Soviet Union smashing the Hungarian rebellion.

When I recognized that quality in the Soviet state, I knew that the ideal of socialism had been betrayed. The betrayal started early, with the victory of the Bolsheviks in the second 1917 Revolution. The Soviets, the local councils of soldiers, sailors, peasants, and workers, in which such lively democratic discussions had taken place, were disbanded and replaced by the rule of the Communist party.

While the tone of Lenin's book, *State and Revolution,* which appeared in 1917, was anti-statist and libertarian, that same year Lenin, at the head of the new government, began the centralization of power in the Communist party and the suppression of all opposition. The revolution against tsarist Russia, for "peace, bread, and land," a genuine popular revolt involving masses of people, grass-roots committees, and local soviets, became transformed into elitist rule.[14]

Maxim Gorky, the great Russian writer, had welcomed and supported the revolution against the tsar in March 1917. But when the Bolsheviks took power in November, Gorky almost immediately saw that they were determined to wipe out any opposition and suppress any criticism. He observed the imprisonment of a number of socialist leaders who disagreed with the Bolsheviks.

Two months after the Bolsheviks took power, in January 1918, they dissolved the Constituent Assembly, which had been elected by popular vote and which included representatives from the various anti-tsarist parties. Gorky was indignant. He became absolutely furious when workers who demonstrated peacefully in Petrograd against the dissolution of the Constituent Assembly were shot down in the streets by government soldiers.

He wrote in the newspaper that he edited in Petrograd, *Novaya Zhizn* (New Life) that the Constituent Assembly represented the century-long Russian dream of democracy. "Rivers of blood have been spilled on the sacrificial altar of this sacred idea, and now the 'People's Commissars' have given orders to shoot the democracy which demonstrated in honor of this idea." Gorky recalled the failed 1905 Revolution:

> On January 9, 1905, when the downtrodden, ill-treated soldiers were firing into unarmed and peaceful crowds of workers by order of the Tsarist regime, intellectuals and workers ran up to the soldiers—the unwilling murderers—and shouted point-blank in their faces: "What are you doing, damn you? Who are you killing? Don't you see they are your brothers, they are unarmed, they bear no malice toward you, they are going to the Tsar to beg his attention to their needs. . . . Come to your senses, what are you doing, idiots!"[22]

Gorky likened this to the shooting that had just taken place in front of the Constituent Assembly:

> And just as on January 9, 1905, people who had not lost their conscience and reason asked those who were shooting: "What are you doing, idiots? Aren't they your own people marching? You see there are red banners everywhere, and not a single poster hostile to the working class, not a single utterance hostile to you!"
>
> And—like the Tsarist soldiers—these murderers under orders answered: "We've got our orders! We've got our orders to shoot."

(The scenes Gorky described seem like the ones that took place in Beijing, the capital of Communist China, in June 1989, when government troops fired at peaceful demonstrators for democracy, mostly students, killing hundreds of them, and bystanders cried out at them: "We are your brothers! Why are you shooting?")

Around the same time, Gorky noted that of all the letters he received "the most interesting are those written by women." He commented on "the psychophysiology of a woman":

> As a being who incessantly replenishes the losses inflicted upon life by death and destruction, she must feel both more deeply and more acutely than I, a man, hatred and aversion for all that reinforces the work of death and destruction.

Yes, Gorky said, this was idealism, but

> I find that social idealism is most necessary precisely in an era of revolution. . . . Without the participation of this idealism, a revolution—and all of life—would turn into a dry, arithmetical problem of distributing material wealth, a problem the solution of which demands blind cruelty and streams of blood, a problem which, arousing savage instincts, kills man's social spirit, as we shall see in our time.[23]

Polish-German revolutionary Rosa Luxemburg (who was executed by the German police after an abortive revolution in 1919) admired the Bolsheviks' seizure of power in 1917. But she criticized the dismissal of the Constituent Assembly. She published a pamphlet, against the advice of some of her comrades, who said it would play into the hands of counter-revolutionaries. This has been the standard excuse given for the failure of revolutionaries to honestly criticize what they see as evil in revolutionary movements: "It will play into the hands of. . . ." Luxemburg wrote,

> Socialism, by its very nature, cannot be dictated, introduced by command. . . . Without general elections, without unrestricted freedom of press and assembly, without a free exchange of opinions, life dies out in every public institution and only bureaucracy remains active.[24]

Emma Goldman and Alexander Berkman, American anarchists who were deported from the United States to their native Russia for opposing World War I, became observers of the early years of the Bolshevik regime, from 1919 to 1921, and were soon disillusioned with what they saw. The culmination of their despair at the betrayal of revolutionary ideals was the shooting down of the sailors who revolted at Kronstadt, outside Leningrad, asking that the revolution meet their needs. Trotsky and Lenin supervised the operation, and Goldman and Berkman left the Soviet Union in disgust.

The extent of Stalin's crimes was finally admitted, three years after his death, when Nikita Khrushchev shook the several thousand delegates to the Twentieth Congress of the Communist party, by giving the bloody details of Stalin's murders of his own revolutionary colleagues. In the Khrushchev era, some reforms took place in the Soviet Union. Censorship of the foreign press stopped. Many political prisoners were set free. But repression persisted, the atmosphere of the police state was still evident. Dissident writers and speakers were put in prison or given psychiatric examinations, pronounced psychotic, and put into mental institutions. A news item from Moscow in late 1986:

> Anatoly T. Marchenko, one of the best known Soviet political dissidents, has died in prison, his wife learned here today. He was 48 years old.
>
> Mr. Marchenko had spent more than 20 years of his life in prisons, labor camps and internal exile. He died in the prison at Chistopol, in the Tatar republic, while serving a 10-year term for "anti-Soviet agitation and propaganda."[25]

A recognition of the terrible things that have happened in the Soviet Union should not lead us—as it has done to certain ex-Communists—to rush from one pole of fanaticism to another, to embrace the anti-communism of the U.S. government, to justify its wars, its control over other countries, its buildup of a genocidal nuclear arsenal at the expense of the American taxpayer, at the cost of poverty, sickness, and homelessness for tens of millions of Americans.

It was precisely such a rush from one hysterical end of the spectrum to another that led some people, indignant at the cruelties of capitalism, the bullying of small countries by the imperial powers, to adopt the Soviet Union uncritically as the representative of socialism, of the future. My own refusal to adopt either Soviet socialism or American capitalism as models of justice and freedom led me, while participating in the movements of the sixties, to read more and more about the philosophy of anarchism.

Anarchists, I discovered, did not believe in *anarchy* as it is usually defined—disorder, disorganization, chaos, confusion, and everyone doing as they like. On the contrary, they believed that society should be organized in a thousand different ways, that people had to cooperate in work and in play, to create a good society. But, anarchists insisted, any organization must avoid hierarchy and command from the top; it must be democratic, consensual, reaching decisions through constant discussion and argument.

What attracted me to anarchism was its rejection of any bullying authority—the authority of the state, of the church, or of the employer. Anarchism believes that if we can create an egalitarian society without extremes of poverty and wealth and join hands across all national boundaries, we will not need police forces, prisons, armies, or war, because the underlying causes of these will be gone.

Anarchists are ferocious critics of all governments, whether capitalist or "socialist." They believe governments naturally tend to authoritarianism and think that we can have a society of self-governing, cooperative communities, linked together in ways that will avoid a central bureaucracy, yet get things done on a national and international scale.

This is, of course, a utopian idea; it has existed nowhere, at least not for long. There have been moments in history when something approaching true democracy existed. One of those was the Paris Commune of 1871, when Parisians were involved in daily meetings, discussing every subject under the sun, trying to come to conclusions without governmental force.

Another such moment was the period of six months or so at the start of the Spanish Civil War, when certain parts of Spain were in the hands of anarchist groups. One of those places was Barcelona, where George

Orwell, wounded in the Spanish war, visited in December 1936. He was astonished by what he saw:

> When one came straight from England the aspect of Barcelona was something startling and overwhelming. It was the first time that I had ever been in a town where the working class was in the saddle. . . . Every shop and cafe had an inscription saying that it had been collectivized; even the bootblacks had been collectivized and their boxes painted red and black. Waiters and shopwalkers looked you in the face and treated you as an equal. Servile and even ceremonial forms of speech had temporarily disappeared.[26]

The year 1989 was a historic year in the history of both communism and anticommunism. The Soviet Union, under the leadership of an extraordinary reformer, Mikhail Gorbachev, began a radical transformation of its society, encouraging freedom of the speech and press, sponsoring contested elections, even surrendering the idea of a one-party state and the monopoly of power by the Communist party.

At the same time, in the countries of the Communist bloc in Eastern Europe, mass movements for nonviolent radical change swept through country after country—Poland, Czechoslovakia, East Germany, Bulgaria, and Rumania. Huge demonstrations led to the toppling of the old Communist leaders. Communists and noncommunists seemed to be joining in new coalitions looking to a society with democratic freedoms and mixed socialist-capitalist economies.

It seems that the human costs of fanaticism, of holding on to a rigid ideology, had become too great, and that Gorbachev and other Communist leaders recognized this. They saw it in bottled-up discontent and in failing economies. They saw the enormous waste in military expenditures burdening their countries. The surrender of fanaticism became a practical necessity for leadership, a powerful demand by the people of the Communist countries.

As the United States entered the 1990s, its leadership was lagging behind the dramatic changes in Eastern Europe. President George Bush and the members of Congress were still voting for an enormous military budget of $290 billion dollars, while millions were homeless and more millions lived in slums and while poverty led to crime, drug addiction, alcoholism, and violence. The environment was deteriorating—the pollution of air and water and the poisoning of the planet—but this was being met with the most puny of measures, while enormous resources were still being expended for space ships and nuclear delivery systems.

On the very day (February 6, 1990) that the Central Committee of the Communist party was meeting in the Soviet Union and deciding to give up its monopoly of power—an astounding event—President Bush was watching American soldiers fight "Soviet tanks" in a mock battle of World War III in California. His speech that day to the troops was a sign that the old anti-communist fanaticism was still alive in the highest circles of American politics. The *New York Times* reported:

> "It's especially encouraging to see anything which might bring the day of true democracy a bit closer for the Soviet people," Mr. Bush said. But he cautioned, "It is important not to let these encouraging changes, political or military, lull us into a sense of complacency. . . . God bless our country. Thank you colonel, and now, back to war."[27]

Even if the political leadership of the United States has been slow to give up the long-held anti-communist fanaticism, so costly to human rights in this country and abroad, perhaps the American people have learned something from the events of the past decades, from watching the extraordinary developments in the Communist world, watching people there surrender *their* fanaticism.

The coming of a new century may be the right time for people all over the world to discard old orthodoxies, frozen dogmas, simple definitions. It may be a time to welcome thinking outside the customary boundaries; to look with fresh eyes at communism, socialism, capitalism, liberalism, and anarchism; and to seek out good ideas wherever they are, because we desperately need them.

When Bertolt Brecht was called before the House Committee on Un-American activities, he wanted to read a statement, but they would not let him do so. Part of his statement was as follows:

> We are living in a dangerous world. Our state of civilization is such that mankind already is capable of becoming enormously wealthy but as a whole is still poverty-ridden. Great wars have been suffered. Greater ones are imminent, we are told. Do you not think that in such a predicament every new idea should be examined carefully and freely?[28]

CHAPTER

ELEVEN

The Ultimate Power

As the twentieth century draws to a close, a century packed with history, what leaps out from that history is its utter unpredictability. Who could have predicted, not just the Russian Revolution, but Stalin's deformation of it, then Khrushchev's astounding exposure of Stalin, and in recent years Gorbachev's succession of surprises?

Or that in Germany, the conditions after World War I that might have brought socialist revolution—an advanced industrial society, with an educated organized proletariat, and devastating economic crisis—would lead instead to fascism? And who would have guessed that an utterly defeated Germany would rise from its ashes to become the most prosperous country in Europe?

Who foresaw the shape of the post–World War II world: the Chinese Communist revolution, and its various turns—the break with the Soviet Union, the tumultuous cultural revolution, and then post–Mao China making overtures to the West, adopting capitalist enterprise, perplexing everyone?

No one foresaw the disintegration of the old Western empires happening so quickly after the war, in Asia, Africa, and the Middle East, or the odd array of societies that would be created in the newly independent nations, from the benign socialism of Nyerere's Tanzania to the madness of Idi Amin's Uganda.

Spain became an astonishment. A million had died in the Spanish Civil War and Franco's fascism lasted forty years, but when Franco died, Spain was transformed into a parliamentary democracy, without bloodshed. In other places too, deeply entrenched regimes seemed to suddenly disintegrate—in Portugal, Argentina, the Philippines, and Iran.

The end of the war left the United States and the Soviet Union as superpowers, armed with frightening nuclear arsenals. And yet these superpowers have been unable to control events, even in those parts of the world considered to be their spheres of influence. The United States could not win wars in Vietnam or Korea or stop revolutions in Cuba or Nicaragua. The Soviet Union was forced to retreat from Afghanistan and could not crush the Solidarity movement in Poland.

The most unpredictable events of all were those that took place in 1989 in the Soviet Union and Eastern Europe, where mass movements for liberty and democracy, using the tactic of nonviolent mass action, toppled long-lasting Communist bureaucracies in Poland, Czechoslovakia, Hungary, Rumania, Bulgaria, and East Germany.

Uncertain Ends, Unacceptable Means

To confront the fact of unpredictability leads to two important conclusions:

The first is that the struggle for justice should never be abandoned on the ground that it is hopeless, because of the apparent overwhelming power of those in the world who have the guns and the money and who seem invincible in their determination to hold on to their power. That apparent power has, again and again, proved vulnerable to human qualities less measurable than bombs and dollars: moral fervor, determination, unity, organization, sacrifice, wit, ingenuity, courage, and patience—whether by blacks in Alabama and South Africa; peasants in El Salvador, Nicaragua, and Vietnam; or workers and intellectuals in Eastern Europe and the Soviet Union. No cold calculation of the balance of power should deter people who are persuaded that their cause is just.

The second is that in the face of the obvious unpredictability of social phenomena all of history's excuses for war and preparation for war—self-defense, national security, freedom, justice, and stopping aggression—can no longer be accepted. Massive violence, whether in war or internal upheaval, cannot be justified by any end, however noble, *because no outcome is sure.* Any humane and reasonable person must conclude that if the ends, however desirable, are uncertain, and the means are horrible and certain, those means must not be employed.

We have had too many experiences with the use of massive violence for presumably good reasons to willingly continue accepting such reasons. In this century there were 10 million dead in World War I, the war "to end all wars"; 40 to 50 million dead in World War II to "stop aggression" and "defeat fascism"; 2 million dead in Korea and another 1 to 2 million dead in Vietnam, to "stop communism"; 1 million dead in the Iran-Iraq war, for "honor" and other indefinable motives. Perhaps a million dead in Afghanistan, to stop feudalism or communism, depending on which side was speaking.[1]

None of those ends was achieved: wars did not end, aggression continued, fascism did not die with Hitler, communism was not stopped, there was no honor for anyone. In short (as I argued earlier in the book) the traditional distinction between "just" and "unjust" war is now obsolete. The cruelty of the means today exceeds all possible ends. No national boundary, no ideology, no "way of life" can justify the loss of millions of lives that modern war, whether nuclear or conventional, demands. The standard causes are too muddy, too mercurial, to die for. Systems change, policies change. The distinctions claimed by politicians between good and evil are not so clear that generations of human beings should die for the sanctity of those distinctions.

Even a war for defense, the most morally justifiable kind of war, loses its morality when it involves a sacrifice of human beings so massive it amounts to suicide. One of my students, a young woman, wrote in her class journal in 1985, "Wars are treated like wines—there are good years and bad years, and World War II was the vintage year. But wars are not like wines. They are more like cyanide; one sip and you're dead."

Internal violence has been almost as costly in human life as war. Millions were killed in the Soviet Union to "build socialism." Countless lives were taken in China for the same reason. A half million were killed in Indonesia for fear of communism; at least a million dead in Cambodia and a million dead in Nigeria in civil wars. Hundreds of thousands killed in Latin America by military dictatorships to stop communism, or to "maintain order." There is no evidence that any of that killing did any good for the people of those nations.

Preparation for war is always justified by the most persuasive of purposes: to prevent war. But such preparation has not prevented a series of wars that since World War II have taken more lives than World War I.

As for the claim that massive nuclear armaments have prevented World War III, that is not at all certain. World War III has certainly not taken place, but it is not clear that this is because of the massive arms race. The logic of that claim is the logic of the man, living in New York City, who sprinkled yellow powder all over his house, explaining to his friends that this was to keep elephants out, and the proof of his success was that no elephant had ever appeared in his house.[2]

There are many reasons why an all-out war between the Soviet Union and the United States has not taken place. Neither nation has anything to gain from such a war. Neither nation can possibly invade and occupy the other. Why would the Soviet Union want to destroy its great source of wheat? The atomic bombs necessary to annihilate the other superpower would create an enormous danger, through radioactivity and nuclear winter, to the attacking power. The conflicts between the United States and the USSR have, therefore, been in *other* places, and those places *have* had wars, undeterred by the arms buildup of the superpowers.

Deterrence is the favorite word of those who urge the buildup of weapons, both in the United States and in the Soviet Union. But it seems that the only thing that has been "deterred" (World War III) is deterred by other factors, which makes the enormous buildup of weapons on both sides a total waste. No politician on either side of the cold war has had the courage to make this statement, which is a matter of the most ordinary common sense.

The chief reason consistently given for spending thousands of billions of dollars on weapons has been that this prevents a Soviet invasion of Western Europe. Probably no American knows more about Soviet policy than veteran diplomat and historian George Kennan, former ambassador to the Soviet Union and one of the theoreticians of the cold war. Kennan insists that the fears of Soviet invasion of Western Europe are based on myth. This is corroborated by a man who worked for the CIA for twenty-five years, Harry Rositzke. Rositzke was at one time the CIA director of espionage operations against the Soviet Union. He wrote, in the 1980s; "In all my years in government and since I have never seen an intelligence estimate that shows how it would be profitable to Soviet interests to invade Western Europe or to attack the United States."[3]

Common sense suggests that the Soviet Union has enough problems at home, has had trouble controlling Eastern Europe, and was unable to defeat Afghanistan, a small backward nation on its border. Would it invade Western Europe and face the united opposition of 200 million people who would never submit, would make it an endless war?

It appears that the citizens of the United States have been taxed several trillion dollars because of an irrational fear. An irrational fear is, by definition, inconsolable and yet infinite in its demands.

So we have this irony. That the arms race has deterred what would not take place anyway. And it has not deterred what *has* taken place: wars all over the world, some involving the superpowers directly (Korea, Vietnam, and Afghanistan), others involving them indirectly (the Israeli-Arab wars, the Iran-Iraq war, the Indonesian war against East Timor, the contra war against Nicaragua).

While the supposed *benefits* of the arms race are very dubious, the human *costs* are obvious, immediate, and awful. In 1989 about a trillion dollars—a thousand billion dollars—were spent for arms all over the world, the United States and the Soviet Union accounting for more than half of this. Meanwhile, about 14 million children die every year from malnutrition and disease, which are preventable by relatively small sums of money.

The new-style *Trident* submarine, which can fire hundreds of nuclear warheads, costs $1.5 billion. It is totally useless, except in a nuclear war, in which case it would also be totally useless, because it would just add several hundred more warheads to the thousands already available. (Its only use might be to *start* a nuclear war by presenting a first-strike threat to the Soviet Union.) The $1.5 billion could finance a five-year program of universal child immunization against certain deadly diseases, preventing 5 million deaths.[4]

The B-2 bomber, the most expensive military airplane in history, approved by the Reagan and Bush administrations, and by many members of Congress in both parties, was scheduled to cost over a half billion dollars for each of 132 bombers. A nuclear arms analyst with the Congressional Budget Office estimated that the total cost would run between $70 billion and $100 billion. With this money the United States could build a million new homes.[5]

Over the past decade, several trillions of dollars have been spent for military purposes—to kill and to prepare to kill. One can only begin to imagine what could be done with the money in military budgets to feed the starving millions in Africa, Asia, and Latin America; to provide health care for the sick; to build housing for the homeless; and to teach reading, writing, and arithmetic to millions of people crippled by their inability to read or write or count.

There have been hundreds of nuclear weapons tests by the Soviet Union and the United States over the years. (News item, 1988: "The

United States has concealed at least 117 nuclear explosions at its underground test site in the Nevada desert over the past quarter-century, a group of private scientists reported yesterday."[6]) The $12 million used for one of these tests would train 40,000 community health workers where they are desperately needed in the Third World.

The United States spent about $28 billion to build 100 B-1 bombers, which turned out to be an enormous waste, even from the standpoint of the military, involving stupidity, greed, and fraud (critics said the B-1 would not survive a collision with a pelican).[7] Imagine what could be done for human health with that $28 billion.

Health and education in the eighties were starved for resources. But in 1985 it was disclosed that $1.8 billion dollars had been spent on sixty-five antiaircraft guns called the Sergeant York, all of which had to be scrapped as useless.[8]

Imagine what could be done to stop the most frightening fact of our time, the steady poisoning of the world's environment—the rivers, the lakes, the oceans, the beaches, the air, the drinking water, and the soil that grows our food—the depletion of the protective ozone layer that covers the entire earth, and the erosion of the world's forests. The money, technology, and human energy now devoted to the military could perform miracles in cleaning up the earth we live on.

But the cost of the arms race is not only the enormous waste of resources. There is a psychic cost—the creation of an atmosphere of fear all over the world. There is no accurate way of measuring that fear in generations of young people who have grown up in the shadow of the bomb. One can only imagine the effect on all those little schoolchildren in the United States, who, in the 1950s, were taught to crouch under their desks when they heard a siren, signifying a bombing attack.

And what is the effect on the 10 or 20 million young men (and women) who are either conscripted or enticed into the armed forces of nations, and then taught to kill, to obey orders, to stop thinking like free human beings?

These are the certainties of evil in the arms race. There are other things that are not certainties, but probabilities, and that is nuclear accidents. When thousands of nuclear weapons are stockpiled, when tests are taking place, and when bombing planes are sent aloft with hydrogen bombs, there is a strong probability that accidents will take place involving those bombs.

In fact there have been over a hundred of those accidents. The military calls them, in its quaint language, "Broken Arrows." One of the first of these was the loss of four hydrogen bombs over Spain in 1966. They didn't explode, but there was radioactive fallout. Lies were told by both the Spanish and American governments for a long time, undoubtedly to try to cool public resentment against the U.S. military presence. Nevertheless, there was a demonstration of a thousand people in front of the U.S. Embassy in Madrid. It was charged by the police, who beat demonstrators with clubs. It seems that many Spanish citizens resented the fact that hydrogen bombs were being flown, like bales of cotton, over their land.[9]

It should be noted that a *hydrogen* bomb—also called a thermonuclear bomb—is the superbomb, developed after the original atomic bomb. Instead of *fission,* splitting a uranium atom, or a plutonium atom, to release the amounts of explosive energy that were released over Hiroshima and Nagasaki, the hydrogen bomb works by *fusion,* in which two hydrogen atoms are put together to release far more explosive energy. Indeed, 1,000 times as much, so we must imagine a bomb 1,000 times as powerful as the bomb dropped on Hiroshima.

The one dropped on Hiroshima was equivalent in its destructive power to 14 kilotons (14,000 tons) of TNT, the material used in ordinary bombs. There are hydrogen bombs with the power of 14 *megatons* (14 million tons) of TNT. And it is these bombs (called "strategic nuclear weapons" to differentiate them from the smaller "tactical nuclear weapons") of which both the United States and the Soviet Union have accumulated 10,000 each.

Two of these superbombs were involved in an accident in North Carolina in 1961. A Defense Department document obtained nineteen years later by the Reuters news agency revealed what the Pentagon at the time refused to confirm or deny. The Reuters article said:

> On January 24, 1961, a crashing B-52 bomber jettisoned two nuclear bombs over Goldsboro, North Carolina, according to the document. A parachute deployed on one bomb, while the other broke apart on impact.
>
> The bomb with the parachute was jolted when the parachute caught in a tree and five of the six interlocking safety switches were released, said the former officials. Only one switch prevented the explosion of a 24 megaton bomb, 1800 times more powerful than the one dropped on Hiroshima in 1945, they said.[10]

That should give anyone pause. The superpowers have in their arsenals the equivalent of a million Hiroshima-type bombs. Only people who were both saints and geniuses might possibly be trusted with such weapons. This does not seem an accurate description of the leaders of the United States and the Soviet Union. Consider an item like the following, shortly after Ronald Reagan took office as president:

> President Ronald Reagan and his top three aides flew to Washington yesterday aboard the so-called "Doomsday Plane", a $117-million jumbo jet equipped to serve as an airborne command post in a nuclear war. . . . Deputy White House secretary Larry Speakes quoted Reagan as saying he was highly impressed and as adding, "It gives me a sense of confidence."[11]

The very possession of nuclear weapons endangers the possessor. The chance of blowing ourselves up by accident is greater than the chance of invasion by a foreign power, just as a homeowner who keeps a rifle handy is (as statistics show) more likely to kill a member of the family with it than to shoot an outside intruder.

We would need an extraordinary faith in technology to believe that we can have 10,000 thermonuclear weapons, some of them in airplanes flying overhead, and perhaps 20,000 smaller nuclear weapons in various places, and not have accidents.

There is an even more awesome prospect than "Broken Arrow" accidents. That is, a radar error that will signal an enemy bombing attack and thus trigger off, perhaps automatically without human intervention, a genuine attack that would be the beginning of the end for everybody.

In fact, there have been many computer errors, over 100 of them in 1980–1981. One of them led to a "red alert," that is, the radar announced an imminent Soviet attack, and planes with hydrogen bombs were about to be sent aloft when the error was discovered. A news dispatch of June 18, 1980:

> On June 3 and June 6, errors in a computer at the North American Air Defense Command headquarters inside Cheyenne Mountain, near Colorado Springs, caused the system to warn erroneously that Soviet intercontinental missiles had been fired at the United States. The alert sent nearly 100 bomber crews to start their planes' engines.[12]

A few months before that incident, there was an Associated Press dispatch:

> The worldwide computer system built to warn the President of an enemy attack or international crisis is prone to break down under pressure, according to informed sources who have worked on or examined the system.
>
> A Pentagon document defending the system said that generally the "computers render effective support; the principal exception occurs in crisis situations."[13]

It will only fail in "crisis situations"!

There have been enough disasters with advanced technology to persuade us not to believe those "experts" who assure us blandly that some device is "foolproof" or "fail-safe" or has quadruple guards, or whatever. There was the near meltdown of the nuclear reactor at Three Mile Island, which came frighteningly close to a major catastrophe and which let loose enough contamination to cause sickness in humans and animals years later. Then came the even worse disaster at the Chernobyl nuclear plant in the Soviet Union. And shortly after that, the failure of the U.S. space shuttle *Challenger,* which killed all those aboard.

Those events were accompanied by official lies to cover up the true nature of what had happened. Indeed, nuclear technology, because its failures have cataclysmic consequences, encourages political leaders to deceive the public, as happened right from the beginning of the atomic tests in the Nevada desert. The Atomic Energy Commission lied to the GIs who participated in those tests and who later developed cancer far beyond the normal statistical expectations.

There is still another cost of the arms buildup, and that is the fact that the possession of superweapons tempts the possessor to use it as a threat in any international crisis. Once the threat is made, it is very difficult, given the traditional concern of political leaders with "credibility," "saving face," "maintaining our image," etc., to back down.

That is why the world came close to nuclear war during the Cuban missile crisis in the fall of 1962, when the discovery of the presence of Russian missiles on Cuba led to an American ultimatum to Khrushchev, where both nations needed to "save face" by bulling it out. As Kennedy's adviser Theodore Sorensen put it, the president "was concerned less about the missiles' military implications than with their effect on the global political balance."[14]

Only Khrushchev's decision to back down enabled an agreement on removal of the missiles, in return for a pledge not to try again to invade Cuba. President Kennedy estimated that there was a one in three chance of nuclear war in that situation and yet he went ahead with his threats.[15]

And what provoked it was that the Soviets did in Cuba what the United States had already done in Turkey and other countries, to place missiles very close to the borders of the other superpower.

Recently, a researcher asked some of the top military and strategic leaders of the United States the commonsense question: Why in the world do we need tens of thousands of nuclear bombs for deterrence? Suppose we assume (what I believe to be false), that nuclear weapons are needed to deter a Soviet invasion or attack, surely a few hundred bombs—enough to destroy every major Soviet city (and which could be carried on *two* submarines)—would be a sufficient deterrent.

The answers of these policymakers were startling; they acknowledged that the weapons were unnecessary from a military point of view, but claimed they served a "political" purpose in that they conveyed a certain *image* of American power. One analyst with the Rand Corporation (a government think tank) told him:

> If you had a strong president, a strong secretary of defense they could temporarily go to Congress and say, "We're only going to build what we need. . . . And if the Russians build twice as many, tough." But it would be unstable politically. . . . And it is therefore better for our own domestic stability as well as international perceptions to insist that we remain good competitors even though the objective significance of the competition is . . . dubious.[16]

In short, hundreds of billions have been spent to maintain an *image.* The image of the United States is that of a nation possessed of a frightening nuclear arsenal. What good has that image done, for the American people, or for anyone in the world? Has it prevented revolutions, coups, wars? Even from the viewpoint of those who want to convey an image of strength—for some mysterious psychic need of their own, perhaps—what image is conveyed when a nation so overarmed is unable to defeat a tiny country in Southeast Asia, or to prevent revolutions in even tinier countries in the Caribbean?

The weapons addiction of all our political leaders, whether Republican or Democrat, has the same characteristics as drug addiction. It is enormously costly, very dangerous, provokes ugly violence, and is self-perpetuating—all on a scale far greater than drug addiction.

Aside from its uselessness for military and political purposes, its colossal waste of human resources, its dangers to the survival of us all, nuclear deterrence is profoundly immoral. It means that the United

States is holding hostage the entire population of the Soviet Union—the very people it claims are suffering under communism—and stands ready to kill them all if the Soviet government makes the wrong move. And the Soviet Union is doing the same to the American population. If we think holding hostage the passengers of an airliner is unspeakably evil and call it terrorism, what name shall we give for holding hostage the entire human race?

The arms race is sustained by a fanatical righteousness that sees international conflict as total good versus total evil, and is willing to sacrifice hundreds of millions of lives in a nuclear war. William Buckley wrote in the mid 1980s:

> The suggestion that . . . no use of nuclear weapons is morally defensible, not even the threat of their use as a deterrent, is nothing less than an eructation in civilized thought, putting, as it does, the protraction of biological life as the fit goal of modern man.[17]

Not only are we supposed to feel intellectually inferior if we have to look up the word *eructation* (which means belching, and Buckley, intent on showing off, is not using it accurately); but we are supposed to feel morally inferior if we oppose nuclear deterrence because of some cowardly feeling that *life* is more precious than political victory. Buckley is a Catholic, and we might contrast his statement with that of Vatican II: "Any act of war aimed indiscriminately at the destruction of entire cities or of extensive areas along with their population is a crime against God and man himself. It merits unequivocal and unhesitating condemnation."[18]

It is sad to see how, in so many countries, citizens have been led to war by the argument that it is necessary because there are tyrannies abroad, evil rulers, murderous juntas. But to make war is not to destroy the tyrants; it is to kill their subjects, their pawns, their conscripted soldiers, their subjugated civilians.

War is a *class* phenomenon. This has been an unbroken truth from ancient times to our own, when the victims of the Vietnam War turned out to be working-class Americans and Asian peasants. Preparations for war maintains swollen military bureaucracies, gives profits to corporations (and enough jobs to ordinary citizens to bring them along). And they give politicians special power, because fear of "the enemy" becomes the basis for entrusting policy to a handful of leaders, who feel bound (as we have seen so often) by no constitutional limits, no constraints of decency or commitment to truth.

Justice Without Violence

Massive violence has been accepted historically by citizens (but not by all; hence desertions, opposition, and the need for bribery and coercion to build armies) because it has been presented as a means to good ends. All over the world there are nations that commit aggression on other nations and on their own people, whether in the Middle East, or Latin America, or South Africa—nations that offend our sense of justice. Most people don't really want violence. But they do want justice, and for that sake, they can be persuaded to engage in war and civil war.

All of us, therefore, as we approach the next century, face an enormous responsibility: How to achieve justice without massive violence. Whatever in the past has been the moral justification for violence—whether defense against attack, or the overthrow of tyranny—must now be accomplished by other means.

It is the monumental moral and tactical challenge of our time. It will make the greatest demands on our ingenuity, our courage, our patience, and our willingness to renounce old habits—but it must be done. Surely nations must defend themselves against attack, citizens must resist and remove oppressive regimes, the poor must rebel against their poverty and redistribute the wealth of the rich. But that must be done without the violence of war.

Too many of the official tributes to Martin Luther King, Jr., have piously praised his nonviolence, the praise often coming from political leaders who themselves have committed great violence against other nations and have accepted the daily violence of poverty in American life. But King's phrase, and that of the southern civil rights movement, was not simply "nonviolence," but *nonviolent direct action.*

In this way, nonviolence does not mean acceptance, but resistance—not waiting, but acting. It is not at all passive. It involves strikes, boycotts, noncooperation, mass demonstrations, and sabotage, as well as appeals to the conscience of the world, even to individuals in the oppressing group who might break away from their past.

Direct action does not deride using the political rights, the civil liberties, even the voting mechanisms in those societies where they are available (as in the United States), but it recognizes the limitations of those controlled rights and goes beyond.

Freedom and justice, which so often have been the excuses for violence, are still our goals. But the means for achieving them must change, because violence, however tempting in the quickness of its action, undermines those goals immediately, and also in the long run. The means for achieving social change must match, morally, the ends.

It is true that human rights cannot be defended or advanced without *power.* But, if we have learned anything useful from the carnage of this century, it is that true power does not—as the heads of states everywhere implore us to believe—come out of the barrel of a gun, or out of a missile silo.

The possession of 10,000 thermonuclear weapons by the United States did not change the fact that it was helpless to stop a revolution in Cuba or another in Nicaragua, that it was unable to defeat its enemy either in Korea or in Vietnam. The possession of an equal number of bombs by the Soviet Union did not prevent its forced withdrawal from Afghanistan nor did it deter the Solidarity uprising in Poland, which was successful enough to change the government and put into office a Solidarity member as prime minister. The following news item from the summer of 1989 would have been dismissed as a fantasy two years earlier: "Solidarity, vilified and outlawed for eight years until April, jubilantly entered Parliament today as the first freely elected opposition party to do so in a Communist country."[19]

The power of massive armaments is much overrated. Indeed, it might be called a huge fake—one of the great hoaxes of the twentieth century. We have seen heavily armed tyrants flee before masses of citizens galvanized by a moral goal. Recall those television images of Somoza scurrying to his private plane in Managua; of Ferdinand and Imelda Marcos quickly assembling their suitcases of clothes, jewels, and cash and fleeing the Philippines; of the Shah of Iran searching desperately for someone to take him in; of Duvalier barely managing to put on his pants before escaping the fury of the Haitian people.

In the United States we saw the black movement for civil rights confront the slogan of "Never" in a South where blacks seemed to have no power, where the old ways were buttressed by wealth and a monopoly of political control. Yet, in a few years, the South was transformed.

I recall at the end of the great march from Selma to Montgomery in 1965 when, after our twenty-mile trek that day, coming into Montgomery, I had decided to skip the speeches at the capitol and fly back to Boston. At the airport I ran into my old Atlanta colleague and friend,

Whitney Young, now head of the Urban League, who had just arrived to be part of the celebration in Montgomery. We decided to have coffee together in the recently desegregated airport cafeteria.

The waitress obviously was not happy at the sight of us. Aside from the *integration* of it, she might have been disconcerted by the fact that the white man was still mud-splattered, disheveled, and unshaven from the march, and the black man, tall and handsome, was impeccably dressed with suit and tie. We noticed the big button on her uniform. It said "Never!" but she served us our coffee.

Racism still poisons the country, north and south. Blacks still mostly live in poverty, and their life expectancy is years less than that of whites. But important changes have taken place that were at one time unimaginable. A consciousness about the race question exists among blacks and whites that did not exist before. The nation will never be the same after that great movement, will never be able to deny the power of nonviolent direct action.

The movement against the Vietnam War in the United States too was powerful, and yet nonviolent (although, like the civil rights movement, it led to violent scenes whenever the government decided to use police or National Guardsmen, against peaceful demonstrators). It seemed puny and hopelessly weak at its start. In the first years of the war, no one in public life dared to speak of unilateral withdrawal from Vietnam. When my book *Vietnam: The Logic of Withdrawal* was published in 1967, the idea that we should simply leave Vietnam was considered radical. But by 1969 it was the majority sentiment in the country. By 1973 it was in the peace agreement, and the huge U.S. military presence in Vietnam was withdrawn.

President Lyndon Johnson had said; "We will not turn tail and run." But we did, and it was nothing to be ashamed of. It was the right thing to do. Of course, the military impasse in Vietnam was crucial in bringing the war to an end, but it took the movement at home to make American leaders decide not to try to break that impasse by a massive escalation, by more death and destruction. They had to accept the limits of military power.

In that same period, cultural changes in the country showed once again the power of apparently powerless people. Women, a century before, had shown their power and won the right to go to college, to become doctors and lawyers, and to vote. And then in the sixties and seventies the women's liberation movement began to alter the nation's perception of women in the workplace, in the home, and in relationships with men, other women, and children. The right to abortion was established by the Supreme Court against powerful opposition by religious conservatives (although that decision is still under heavy attack).

Another apparently powerless group—homosexual men and lesbian women—encouraged perhaps by what other movements had been able to accomplish against great odds, took advantage of the atmosphere of change. They demanded, and in some places received, acceptance for what had before been unmentionable.

These last decades have shown us that ordinary people can bring down institutions and change policies that seemed entrenched forever. It is not easy. And there are situations that seem immovable except by violent revolution. Yet even in such situations, the bloody cost of endless violence—of revolt leading to counterrevolutionary terror, and more revolt and more terror in an endless cycle of death—suggests a reconsideration of tactics.

We think of South Africa, which is perhaps the supreme test of the usefulness of nonviolent direct action. It is a situation where blacks have been the victims of murderous violence and where the atmosphere is tense with the expectation of more violence, perhaps this time on both sides. But even the African National Congress, the most militant and most popular of black organizations there, clearly wants to end apartheid and attain political power without a bloodbath that might cost a million lives. Its members have tried to mobilize international opinion, have adopted nonviolent but dramatic tactics: boycotts, economic sanctions, demonstrations, marches, and strikes. There will undoubtedly be more cruelty, more repression, but if the nonviolent movement can grow, perhaps one day a general strike will paralyze the economy and the government and compel a negotiated settlement for a multiracial, democratic South Africa.

The Palestinians in the West Bank and Gaza Strip, under the military occupation of the Israelis since the war of 1967, began around 1987 to adopt nonviolent tactics, massive demonstrations, to bring the attention of the world to their brutal treatment by the Israelis. This brought more brutality, as hundreds of Palestinians, unarmed (except for clubs and rocks), were shot to death by Israeli soldiers. But the world did begin to pay attention and if there is finally a peaceful arrangement that gives the Palestinians their freedom and Israel its security, it will probably be the result of nonviolent direct action.

Certainly, the use of terrorist violence, whether by Arabs placing bombs among civilians or by Jews bombing villages and killing large numbers of noncombatants, is not only immoral, but gains nothing for anybody. Except perhaps a spurious glory for macho revolutionaries or ruthless political leaders puffed up with their "power" whenever they succeed in blowing up a bus, destroying a village, or (as with Reagan) killing a hundred people by dropping bombs on Tripoli.

People made fearful by politicians but also by real historical experience worry about invasion and foreign occupation. The assumption has always been that the only defense is to meet violence with violence. We have pointed out that, with the weaponry available today, the result is only suicidal (South Korea against North Korea, Iran against Iraq, even Vietnam against the United States).

A determined population can not only force a domestic ruler to flee the country, but can make a would-be occupier retreat, by the use of a formidable arsenal of tactics: boycotts and demonstrations, occupations and sit-ins, sit-down strikes and general strikes, obstruction and sabotage, refusal to pay taxes, rent strikes, refusal to cooperate, refusal to obey curfew orders or gag orders, refusal to pay fines, fasts and pray-ins, draft resistance, and civil disobedience of various kinds.[20] Gene Sharp and his colleagues at Harvard, in a study of the American Revolution, concluded that the colonists were hugely successful in using nonviolent tactics against England. Opposing the Stamp Tax and other oppressive laws, the colonists used boycotts of British goods, illegal town meetings, refusal to serve on juries, and withholding taxes. Sharp notes that "in nine or ten of the thirteen colonies, British governmental power had already been effectively and illegally replaced by substitute governments" before military conflict began at Lexington and Concord.[21]

Thousands of such instances have changed the world, but they are nearly absent from the history books. History texts feature military heroes, lead entire generations of the young to think that wars are the only way to solve problems of self-defense, justice, and freedom. They are kept uninformed about the world's long history of nonviolent struggle and resistance.

Political scientists have generally ignored nonviolent action as a form of power. Like the politicians, they too have been intoxicated with *power.* And so in studying international relations, they play games (it's called, professionally, "game theory") with the strategic moves that use the traditional definitions of power—guns and money. It will take a new movement of students and faculty across the country to turn the universities and academies from the study of war games to peace games, from military tactics to resistance tactics, from strategies of "first-strike" to those of "general strike."

It would be foolish to claim, even with the widespread acceptance of nonviolent direct action as *the* way of achieving justice and resisting tyranny, that all group violence will come cleanly to an end. But the gross instances can be halted, especially those that require the cooperation of the citizenry and that depend on the people to accept the legitimacy of the government's actions.

Military power is helpless without the acquiescence of those people it depends on to carry out orders. The most powerful deterrent to aggression would be the declared determination of a whole people to resist in a thousand ways.

When we become depressed at the thought of the enormous power that governments, multinational corporations, armies and police have to control minds, crush dissents, and destroy rebellions, we should consider a phenomenon that I have always found interesting: Those who possess enormous power are surprisingly nervous about their ability to hold on to their power. They react almost hysterically to what seem to be puny and unthreatening signs of opposition.

For instance, we see the mighty Soviet state feeling the need to put away, out of sight, handfuls of disorganized intellectuals. We see the American government, armored with a thousand layers of power, work strenuously to put a few dissident Catholic priests in jail or keep a writer or artist out of this country. We remember Nixon's hysterical reaction to a solitary man picketing the White House: "Get him!"

Is it possible that the people in authority know something that we don't know? Perhaps they know their own ultimate weakness. Perhaps they understand that small movements can become big ones, that if an idea takes hold in the population, it may become indestructible.

It is one of the characteristics of complex and powerful machines that they are vulnerable to tiny unforeseen developments. The disaster of the giant space vessel *Challenger* was due to the failure of a small ring that was affected by cold. Similarly, huge organizations can be rendered helpless by a few determined people. A headline in the *New York Times* in the summer of 1989 read: "Environmentalists' Vessels Sink Navy Missile Test." The story began,

> The Navy was forced to cancel a test launching of its newest missile today when four vessels manned by protesters sailed into a restricted zone 50 miles

off the Atlantic coast of Florida and attached an antinuclear banner on the side of the submarine that was to fire the missile.[22]

As all-controlling a government as that in the Soviet Union must still worry about its citizens' protest, especially when large numbers of people are involved. The Soviet Union, after unilaterally halting its nuclear tests for a year and a half and finding that the United States did not respond, announced in February 1987, that it would now resume testing. And it did. But suddenly, it mysteriously halted testing for five months in 1989. Why?

According to two American physicians connected with "Physicians for Social Responsibility," and in touch with Soviet doctors, the mysterious five-month absence of nuclear testing may well have been due, in their words, to "the rapid growth of a grassroots environmental movement in Kazakhstan." It seems that two underground tests had released radioactive gases into the atmosphere. This led a prominent Kazakh poet to call a meeting of concerned citizens. Five thousand people assembled and made a public appeal to close the test site in Kazakhstan. They said; "We cannot be silent. In the process of our growing democracy, the people's opinion gains power and range. Everything happening on this earth applies to all of us. Only by uniting our efforts . . . will we help ourselves survive in this still green world."[23]

Whether or not their protest stopped the testing is not certain. But the fact that in the Soviet Union such a meeting could take place and boldly call for a change in national policy was a sign of a new power developing to contest the power of the government.

Nonviolent direct action is inextricably related to democracy. Violence to the point of terrorism is the desperate tactic of tiny groups who are incapable of building a mass base of popular support. Governments much prefer violence committed by disciplined armies under their control, rather than adopt tactics of nonviolence, which would require them to entrust power to large numbers of citizens, who might then use it to threaten the elites' authority.

A worldwide movement of nonviolent action for peace and justice would mean the entrance of democracy for the first time into world affairs. That's why it would not be welcomed by the governments of the world, whether "totalitarian" or "democratic." It would eliminate the dependence on *their* weapons to solve problems. It would bypass the official makers of policy and the legal suppliers of arms, the licensed dealers in the most deadly drug of our time: violence.

It was 200 years ago that the idea of democracy was introduced into modern government, its philosophy expressed in the American Declaration of Independence: Governments derive their powers from the consent of the governed and maintain their legitimacy only when they answer the needs of their citizens for an equal right to life, liberty, and the pursuit of happiness.

It is surely time to introduce that basic democratic concept into international affairs. The terrifying events of this century make it clear that the political leaders of the world and the experts who advise them are both incompetent and untrustworthy. They have put us all in great danger.

We recall the British historian Arnold Toynbee, surveying thousands of years of human history, and despairing of what he saw in the atomic age. He cried out: "No annihilation without representation!"

The New Realism

Those of us who call for the repudiation of massive violence to solve human problems must sound utopian, romantic. So did those who demanded the end of slavery. But utopian ideas do become realistic at certain points in history, when the moral power of an idea mobilizes large numbers of people in its support. This may then be joined to the realization, by at least some of those in authority, that it would be *realistic* for them to change their policy, even perhaps share power with those they have long controlled.

It is becoming more and more clear that "military victory," that cherished goal of generals and politicians, may not be possible any more. Wars end in stalemates, as with the United States in Korea, or with Iran and Iraq, or in forced withdrawals, as the United States in Vietnam, the Soviet Union in Afghanistan. So called "victories," as Israel in the 1967 war, bring no peace, no security. Civil wars become endless, as in El Salvador, and after rivers of blood the participants must turn to negotiated settlements. The contras in Nicaragua could not win militarily, and finally had to negotiate for a political solution.

The economic costs of war and preparations for war threaten the stability of the great powers. One of the reasons the United States withdrew from Vietnam was the drain on its budget, which required the neglect of social problems at home, bringing on the black riots of 1967 and 1968, throwing a scare into the establishment. The Soviet

Union undertook bold initiatives for disarmament in the mid-1980s when it recognized that its economy was overmilitarized and failing. Both superpowers must be reminding themselves more and more of all those empires in history that became arrogant with power, overburdened with armies, impoverished by taxes, and collapsed.[24]

Heads of governments become nervous when public opinion begins to veer away from their control. This happened in the 1980s, when dramatic changes took place in the public's views on war and militarism. In the United States in 1981 public opinion surveys showed that 75 percent of those polled said more money was needed for the military. But by the beginning of 1985, only 11 percent favored an increase in military spending, and 46 percent favored a decrease.[25]

When military bureaucrats worry about the growth of peace signs, the rest of the world might well be pleased. Caspar Weinberger, leaving his job as secretary of defense for seven years under Reagan, was alarmed: "A recent, rather startling poll indicated that 71% of Republicans and 74% of Democrats believe that the United States can trust the general secretary of the Soviet Union, Mikhail Gorbachev."[26]

In 1983 in West Germany, so close to the Soviet bloc, 55 percent saw the Soviet Union as a military threat; by 1988, only 24 percent saw such a threat, and half of those polled were in favor of unilateral disarmament.[27] In 1984 a quarter of a million West Germans gathered in Kassel to protest the installation of Pershing and cruise missiles. They erected ninety-six crosses in a field outside the U.S. Air Force Station, one for each cruise missile deployed there.

With both the United States and the Soviet Union facing severe economic problems—stagnation and budget deficits—there is suddenly a *realistic* incentive to cut back on military spending. Indeed, the forbidden phrase *unilateral disarmament* may become very practical.

Unilateral actions are the best way; they avoid endless negotiations, as was seen in 1963 when John F. Kennedy took the initiative to stop atmospheric nuclear testing and the Soviet Union followed suit.[28] There had been an earlier "moment of hope" (the phrase of Nobel Prize winner Philip Noel-Baker), when Khrushchev became the Soviet leader and his government withdrew Soviet forces from Austria and returned a naval base to Finland. But that didn't lead to anything significant, and, according to Soviet specialist Walter Clemens; "Washington never tested Moscow's offer to join both Germanys in a neutral and demilitarized Central Europe."[29]

The nation that takes the first initiatives to disarm will be at a great advantage. First, in world prestige, that much-desired *image.* Note how Gorbachev, after his initiatives, became the most popular political figure in West Germany, the United States' strongest ally. Second, in freeing huge resources for economic development. The obvious benefits to the nation that first disarms might well lead to a disarmament race.

Statistics indicate that, of the industrialized nations, those that spend the least for military purposes show the greatest economic progress. The United States between 1982 and 1986 spent 6 percent of its gross national product for the military while Japan spent about 1 percent. Japan's economy, everyone agreed, was more efficient, more dynamic, and healthier.

Of course, those realistic incentives are not enough by themselves to alter the habits of governments so deeply dug into old policies of militarism and war. But they create the possibility, *if* a great popular movement should develop to insist on change. Such a movement, if it became large enough and strong enough to threaten the political power of the government, would create an additional incentive for change.

A great movement must be driven by a vision, as the civil rights movement was driven by the dream of equality and the antiwar movement by the prospect of peace. The vision in this case, for people all over the world, is the most inspiring of all, that of a world without war, without police states nourished by militarism, and with immense resources now free to be used for human needs. It would be a tremendous shift of resources from death to life. It would mean a healthy future for ourselves, our children, and our grandchildren.

The vision would be of a trillion dollars (the annual military costs around the world) made available to the coming generation, to the young, who could use their energy, their talents, their idealism, and their love of adventure to rebuild the cities, feed the hungry, house the homeless, clean the rivers and lakes, refresh the air we breathe, and revitalize the arts. Imagine the 30 million young men now in uniform, imagine those several hundred million people in the world either unemployed or underemployed (the International Labor Office estimates over 400 million people in the 1980s)—imagine all that wasted energy mobilized to make their lives useful and exciting and to transform the planet.[30]

If the U.S. government can give several hundred billion dollars in contracts to corporations to build weapons, why can it not (by powerful public demand) give that valuable money to public-service corporations whose contracts will require them to employ people, young and old, to make life better for everyone? The conversion of resources requires a conversion of language. New definitions of old terms could become a part of the common vocabulary. The old definitions have misled us and caused monstrous harm.

The word *security*, for instance, would take on a new meaning: the health and well-being of people, which is the greatest strength and the most lasting security a nation can have. (A simple parable makes this clear: Would a family living in a high-crime city feel more "secure" if it put machine guns in its windows, dynamite charges in the yard, and tripwires all around the house, at the cost of half the family income and less food for the children? The analogy is not far-fetched. It is an understatement of what nations do today.)

The word *defense* would mean, not the waging of war and the accumulation of weapons, but the united actions of people against tyranny, using every ingenious device of nonviolent resistance.

Democracy would mean the right of people everywhere to determine for themselves, rather than have political leaders decide for them, how they will defend themselves, how they will make themselves secure, and how they will achieve justice and freedom.

Patriotism would mean not blind obedience to a nation's leaders, but a commitment to help one's neighbors and to help anyone, regardless of race or nationality, achieve a decent life.

It is impossible to know how quickly or how powerfully such new ways of thinking, such reversals of priorities, can take hold, can excite the imagination of millions, can cross frontiers and oceans, and can become a world force. We have never had a challenge of this magnitude, but we have never had a need so urgent, a vision so compelling.

History does not offer us predictable scenarios for immense changes in consciousness and policy. Such changes have taken place, but always in ways that could not be foretold, starting often with imperceptibly small acts, developing along routes too complex to trace. All we can do is to make a start, wherever we can, to persist, and let events unfold as they will.

On our side are colossal forces. There is the desire for survival of 5 billion people. There are the courage and energy of the young, once their adventurous spirit is turned toward the ending of war rather than the waging of war, creation rather than destruction, and world friendship rather than hatred of those on the other side of the national boundaries.

There are artists and musicians, poets and actors in every land who are ready to make the world musical and eloquent and beautiful for all of us, if we give them the chance. They, perhaps more than anyone, know what we are all missing by our infatuation with violence. They also know the power of the imagination and can help us to reach the hearts and souls of people everywhere.

The composer Leonard Bernstein a few years ago spoke to a graduating class at John Hopkins University; "Only think: if all our imaginative resources currently employed in inventing new power games and bigger and better weaponry were re-oriented toward disarmament, what miracles we could achieve, what new truths, what undiscovered realms of beauty!"[31]

There are teachers in classrooms all over the world who long to talk to their pupils about peace and solidarity among people of all nations and races.

There are ministers in churches of every denomination who want to inspire their congregations as Martin Luther King, Jr., did, to struggle for justice in a spirit of joy and love.

There are people, millions of them, who travel from country to country for business or pleasure, who can carry messages that will begin to erase, bit by bit, the chalk marks of national boundaries, the artificial barriers that keep us apart.

There are scientists anxious to use their knowledge for life instead of death.

There are people holding ordinary jobs of all kinds who would like to participate in something extraordinary, a movement to beautify their city, their country, or their world.

There are mothers and fathers who want to see their children live in a decent world and who, if spoken to, if inspired, if organized, could raise a cry that would be heard on the moon.

It is, of course, an enormous job to be done. But never in history has there been one more worthwhile. And it needn't be done in desperation, as if it had to be done in a day. All we need to do is make the first moves, speak the first words.

One of the scientists who worked on the atomic bomb, who later was a scientific adviser to President Eisenhower, chemist George Kistiakowsky, devoted the last years of his life, as he was dying of cancer, to speaking out against the madness of the arms race in every public forum he could find. Toward the very end, he wrote, in the *Bulletin of the Atomic Scientists:* "I tell you as my parting words. Forget the channels. There simply is not enough time left before the world explodes.

Concentrate instead on organizing, with so many others of like mind, a mass movement for peace such as there has not been before."

He understood that it was not the bomb he had worked on, but the people he had come to work with, on behalf of peace, that were the ultimate power.

Notes

CHAPTER ONE *Introduction: American Ideology*

1. When Gorbachev came to power in the Soviet Union, he clearly grasped this idea, one that the giant American corporations had learned long ago; one did not have to monopolize the field to maintain control and allowing for a bit of competition was the most ingenious way to dominate. And so he initiated some socialist "pluralism."

2. Edward Herman and Noam Chomsky, in their book *Manufacturing Consent* (South End Press, 1989) argue powerfully that the function of the media in the United States (and, of course, not *only* in the United States) "is to inculcate and defend the economic, social, and political agenda of privileged groups that dominate the domestic society and the state." They document this with examples of how the press treated certain historical events: the Tet offensive during the Vietnam War, the Watergate scandals of the Nixon era, and the Iran-Contra affair of the Reagan years.

3. Ched Noble, in a remarkable essay, "Ethics and Experts," *Working Papers* (July–Aug. 1980), rebels against the field in which she received her Ph.D. (philosophy), as she finds in it a "new philosophical sub-discipline, applied ethics." She challenges the assumption she finds in this new area, that "in order to think properly about moral issues . . . one needs a background in classical moral theories and modern theory of value." While she does not believe common sense alone can solve the profound moral problems, she insists that "contemporary theoretical ethics cannot supply the deficiencies of common sense." She resents the arrogance of philosophers "who believe that philosophy is the proper academic discipline to assume responsibility for solving today's moral problems."

A similar view is expressed by a veteran philosopher, Bernard Williams, in his book *Ethics and the Limits of Philosophy* (Harvard University Press, 1986), who argues that philosophy cannot do much to guide ethical actions.

4. John Le Carré, *The Russia House* (Knopf, 1989), 207.

5. The German scientist Werner Heisenberg became famous for, among other things, his "principle of uncertainty," which makes this point. Heisenberg, in his book *From*

Plato to Planck, said that "in science we are not dealing with nature itself but with the science of nature—that is, with nature which has been thought through and described by man." Quoted in Paul Mattick, "Marxism and the New Physics," *Philosophy of Science* (Oct. 1962): 360.

CHAPTER TWO *Machiavellian Realism and U.S. Foreign Policy: Means and Ends*

1. Ralph Roeder, *The Man of the Renaissance* (Viking, 1933), 120–130, gives us a dramatic description of Savonarola's arrest and execution.

2. Niccolò Machiavelli, *The Prince and The Discourses,* Introduction by Max Lerner, (Modern Library College Edition, 1950), Chapt. 6, p. 22. All citations are from this edition unless otherwise specified.

3. Ibid., Chapt. 15, p. 56.

4. In the period after World War II, the term *realism* became known among theorists of international relations as meaning a recognition that "national interest" and "power" predominated in the foreign policy of nations. Political scientist Hans Morgenthau made this the center of his theory, explained in his book *Politics among Nations* (Knopf, 1948), which became the most influential textbook of the postwar period. The "realist paradigm" is discussed at length and criticized in John Vasquez, *The Power of Power Politics* (Rutgers, 1983).

5. Editorial, *Wall Street Journal,* Apr. 6, 1989. That same week, the Supreme Judicial Court in Massachusetts rejected such "realism" when it overturned the arrest of a protester against nuclear weapons for "trespassing," saying that the traditional police practice of using "disorderly conduct" or "loitering" charges as a catch-all for arresting undesirable persons violated rights of free speech and assembly.

6. Scholars, as is their habit, have always argued about Machiavelli and what he "really meant," although the language of *The Prince* is quite simple and direct. Political philosopher Leo Strauss, in his *Thoughts on Machiavelli* (Free Press, 1958), believes that we cannot read Machiavelli directly, that we must look for hidden meanings. This approach is strongly criticized by Robert McShea in "Leo Strauss on Machiavelli," *Western Political Quarterly,* (Dec. 1963), who says, "The theory of concealed teaching and the rules for reading as used by Strauss in the explication of Machiavelli's text seem less a means for finding what that thinker purports to say than for reading preconceived notions into his writing."

7. The British political philosopher Isaiah Berlin has written about Machiavelli in the *New York Review of Books,* March 17, 1988, that he believed "one needed a ruling class of brave, resourceful, intelligent, gifted men who knew how to seize opportunities and use them, and citizens who were adequately protected, patriotic, proud of their state, epitomes of manly, pagan virtues. That is how Rome rose to power and conquered the world. . . . Decadent states were conquered by vigorous invaders who retained those virtues." Berlin takes a kindly view of Machiavelli, saying that Machiavelli recognizes the Christian virtues, which are different, but "leaves you to choose." This seems naive to me. A writer who argues so powerfully for those "pagan virtues" hardly leaves it to us to choose. Of course, we can still choose, but he has loaded the argument so as to push our choice his way. J. H. Hexter, a Yale historian, noted Machiavelli's chief concern as *lo stato* (roughly, "the state"), as "an instrument of exploitation, the mechanism the prince uses to get what he wants." J. H. Hexter, "The Loom of Language and

the Fabric of Imperatives: The Case of *Il Principe* and *Utopia,*" *The American Historical Review* (July 1964).

8. *Parade Magazine,* Apr. 13, 1986.

9. Robert W. Tucker, "The Purposes of American Power," *Foreign Affairs* (Winter 1980).

10. *New York Times,* Mar. 10, 1983.

11. For a history of U.S. involvement in Central America, see Walter LaFeber, *Inevitable Revolutions* (Norton, 1983). For a more specific look at the policy of the Reagan administration in El Salvador, read Raymond Bonner, *Weakness and Deceit* (Times Books, 1984).

12. Robert W. Tucker, "The Purposes of American Power," *Foreign Affairs* (Winter 1980).

13. Machiavelli, *The Prince,* Chapt. 7, p. 27.

14. The Select Committee to Study Governmental Operations with Respect to Intelligence Activities, *Interim Report,* Senate, Nov. 20, 1975, 11.

15. In the trial of former National Security Council staff member Colonel Oliver North, who had been in charge of the covert operation to deliver military supplies to the counterrevolutionary group in Nicaragua (the "contras") after Congress had made such aid illegal, it was disclosed that Vice President George Bush had made a visit to the president of Honduras in the spring of 1985 to discuss giving additional military supplies to Honduras in return for Honduras's continuing aid to the contras, who were based on their territory. *Boston Globe,* Apr. 7, 1989.

16. Machiavelli, *The Prince,* Chapt. 18, p. 64.

17. Ibid., Chapt. 6, p. 22; Chapt. 25, p. 94.

18. Ibid., Chapt. 18, p. 64.

19. We are told this by Christian Gauss, a Princeton philosopher, in his introduction to the Mentor edition of *The Prince.* Max Lerner, in his introduction to the Modern Library College Edition, names Lenin and Stalin as having read Machiavelli. But we get no details or sources.

20. Antonio Gramsci, *The Modern Prince* (International Publishers, 1959), 142.

21. Machiavelli, *The Prince* xxxvii.

22. This was Senator Hayakawa, of California, at the time of the renegotiation of the Panama Canal treaty in the late 1970s.

23. James M. Burns, *Roosevelt: The Lion and the Fox* (Harcourt Brace, 1956).

24. The incidents are described and the historian Thomas A. Bailey is quoted in Bruce M. Russett, *No Clear and Present Danger* (Harper & Row, 1972), 79–82.

25. For a detailed account of the Gulf of Tonkin incidents, see Anthony Austin, *The President's War* (New York Times Books, 1971).

26. More did not follow his own advice. He accepted the job of adviser to Henry VIII, and according to a recent biographer, he served Henry obediently, only dissenting on the issue of Henry's divorce and remarriage. Richard Marius, *Thomas More: A Biography* (Knopf, 1984).

27. Thomas More, *Utopia* (Appleton-Century-Crofts, 1948), 18.

28. Arthur Schlesinger, Jr., *A Thousand Days* (Houghton Mifflin, 1965), 254–255.

29. The Schlesinger memo is analyzed critically in Ronald Radosh, "Historian in the Service of Power," *The Nation,* Aug. 6, 1977.

30. *The Nation,* Aug. 6, 1977. A Harvard colleague of Schlesinger, Richard Neustadt, wrote *Presidential Power: The Politics of Leadership* (John Wiley, 1960). This book is a kind of modern version of *The Prince.* Neustadt wrote in his preface, "My theme is personal power and its politics; what it is, how to get it, how to keep it, how to use it." He understands that whatever the personal scruples of the expert adviser, he will not stray far from the policies of *The Prince.* Speaking of the expert, "His expertise assures a contribution to the system and it naturally commits him to proceed within the system."

31. Victor Bernstein and Jesse Gordon, "The Press and the Bay of Pigs," *Columbia University Forum* (Fall 1967).

32. According to investigative journalist Seymour Hersh, who did years of research on Kissinger's role in Washington, "By the early summer of 1969, Nixon and Kissinger had reached agreement in secret on a Gotterdammerung solution to the Vietnam War. North Vietnam would be threatened with a 'savage, decisive blow,' a phrase Kissinger now used openly and repeatedly in meetings with the NSC staff that summer and fall—if it did not begin serious negotiations in Paris. . . . All in all, twenty-nine major targets in North Vietnam were targeted for destruction in a series of attacks planned to last four days and to be renewed, if necessary, until Hanoi capitulated." Seymour Hersh, *The Price of Power* (Summit Books, 1983), 118–120.

33. Marvin Kalb and Bernard Kalb, *Kissinger* (Little, Brown, 1974), 161–162.

34. *New York Times,* Dec. 30, 1972.

35. Machiavelli, *The Prince,* Chapt. 18, p. 65.

36. Words and music by Tom Lehrer, reproduced in his book *Too Many Songs* (Pantheon, 1981).

37. Richard Rhodes, *The Making of the Atomic Bomb* (Simon & Schuster, 1986), 770, 776.

38. Robert Jungk, *Brighter Than a Thousand Suns* (Harcourt, Brace, 1956), 209.

39. Rhodes, *The Making of the Atomic Bomb,* 696.

40. U.S. Strategic Bombing Survey, *Japan's Struggle to End the War* (U.S. Government Printing Office, 1945).

41. Robert Butow, *Japan's Decision to Surrender* (Stanford University Press, 1954).

42. Barton J. Bernstein, "Hiroshima and Nagasaki Reconsidered: The Atomic Bombings of Japan and the Origins of the Cold War, 1941–1945" (University Program Modular Studies, General Learning Press, 1945).

43. Ibid.

44. Rhodes, *The Making of the Atomic Bomb,* 698.

45. Herbert Feis, *Japan Subdued* (Princeton University Press, 1962).

46. Gar Alperovitz, *Atomic Diplomacy: Hiroshima and Potsdam* (Simon & Schuster, 1965).

47. Quoted in Gar Alperovitz, "More on Atomic Diplomacy," *Bulletin of Atomic Scientists* (Dec. 1985). James Forrestal, secretary of the navy, wrote in his diary: "Byrnes said he was most anxious to get the Japanese affair over with before the Russians got in."

48. Atomic radiation is not the only horrible way of killing people. Chemical weapons, of which we had a brief and unforgettable glimpse in the use of mustard gas in World War I, were being produced by at least twenty nations in 1989, including the United States. In 1983, two years after he became president, Reagan ordered the Pentagon to end a fourteen-year cessation of the production of these weapons (Nixon had called a halt when he became president), and soon the country had thousands of tons of chemical

warfare material stored in various places around the country. Three times there was a tie vote in the Senate on the resumption of production of chemical weapons, and each time Vice President Bush, as president of the Senate, cast the deciding vote in favor of resumption. *The Nation*, Jan. 23, 1989.

49. Rhodes, *The Making of the Atomic Bomb*, 720–721.

50. Ibid., 452. Rabi did, however, come to Los Alamos occasionally as a consultant.

51. Daniel J. Kevles, *The Physicists* (Random House, 1979), 334.

52. Martin Sherwin, *A World Destroyed* (Vintage, 1977), 210–213.

53. Rhodes, *The Making of the Atomic Bomb*, 749.

54. Fletcher Knebel and Charles W. Bailey II, *No High Ground* (Bantam, 1960), 84.

55. Interview of November 11, 1963, quoted by Rhodes, *The Making of the Atomic Bomb*, 688.

56. Schlesinger, *A Thousand Days*, 252.

57. James Thomson, "How Could Vietnam Happen: An Autopsy," *Atlantic Monthly* (April 1968).

58. Marvin Kalb and Bernard Kalb, *Kissinger*, 161–162.

CHAPTER THREE *Violence and Human Nature*

1. This statement is in my files as from John Stuart Mill. I have not been able to find the exact source even after consulting a leading Mill scholar and searching the Intelex collection of Mill's major works. I will be pleased to hear from anyone who can locate the source.

2. For instance, a recent survey of students at Wesleyan University conducted by psychologist David Adams found that nearly half of them believed war was built into human nature. *U.S. News & World Report*, Apr. 11, 1988.

3. Machiavelli, *The Prince*, 56.

4. Thomas Hobbes, *The Leviathan*. How many political scientists today accept this view of human nature is hard to say. Certainly some do. For instance, Professor Stephen J. Andriole, writing in the *American Political Science Review* (1975), criticizes author Robert Nye because "Nye rejects the hypothesis that aggressive behavior is in any way innate, or instinctual, in the human species."

5. The exchange of letters between Einstein and Freud was printed in *Why War?* a pamphlet published by the International Institute of Intellectual Co-operation of the League of Nations in 1933. Freud's response appears also in *Sigmund Freud, Collected Papers*, Vol. 5 (Basic Books, 1959).

6. E. O. Wilson, *Sociobiology, The New Synthesis* (Harvard University Press, 1975).

7. E. O. Wilson, *On Human Nature* (Harvard University Press, 1978).

8. The great Goya, who painted the horrors of the Napoleonic wars, came to a pessimistic view. As one art critic said: "The liberal message was that human nature is naturally good, but is deformed by corrupt laws and bad customs. . . . Goya's message, late in his life, is different. The chains are attached to something deep inside human nature." Robert Hughes, "The Liberal Goya," *New York Review of Books*, June 29, 1989.

9. *New York Review of Books*, Mar. 8, 1973.

10. Freud's essay *Civilization and Its Discontents* (Hogarth Press, 1930) refers to aggression as an "indestructible feature of human nature," and in it he talks of the "constitutional inclination of people to be aggressive to one another." But Freud is not consistent

or clear. In his lecture "Anxiety and Instinctual Life" (*New Introductory Lectures on Psychoanalysis,* #32 [Norton, 1933]) he talks in one paragraph about "what history tells us . . . that belief in the 'goodness' of human nature is one of those evil illusions" and in the same paragraph says "We have argued in favor of a special aggressive and destructive instinct in men not on account of the teachings of history or of our experience in life but on the basis of general considerations to which we were led by examining the phenomena of sadism and masochism."

11. The story of the experiment is told in detail by Stanley Milgram, *Obedience to Authority* (Harper & Row, 1974).

12. Quoted in Milgram, *Obedience to Authority,* 2.

13. Colin Turnbull, *The Mountain People* (Simon & Schuster, 1972).

14. Konrad Lorenz, *On Aggression* (Harcourt Brace & World, 1966).

15. Erik Erikson, "Psychoanalysis and Ongoing History," *American Journal of Psychiatry* (Sept. 1965).

16. J. Glenn Gray, *The Warriors* (Harper & Row, 1970).

17. For a summary of antiwar feeling during World War I, see Howard Zinn, "War Is the Health of the State," *A People's History of the United States* (Harper & Row, 1981). For the specifics of draft resistance and other forms of opposition to the war, see H. C. Peterson and Gilbert C. Fite, *Opponents of War, 1917–1918* (University of Washington Press, 1957).

18. John Ketwig's letter became a book, *. . . and a hard rain fell* (Macmillan, 1985). Part of it is reprinted in *Unwinding the Vietnam War* (Real Comet Press, 1987).

19. *Boston Globe,* Mar. 20, 1983. The story of the massacre is vividly described in detail by Seymour Hersh, *Mylai 4* (Random House, 1970). Hersh was the first American journalist to write about it; he wrote for a small antiwar news agency in Southeast Asia called Dispatch News Service.

20. *Boston Globe,* Mar. 20, 1983.

21. Emma Goldman, "Anarchism: What It Really Stands For," in *Red Emma Speaks,* ed. Alix Kates Shulman (Schocken Books, 1983), 73.

22. *New York Review of Books,* Feb. 7, 1974.

23. Terence des Pres, *The Survivor* (Oxford University Press, 1976).

24. Howard Zinn, *S.N.C.C.: The New Abolitionists* (Greenwood Press, 1985), 85–86.

CHAPTER FOUR *The Use and Abuse of History*

1. Noam Chomsky, who began exploring human potential in his theories of language acquisition, concluded that humans have a built-in grammar that gives them a universal ability to learn language, even though the specific form of the language depends on history and culture. He seems also to believe that there is an innate human desire for freedom, which can be suppressed or distorted, but which continually strives to express itself. See his book *Reflections on Language* (Pantheon, 1975). The philosopher Bernard Williams, in his book *Ethics and the Limits of Philosophy* (Harvard University Press, 1986), is dubious that we can derive our values from the pure thought of philosophy, that there is some rational system of thought to tell us what is right and what is wrong. But this does not leave us hanging helplessly in an amoral atmosphere. There is something *inside* us that is a better guide than cool philosophical analysis. And we can help this along, he thinks, not through abstract ethical theories, but by taking a closer, deeper look at the world about us, its history and its present characteristics.

2. Oddly enough, E. O. Wilson, who sometimes speaks as if he finds aggressiveness in human nature, also finds cooperation in it. He talks about (in his book *On Human Nature*) "the mammalian imperative." He finds it a basic characteristic of mammals (which includes human beings and other species) to seek, among other things, "grudging cooperation . . . to enjoy the benefits of group membership." This "fact" about people is, therefore, a basis for the value of "universal rights," he says. The philosopher Peter Singer disputes this, saying, "Human beings have been mammals at all times and in all places." And yet they have not always supported universal rights. We do not need a biological justification for universal rights, Singer says. It is a good in itself. Peter Singer, "Ethics and Sociobiology," *Philosophy and Public Affairs* (Winter 1982).

3. Samuel Yellen, *American Labor Struggles* (Harcourt Brace, 1936).

4. The entire speech is reprinted in the report of the House Mines and Mining Committee, *Conditions in the Coal Mines of Colorado* (1914), 2631–2634.

5. Commission on Industrial Relations, Senate, *Report and Testimony* (1915), 8607.

6. A Yale law professor named William Brewster compiled a 600-page document of eyewitness reports of National Guard brutality, titled *Militarism in Colorado* (1914).

7. Glenn M. Linden, Dean C. Brink, and Richard H. Huntington, *Legacy of Freedom*, Vol. II (Laidlaw Brothers, 1986) 15.

8. *New York Times*, Apr. 21, 1914.

9. Howard Zinn, *The Politics of History* (Illinois University Press, 1990).

10. I am drawing this account of Columbus from the first chapter of my book, *A People's History of the United States* (Harper & Row, 1980).

11. The *Jewish Advocate* in Boston, Oct. 5, 1989, carried an article by Judea B. Miller, who asked: "Why would American Jews gather in a cathedral in Spain to honor Columbus? The reason is that Columbus was of Jewish origin." How ironic that Jews, remembering their own Holocaust, would honor the perpetrator of an earlier one. I would guess that the writer, like most Americans, did not know the story of Columbus's treatment of the Indians.

12. B. de las Casas, *History of the Indies* (Harper & Row, 1971).

13. Samuel Eliot Morison, *Admiral of the Ocean Sea* (Little Brown, 1942).

14. Samuel Eliot Morison, *Christopher Columbus, Mariner* (Little Brown, 1955).

15. Zinn, *A People's History of the United States*, 8.

16. *Boston Globe*, Oct. 7, 1986.

17. Robert Lynd and Helen Lynd, *Middletown, A Study in American Culture* (Harcourt Brace, 1929), 85.

18. Merle Curti, *The Growth of American Thought* (Harper, 1933), 692–693.

19. *LaGuardia Papers*, New York Public Library. Quoted in Howard Zinn, *LaGuardia in Congress* (Cornell University Press, 1959).

20. Karl Marx, *The Eighteenth Brumaire of Louis Bonaparte*

21. *New York Times*, Aug. 31, 1974.

22. A useful corrective to the orthodox treatment of the American revolution is a set of essays edited by Alfred Young, *The American Revolution* (Northern Illinois University Press, 1976). There is also an excellent overview of the revolution in Edward Countryman, *The American Revolution* (Hill & Wang, 1985).

23. For an excellent treatment of this, see James McPherson, *The Negro's Civil War* (Pantheon, 1965).

24. See Jeremy Brecher's book *Strike!* (South End Press, 1979), for the labor actions of the 1930s. For a specific example of the effect of strikes on the passage of the National Labor Relations Act, see Peter Irons, *New Deal Lawyers* (Princeton University Press, 1982).

25. For a study of the anti-imperialist movement during the war in the Philippines, see Daniel B. Schirmer, *Republic or Empire* (Schenkman, 1972).

26. Patrick S. Washburn, reviewing Richard Drinnon, *Keeper of Concentration Camps: Dillon S. Myer and American Racism,* (University of California Press, 1987) in *New York Times Book Review,* Feb. 22, 1988.

27. The argument against "presentism" is looked on skeptically by the historian Immanuel Wallerstein, who writes, in his book *The Modern World-System* (Academic Press, 1974), that "recounting the past is a social act of the present done by men of the present and affecting the social system of the present."

28. Peter Novick, *That Noble Dream* (Cambridge University Press, 1988).

29. Ibid., 121.

30. See Richard Polenberg's review of George T. Blakey, *Historians on the Homefront: American Propagandists for the Great War* (University Press of Kentucky, 1970), which appeared in *American Political Science Review* (Sept. 1973).

31. Quoted by Novick, *That Noble Dream.*

32. Quoted by Novick, Ibid.

33. Quoted by Novick, Ibid.

34. Novick, *That Noble Dream.*

35. Ibid.

36. Two best-selling books played a part in this attack. Allen Bloom's *The Closing of the American Mind* (Simon & Schuster, 1987) deplored the sixties' students leaving the seminar room to take part in demonstrations for racial equality or against the war. E. D. Hirsch's *Cultural Literacy* (Vintage, 1988) drew up lists of facts that he believed all educated people should know.

37. *New York Times,* May 3, 1976.

38. Ibid.

CHAPTER FIVE *Just and Unjust War*

1. *NBC White Paper,* broadcast Jan. 30, 1972.

2. Christian Gauss, of Princeton University, in an introduction to the Mentor edition of *The Prince,* says, "Machiavelli was not interested in peace and did not believe it was necessary."

3. The idea persists that military service and war itself "build character." Poet Michael Blumenthal (is he trying to destroy my notion about the pacifism of poets?) wrote in the 1980s about his success in evading the draft call in the Vietnam War. He felt now that he had missed something positive and strengthening in the experience of the army, the testing ground of war. "And maybe, short of violating one's most deeply held moral principles, serving in the armed forces or, for that matter, being in a war, isn't the greatest tragedy that can occur in life" (Op-ed essay, *New York Times*). But if "one's most deeply held moral principles" include refusing to kill another human being, what is left of Blumenthal's point?

4. Paul Seabury and Angelo Codevilla, *War: Ends and Means* (Basic Books, 1989), quoted in a review article by Gordon A. Craig, *New York Review of Books*, Aug. 17, 1989.

5. Michael Walzer's book, *Just and Unjust Wars* (Basic Books, 1977), engages thoughtfully in this discussion, and there have been many articles on the subject in the pages of the periodical *Philosophy and Public Affairs.*

6. Erasmus is quoted in Michael Howard, *War and the Liberal Conscience* (Rutgers University Press, 1978). Erasmus's major work is *In Praise of Folly.*

7. Ronald Clark, *Einstein: The Life and Times* (World, 1971), 370. Clark, a "realist," deplored Einstein's action, saying it "offered a useful weapon to those only waiting to claim that outside his own field Einstein was something between crank and buffoon." And what would Clark say about the delegates who sat for over a year scrupulously going over ways to "humanize" war, their work ignored almost immediately in a war that was a long succession of horribly inhuman acts?

8. *New York Times,* Feb. 10, 1990.

9. H. D. Kitto, *The Greeks* (Peter Smith, 1988).

10. Thucydides in his *History of the Peloponnesian War* seemed to understand the injustices committed by Athens against her neighbors and gave full attention to the arguments against Athens. See Christopher Bruell, "Thucydides' View of Athenian Imperialism," *American Political Science Review* (1974).

11. Thucydides, *History of the Peloponnesian War.*

12. Quoted in Michael Howard, *War and the Liberal Conscience.*

13. Tom Paine, *The Rights of Man* (1791–1792).

14. Helen Keller's speech was reprinted in a Socialist newspaper, the *New York Call,* Jan. 6, 1916.

15. Barbara Tuchman, *The Guns of August* (Macmillan, 1962).

16. Dalton Trumbo, *Johnny Got His Gun* (Bantam Books, 1983), 230.

17. Arnold Rampersad, *The Life of Langston Hughes* (Oxford University Press, 1986), 322.

18. When the war ended, Trumbo agreed to a new edition of the book. In 1970, he wrote a bitter "addendum": "Numbers have dehumanized us. Over breakfast coffee we read of 40,000 American dead in Vietnam. Instead of vomiting, we reach for the toast. Our morning rush through crowded streets is not to cry murder but to hit that trough before somebody else gobbles our share." Dalton Trumbo, *Johnny Got His Gun* (Lyle Stuart, 1970).

19. James E. Miller was reviewing a book by an Italian historian, Gian Giacomo Migone, *Gli Stati Uniti e il fascismo* (Feltrinelli, 1980), in *American Historical Review* (1981).

20. Akira Iriye, *The Origins of the Second World War* (Longman, 1987)

21. Bruce Russett, in his book on the origins of U.S. entrance into World War II, *No Clear and Present Danger* (Harper & Row, 1972), says: "Throughout the 1930s, the United States government had done little to resist the Japanese advance on the Asian continent." But: "The Southwest Pacific area was of undeniable economic importance to the United States—at the time most of America's tin and rubber came from there, as did substantial quantities of other raw materials."

The cutoff of oil to Japan was crucial in their decision to make war on the United States and try to conquer the oil-rich Dutch East Indies. The cutoff was actually not planned to be total when the United States froze Japanese assets, but Dean Acheson, who was the State Department's representative to the Foreign Funds Control Committee and who was not fully aware of how desperate Japan was about oil nor of the violent reaction it might bring, engineered the policy so as to cut off oil completely to Japan. See Jonathan G. Utley, "Upstairs, Downstairs at Foggy Bottom: Oil Exports and Japan, 1940–41," *Prologue* (Spring 1976).

22. See Arnold Offner, *American Appeasement: U.S. Foreign Policy and Germany, 1933–1938* (Norton, 1976).

23. Raul Hilberg, *The Destruction of the European Jews* (Harper & Row, 1961). Indeed, at certain critical moments, the Allies suppressed information about the operations of extermination camps, according to Walter Laqueur, *The Terrible Secret* (Weidenfeld and Nicolson, 1980).

24. Henry L. Feingold, *The Politics of Rescue: The Roosevelt Administration and the Holocaust, 1938–1945* (Rutgers University Press, 1970). Robert Dallek in a review article on this issue wrote, "The congressional restrictionists, the British, the Vatican, the Latin Americans, the neutrals, the Arabs, the exiled governments, the conquered Europeans, the Committee of the International Red Cross, and American Jewry itself directly and indirectly threw up obstacles to effective rescue that would have inhibited even the most determined administration effort." Robert Dallek, "Franklin Roosevelt as World Leader," *American Historical Review* (1971). But, of course, there was no determined administration effort.

25. Hilberg, *The Destruction of the European Jews.*

26. The quotation is from a summary of the commission report by its chairman, former Supreme Court Justice Arthur Goldberg, and a member of the commission, Rabbi Arthur Hertzberg. For a comprehensive picture of the callousness, incompetence, and self-seeking behavior, by both the U.S. government and Jewish organizations, that stood in the way of saving Jews during the war, see David S. Wyman, *The Abandonment of the Jews* (Pantheon, 1984).

27. Hilberg, *The Destruction of the European Jews,* 723.

28. Arno J. Mayer, *Why Did the Heavens Not Darken?* (Pantheon, 1989). The first of the extermination camps began to operate in December 1941, six months into the invasion of the Soviet Union.

29. Hilberg, *The Destruction of the European Jews,* 258.

30. These and other references to the French control over Indochina appear in *The Pentagon Papers,* Vol. I (Beacon Press, 1971).

31. A. J. P. Taylor, *The Second World War: An Illustrated History* (Putnam, 1975).

32. For a detailed picture of the basic conservatism of the Churchill government, its single-minded pursuit of military victory and its resistance to social change, see Angus Calder, *The People's War: Britain 1939–1945* (Pantheon, 1969).

33. Cordell Hull, *Memoirs* (Macmillan, 1948), 1177.

34. *Life,* Feb. 17, 1941.

35. U.S. economic expansion during the war is described in detail by Lloyd Gardner, *Economic Aspects of New Deal Diplomacy* (University of Wisconsin Press, 1964), and Gabriel Kolko, *The Politics of War* (Random House, 1968).

36. Anthony Sampson, *The Seven Sisters: The Great Oil Companies and the World They Shaped* (Viking, 1975).

37. Quoted in Lloyd Gardner, *Economic Aspects of New Deal Diplomacy* (University of Wisconsin Press, 1964), 264.

38. There is a summary of wartime discrimination in the *Report of the National Advisory Commission on Civil Disorders* (U.S. Government Printing Office, 1968).

39. Lawrence S. Wittner, *Rebels against War* (Columbia University Press, 1969), 46.

40. Ibid., 46.

41. Ibid., 47.

42. Robert L. Allen, *The Port Chicago Mutiny* (Warner Books, 1989).

43. Robert L. Allen, "The Port Chicago Disaster and Its Aftermath," *Black Scholar* (Spring 1982).

44. Representative John Rankin of Mississippi in the *Congressional Record,* Dec. 15, 1941, quoted in Michi Weglyn, *Years of Infamy* (William Morrow, 1976), 54.

45. *Hirabayashi v. United States,* 320 U.S. 81 (1943).

46. Peter Irons, *Justice at War* (Oxford University Press, 1983), 286–292. The footnote appeared in a draft of the Justice Department brief written by John L. Burling, assistant director of the Alien Enemy Control Unit of the department. It said that the reports of the army on espionage by Japanese on the West Coast were "in conflict with information in the possession of the Department of Justice." In view of that conflict, it said, "we do not ask the Court to take judicial notice of the recital of those facts contained in the Report." McCloy objected to the footnote, and then Burling told a colleague: "Presumably at Mr. McCloy's request, the Solicitor General had the printing stopped at about noon." What happened then was that Herbert Wechsler, a former Columbia law school professor and a wartime assistant attorney-general, wrote a substitute footnote that cast no such doubt on the army report and made no such suggestion to the Supreme Court.

47. Weglyn, *Years of Infamy.*

48. See John Dower, *War without Mercy* (Pantheon, 1986) for the way the press treated the incarceration of the Japanese.

49. Studs Terkel, "*The Good War*" (Pantheon, 1984).

50. The story was told in a television documentary by Lavinia Warner, *Jailed by the British,* reviewed in the *Daily Mail,* Feb. 17, 1983.

51. An account of their trial is the book by James P. Cannon, *Socialism on Trial* (Pathfinder Press, 1973).

52. Senate Armed Services Committee hearing on the McCone appointment to the CIA, Jan. 18, 1962, quoted in *I.F. Stone's Weekly,* Jan. 29, 1962.

53. Bruce Catton, *The Warlords of Washington* (Harcourt Brace, 1948).

54. Quoted in Bruce Russett, *No Clear and Present Danger* (Harper & Row, 1972), 73.

55. *U.S. Strategic Bombing Survey (Overall Report, European War)* (U.S. Government Printing Office, 1945).

56. Michael S. Sherry, *The Rise of American Air Power* (Yale University Press, 1987).

57. David Irving, *The Destruction of Dresden* (Ballantine, 1965).

58. These recollections are from a story in the *New York Times,* Jan. 30, 1985.

59. See David Irving, *The Destruction of Dresden*, Chapt. 2. One historian of the Holocaust, Lucy Davidowitz, rightfully enraged at what happened to the Jews, goes on to defend the bombing of Dresden. She says it was a justified retaliation for the Germans sending V-2 rockets, terrible engines of indiscriminate destruction, over London. Telford Taylor, reviewing her book, *The War Against the Jews* (Holt, Rinehart & Winston, 1975), points out that the V-2 bombardment was already five months old, so that "the explanation is implausible on its face." Taylor notes that "the announced justification for the raid was to stop German reinforcement of the Russian front, and that it was carried out despite plain evidence from *Ultra* sources (counterintelligence) that no such reinforcement was in process." *New York Times,* Mar. 10, 1985.

60. Barton J. Bernstein, "Churchill's Secret Biological Weapons," *Bulletin of Atomic Scientists* (Jan./Feb., 1987).

61. *New York Times,* Nov. 23, 1974.

62. The item on the BBC reaction to Burton appeared in the *New York Times,* Nov. 30, 1974.

63. *New York Times,* Mar. 10, 1985.

64. Committee for the Compilation of Materials on Damage Caused by the Atomic Bombs in Hiroshima and Nagasaki, *Hiroshima and Nagasaki: The Physical, Medical, and Social Effects of the Atomic Bombings* (Basic Books, 1981).

65. In his private diary, Truman wrote: "The target will be a purely military one." It was an astounding self-deception. The quote is from Richard Rhodes, *The Making of the Atomic Bomb,* 691.

66. Martin Sherwin, *A World Destroyed* (Knopf, 1975).

67. This recollection appeared in August 1966 in *The Journal of Social and Political Ideas in Japan,* and is quoted by Noam Chomsky in his essay, "The Revolutionary Pacifism of A. J. Muste: On the Backgrounds of the Pacific War," *Liberation* (Sept.–Oct., 1967).

68. Wesley Craven and James Cate, *The Army Air Forces in World War II,* Vol. 3 (University of Chicago Press, 1948–58).

69. *New York Times,* Apr. 16, 1945.

70. This is from a little book produced by a printer in Royan, a former member of the Resistance named Botton. Botton, *Royan—Ville Martyre.*

71. Quoted by Botton, *Royan—Ville Martyre.* I returned to Royan in 1966 to do research on the bombing and the resulting essay appears in Howard Zinn, *The Politics of History* (Beacon, 1970).

72. Irving, *The Destruction of Dresden.*

73. Lawrence Freedman, *Atlas of Global Strategy* (Facts on File, 1985), 51.

74. Dan Jacobs, *The Brutality of Nations* (Knopf, 1987).

75. The original edition: Zora Neale Hurston, *Dust Tracks on a Road* (Lippincott, 1942). A new edition added the deleted portion as an appendix as well as an introduction by Robert Hemenway describing the censorship incident. Hurston, *Dust Tracks on a Road* (Harper & Row, 1984).

76. Ibid., xxxiii.

77. Ibid., 322–348.

78. Ibid., xxxi.

79. Wittner, *Rebels against War,* 34–61.

80. Simone Weil, "Reflections on War," *Politics* (Feb. 1945). Quoted in Wittner, 95.

81. *Politics,* Aug. 1945. Quoted in Dwight MacDonald, *Politics Past* (Viking, 1957), 169.

82. In 1989 Oxford University Press published a book by Paul Fussell, *Wartime: Understanding and Behavior in the Second World War.* Fussell sees war as madness, much like the lieutenant surveying the corpses at the end of the film *The Bridge on the River Kwai.* A British reviewer, Noel Annan (*New York Review of Books,* Sept. 28, 1989) is unhappy with Fussell's total rejection of war. "The great question you expect Fussell to ask he never does. If war is so bestial, should America have gone to war after Pearl Harbor and should the French and the British have decided to stop Hitler in 1939?" Perhaps Fussell doesn't ask that question. I have asked it and try to answer it in my own way. But Annan does not want to consider any but the stock answer: "Of course." He cannot give any other answer because he is caught up in the orthodox glorification of combat

for "just cause," saying, near the end of his review: "It is not sweet to die for one's country. It is bitter. But it can be noble." If Fussell doesn't answer Annan's question, neither does Annan answer a crucial question: What makes you so sure that to die in war is to die "for one's country," rather than for national aggrandizement, political power, economic greed, and fanaticism?

83. See Hilberg, *The Destruction of the European Jews,* where he discusses the ways in which the Nazi satellite countries in Eastern Europe reacted in different ways to the "Jewish question." Bulgaria, for instance, resisted by procrastination and was always affected by opportunism. Hilberg writes: "For twelve months the Bulgarian Jews remained subject to all the discriminations and persecutions of the disrupted destruction process. Then, on August 30, 1944 . . . the morning newspapers in Sofia displayed in prominent headlines the Cabinet's decision to revoke all of the anti-Jewish laws" (p. 484).

84. Gene Sharp, *Making Europe Unconquerable* (Ballinger, 1985), 3–4.

85. Gene Sharp, *The Politics of Nonviolent Action* (Porter Sargent, 1974).

86. Reprinted in *Seeds of Liberation,* Paul Goodman, ed. (George Braziller, 1965).

87. Terkel, "*The Good War.*"

88. Ibid.

CHAPTER SIX *Law and Justice*

1. *U.S. v. O'Brien* 393 U.S. 900.

2. Some of the material in this chapter is drawn from Howard Zinn, *Disobedience and Democracy* (Random House, 1968).

3. Tommy Trantino, *Lock the Lock* (Knopf, 1974), 6–8.

4. Claudia Koonz, *Mothers in the Fatherland* (St. Martins, 1987).

5. Eichel, Jost, Luskin, and Neustadt, *The Harvard Strike* (Houghton Mifflin, 1970), quoted in Nancy Zaroulis and Gerald Sullivan, *Who Spoke Up?* (Doubleday, 1984), 241.

6. Some of the material in this chapter is drawn from my essay "The Conspiracy of Law," in a book edited by Robert Paul Wolff, *The Rule of Law* (Simon & Schuster, 1971).

7. Michael Walzer, in his book *Obligations* (Harvard University Press, 1970) says "there is very little evidence which suggests that carefully limited, morally serious civil disobedience undermines the legal system or endangers physical security."

8. Palmer and Colton, *A History of the Modern World* (Knopf, 1984).

9. It was distinguished historian Charles Beard, in *An Economic Interpretation of the Constitution* (Macmillan, 1935), who broke through the romanticization of the Founding Fathers with his exploration of their economic interests and their political ideas. Other scholars have claimed to refute him, but I believe his fundamental thesis remains untouched: the relationship between wealth and political power.

10. Jerold S. Auerbach, *Unequal Justice* (Oxford University Press, 1976).

11. Political theorist Michael Walzer writes about "the obligation to disobey." He talks about people having the "obligation to honor the engagements they have explicitly made, to defend the groups and uphold the ideals to which they have committed themselves, even against the state, so long as their disobedience of laws or legally authorized comands does not threaten the very existence of the larger society or endanger the lives of its citizens. Sometimes it is obedience to the state, when one has a duty

to disobey, that must be justified." Michael Walzer, *Obligations* (Harvard University Press, 1970).

12. *Euthyphro, Apology, Crito* (Bobbs-Merrill, 1956).

13. Ibid.

14. Emma Goldman, *Anarchism and Other Essays* (Dover, 1969), 128–129.

15. Op-ed page, *New York Times*, July 2, 1989.

16. Carl Cohen, for instance, in his book *Civil Disobedience* (Columbia University Press, 1971), makes a distinction between "direct disobedience" (disobeying a law that is in itself wrong, like a law drafting you into military service), in which case evading punishment is justified, and "indirect disobedience," where someone is violating something like a trespassing law that is not in itself bad, in which case "it is right for him to be punished." That distinction makes no sense to me, because while the trespass law may be theoretically okay, if it is applied unjustly against a political protester, the punishment for disobeying it is also unjust.

The philosopher Sidney Hook, once a radical, later a supporter of American foreign policy in Vietnam and other military interventions, dealt with this question in his book *The Paradoxes of Freedom* (University of California Press, 1964). He says a *democrat* (emphasis in the original) can defend an unlawful action "*only* [emphasis in the original] if he willingly accepts the punishment entailed by his defiance of the law." Otherwise, Hook says, "he has in principle embarked upon a policy of revolutionary overthrow." This seems silly to me. Sure, a person who evades prison *may* be a revolutionary, but he also may not. Angela Davis was a Communist and presumably a revolutionary. Daniel Berrigan was bitterly opposed to the war, but hardly "embarked upon a policy of revolutionary overthrow." Both evaded prison.

17. The legal philosopher Ronald Dworkin, in his book *Taking Rights Seriously* (Harvard University Press, 1978), argues that people should not be punished when committing civil disobedience when "the law is uncertain, in the sense that a plausible case can be made on both sides." And, he claims, any moral issue can find a plausible basis in the Constitution, even if the Supreme Court has not yet come to that conclusion. He seems to be straining to find a *legal* basis for civil disobedience, as if the morality of the disobedient act is not enough. Granted, the Constitution has enough open-ended rights (the Ninth Amendment, for instance, has endless possibilities for asserting the rights of people) to cover just about anything. But to seek refuge in that gives too much support to the idea that you *must* have a legal cover for your moral act. Dworkin's undue respect for the law shows itself when he says (near the end of his chapter on civil disobedience): "If acts of dissent continue to occur after the Supreme Court has ruled that the laws are valid, or that the political question doctrine applies, then acquittal on the grounds I have described [an 'uncertain' law] is no longer appropriate." In other words, Dworkin is willing to accept punishment—he suggests "minimal or suspended sentences"—for insistent civil disobedience. Dworkin finds himself in the humble position of appealing for leniency to the authorities—to Congress, to the prosecutor, to the judge—because he is constantly addressing, not the citizenry, but the government (the prince).

18. Martin Luther King, Jr., "Letter from Birmingham City Jail," reprinted in a collection edited by Staughton Lynd, *Nonviolence in America* (Bobbs-Merrill, 1966).

19. Pamphlet distributed by the Catonsville Nine Defense Committee in 1968.

20. Produced by Lee Lockwood, West Newton, Mass. 1970.

21. John Rawls, in his book *A Theory of Justice* (Harvard University Press, 1971), has a section on civil disobedience (pp. 333–391) in which he worries about civil disobedience

going so far as to bring about a general disrespect for law, but he does speak strongly about the obligation to resist under certain circumstances. Rawls, however, confines his discussion to the situation in a "nearly just constitutional regime" by which he seems to mean the United States. This exaggerates the justness in our system and, therefore, creates a basis for a more cautious and partial acceptance of civil disobedience. There is an excellent comparison of the views on civil disobedience of Rawls, Dworkin, and myself, in an unpublished paper by Roger Karapin (as part of his National Science Foundation Graduate Fellowship), titled, "The State, Democracy, and the Disobedient Citizen: A Review of Some Recent North American Contributions."

22. Liberals and conservatives often join on this issue. For instance, Irving Kristol, a leading American conservative, wrote during the Vietnam war, "Even were I opposed to the Administration's policy in Vietnam, which I am not, I would not regard this case as one in which civil disobedience is justified. The opportunities for dissent are obviously abundant." What Kristol misses is that citizens may have the opportunity to speak up, but speaking alone may not be effective enough, powerful enough, to get a nation out of a war. *New York Times Magazine,* Nov. 26, 1967. Reprinted in Hugo Bedau, ed., *Civil Disobedience* (Pegasus, 1969).

23. James Birney.

24. Samuel May. See Martin Duberman, ed. *The Anti-Slavery Vanguard* (Princeton University Press, 1965).

25. Abe Fortas, *Concerning Dissent and Civil Disobedience* (Signet, 1968).

26. Quoted by Arthur Schlesinger, Jr., *The Imperial Presidency* (Popular Library, 1974).

27. The nineteenth-century English philosopher T. H. Green, in his 1879 lectures on the *Principles of Political Obligation* (University of Michigan Press, 1967), recognizes the right of civil disobedience, especially in war. He says, "If most wars had been wars for the maintenance or acquisition of political freedom" then disobedience might not be justified. But, in fact, "in most modern wars the issue has not been of this kind at all. The wars have arisen primarily out of the rival ambition of kings and dynasties for territorial aggrandisement."

28. *U.S. v. Curtiss-Wright Export Corp.* 299 U.S. 304.

29. Jerome Barron and C. Thomas Dienes, *Constitutional Law* (West Publishing, 1986), comment on the *Curtiss-Wright* case: "While this declaration of inherent foreign affairs powers, operating independently of the Constitution, represents a questionable interpretation of history, it has never been rejected by the Court, and has, on occasion, been embraced."

30. *Da Costa v. Melvin R. Laird,* 405 U.S. 979 (1972).

31. Joint Meeting of the Committee on Foreign Relations and the Armed Services Committee of the U.S. Senate, Sept. 17, 1962.

32. *Boston Globe,* May 15, 1975.

33. Studies of the effects of civil disobedience on the psyches of those engaged in it do not show that breaking the law for a social purpose will lead to breaking the law for other purposes. A study of 300 young black people who engaged in civil disobedience found "virtually no manifestations of delinquency or anti-social behavior, no school drop-outs, and no known illegitimate pregnancies." Pierce and West, "Six Years of Sit-Ins: Psychodynamics, Cause and Effects," *International Journal of Social Psychiatry* (Winter 1966). The authors conclude, "In any event, the evidence is insufficient to demonstrate that acts of civil disobedience of the more limited kind inevitably lead to an increased disrespect for law or propensity toward crime."

34. There is an eighteen-page summary of the antiwar movement in Howard Zinn, *A People's History of the United States* (Harper & Row, 1971). The quote from the leaflet is in my files, *Wisconsin Historical Society.* Also see Nancy Zaroulis and Gerald Sullivan, *Who Spoke Up?* (Doubleday, 1984).

35. For the background of this speech, see David Garrow, *Bearing the Cross* (William Morrow, 1986), 552–553.

36. The facts about conscription during the Vietnam War come from Lawrence Baskir and William Strauss, *Chance and Circumstance* (Random House, 1978).

37. Philip Supina's letter is in my files.

38. The details on draft evaders, deserters, exiles, less-than-honorable discharges, etc. can be found in Baskir and Strauss, *Chance and Circumstance.* They report 500,000 "desertion incidents," which is probably a multiple of the number of permanent desertions.

39. *Boston Globe,* Apr. 3, 1972.

40. *New York Times,* June 3, 1973.

41. *New York Times,* 1980.

42. Elinor Langer, "The Oakland Seven," *The Atlantic* (Oct. 1969).

43. There are more details on this trial in Howard Zinn, "The Camden Trial," *Liberation* (June 1973).

44. Much of the material on jury nullification comes from an article by Professor Alan W. Scheflin of the Georgetown University Law Center, "Jury Nullification: The Right to Say No," *Southern California Law Review* 45 (no. 1).

45. Quoted in Sidney Hook, *The Paradoxes of Freedom,* from *The Life and Writings of B. R. Curtis* (Little Brown, 1879).

46. Quoted by Scheflin, "Jury Nullification."

47. Roscoe Pound, "Law in Books and Law in Action," *American Law Review* (1910), quoted by Scheflin, "Jury Nullification."

48. Jessica Mitford, *The Trial of Dr. Spock* (Vintage, 1969).

49. *Boston Globe,* Sept. 8, 1968.

50. The Catonsville Nine case was officially *U.S. v. Moylan* (1969). Quoted by Scheflin, "Jury Nullification."

51. *New York Times,* Apr. 16, 1987.

52. Film, *The Holy Outlaw,* produced by Lee Lockwood for *NET Journal,* public television, 1970.

53. *The Pentagon Papers* (Gravel Edition, Beacon Press, 1971), 564.

54. Richard Nixon, *RN: The Memoirs of Richard Nixon* (Grosset & Dunlap, 1978).

55. Dumas Malone, *Jefferson and the Rights of Man* (Little, Brown, 1951).

CHAPTER SEVEN *Economic Justice: The American Class System*

1. *New York Times,* July 13, 1969.

2. *New York Times,* May 16, 1986.

3. *Boston Globe,* Feb. 26, 1985. This Physicians Task Force on Hunger had spent two years traveling to fourteen states and going into hundreds of homes, and made their estimates on the basis of their observations along with official reports from the Census

Bureau and the U.S. Department of Agriculture. They said their estimate was a conservative one.

4. *New York Times,* Aug. 17, 1984.

5. UPI dispatch, Apr. 3, 1984.

6. *New York Times,* Jan. 12, 1990.

7. For more information on the class distinctions of early America, see Howard Zinn, *The Politics of History* (Illinois University Press, 1990), "Inequality."

8. It should be pointed out that the word *property* was defined by John Locke more broadly a century before the adoption of the U.S. Constitution. He said that people leave the state of nature and join in society "for the mutual Preservation of their Lives, Liberties and Estates, which I call by the general Name, Property." *Two Treatises of Government.* But that broader definition of the word was not used in the Constitution or later in the courts of the United States. C. B. Macpherson in his essay "A Political Theory of Property" suggests that the definition of the word be expanded to give not just corporations but individuals "a right of access to the means of labor" and that in some future society of abundance property will also mean "a right to a share in political power to control the uses of the amassed capital and the natural resources of the society, and beyond that, a right to a kind of society, a set of power relations throughout the society, essential to a fully human life." He says that "up to now, property has been a matter of a right to a *material* revenue. With the conquest of scarcity that is now foreseen, property must become rather a right to an *immaterial* revenue, a revenue of enjoyment of the quality of life." *Democratic Theory* (Oxford University Press, 1973), 139.

9. Louis Hartz, *Economic Policy and Democratic Thought,* shows how the state of Pennsylvania played an important role in the development of its economy in the years between the Revolution and the Civil War. Frank Bourgin, *The Great Challenge: The Myth of Laissez-Faire in the Early Republic* (Braziller, 1989), shows in detail how the strong hand of government was involved in the economy.

10. For the relations between government and the wealthy, there are a number of good sources. Charles Beard, *An Economic Interpretation of the Constitution of the United States* (Macmillan, 1935), shows how the Constitution responded to the needs of the upper classes. Thomas Cochran and William Miller, *The Age of Enterprise* (Macmillan, 1942), deal with government aid to business in the nineteenth century, as does Gustavus Myers, *History of the Great American Fortunes* (Modern Library, 1936).

11. Harvey O'Connor, *Mellon's Millions* (John Day, 1933).

12. *Harper's Magazine,* 1931.

13. *New York Times,* Oct. 16, 1984.

14. Morton Horwitz, *The Transformation of American Law* (Harvard University Press, 1977).

15. Philosopher Morris Cohen in the 1920s lectured at Cornell University on property and sovereignty: "The extent of the power over the life of others which the legal order confers on those called owners is not fully appreciated by those who think of the law as merely protecting men in their possession. Property law does more. It determines what men shall acquire. Thus, protecting the property rights of a landlord means giving him the right to collect rent, protecting the property of a railroad or a public-service corporation means giving it the right to make certain charges." Morris Cohen, "Property and Sovereignty," *Cornell Law Quarterly* (1927), reprinted in Virginia Held, ed., *Property, Profits, and Economic Justice* (Wadsworth, 1980).

16. *Lochner v. New York,* 198 U.S. 45 (1905).

17. *West Coast Hotel Co. v. Parrish,* 300 U.S. 379 (1937).

18. *San Antonio Independent School District v. Rodriguez* (411 U.S. 1) 1973. Two years earlier, a court in California *(Serrano v. Priest)* had ruled that the accident of school district wealth was not a legitimate basis for fixing the limits of a child's education.

19. *New York Times*, Sept. 28, 1969.

20. *New York Times*, May 27, 1969.

21. *New York Times*, July 28, 1978.

22. This appears in a speech by Adam Smith made in the 1760s, reprinted in F. L. Meek, ed. *Lectures on Jurisprudence: Adam Smith* (Oxford University Press, 1978)

23. Canadian television film *Struggle for Democracy*, Public Broadcasting Corporation.

24. Quoted from Thomas Malthus, *On Population*, in an article by John Hess, "Malthus, Then and Now," *The Nation*, Apr. 18, 1987. Hess notes that Malthus is still admired by conservatives. He takes note of Gertrude Himmelfarb, who, in her introduction to a new edition of Malthus's book, talks about the benefits of the Poor Law, adopted in England in 1834, which ended home relief but set up workhouses for the poor that were "so appalling as to discourage even the most determined malingerers, and the sexes being separated so as to prevent the population increase Malthus had warned against."

25. Conservative writer Irving Kristol argued this way in his article "About Equality," *Commentary*, Nov. 1972. Michael Walzer tackles his argument in an essay "In Defense of Equality," *Radical Principles* (Basic Books, 1980).

26. Milton Friedman, *Capitalism and Freedom* (Chicago: University of Chicago Press, 1962).

27. This is from a survey conducted by Columbia University's Center for the Social Sciences. *New York Times*, Jan. 23, 1986.

28. Walzer, *Radical Principles*, 240, says, "In a capitalist world money is the universal medium of exchange; it enables the men and women who possess it to purchase virtually every other sort of social good. . . . Now isn't it odd, and morally implausible and unsatisfying, that all these things should be distributed to people with a talent for making money?"

29. George Bernard Shaw, *The Intelligent Woman's Guide to Socialism*, (Brentano's, 1928), 71.

30. A. M. Honore, of Oxford University, in his essay "Property, Title, and Redistribution," points to the many older societies in which property was communal in various degrees. The ways include family or clan ownership, public ownership, rotating individual use, individual ownership coupled with compulsory sharing, etc. He cites these to show there is "nothing unnatural" about these departures from the "private property" idea that many people in capitalist society take to be "natural." Virginia Held, ed., *Property, Profits, and Economic Justice.*

31. Karl Marx, *Critique of the Gotha Program* (1875).

32. Walzer, *Radical Principles*, 242–243.

33. These data are from a study by the Employee Benefit Research Institute, a private group that analyzes government data. *New York Times*, July 30, 1989.

34. Robert Nozick in *Anarchy, State and Utopia* (Basic Books, 1974) does introduce a "principle of rectification" (p. 231). He acknowledges that past injustices may have led to a certain present distribution of wealth. He then says, "These issues are very complex and are best left to a full treatment of the principle of rectification." And then in the two concluding sentences of his chapter on distributive justice, he seems to suddenly throw over the whole truckload of arguments he has given against transfer payments to the poor:

> In the absence of such a treatment applied to a particular society, one *cannot* use the analysis and theory presented here to condemn any particular scheme of transfer payments, unless it is clear that no considerations of rectification of injustice could apply to justify it. Although to introduce socialism as the punishment for our sins would be to go too far, past injustices might be so great as to make necessary in the shortrun a more extensive state in order to rectify them.

This is a surprising retreat from his entitlement theory, but he then goes on as if this were an aside. In the very next chapter, discussing equality of opportunity, he worries that if you are going to increase opportunity for people who haven't had much, you will have to take away something [he means taxes, I suppose] from those who have been "more favored with opportunity." He says, "But holdings to which these people are entitled may not be seized [a scare word like *seized* helps Nozick's argument more than a word like *taxed*] even to provide equality of opportunity for others," p. 235.

35. Quoted by Virginia Held, ed., *Property, Profits, and Economic Justice* (Wadsworth, 1980).

36. Milton Friedman, *Capitalism and Freedom* (University of Chicago Press, 1981).

37. *New York Times,* Aug. 3, 1986. In its feature story on the death of Roy Cohn, who had been counsel to Senator Joseph McCarthy of Wisconsin when McCarthy was on his famous hunt for Communists in government, the *Times* said,

> A lifelong bachelor, he lived extremely well. To avoid high taxes, he drew a comparatively low salary of $100,000 a year from his law firm, which compensated him further, and regally, by giving him a rent-free Manhattan apartment, paying part of the rent on his Greenwich home, supplying him with the use of a chauffeured Rolls-Royce and other fine cars and paying all his bills at expensive restaurants such as Le Cirque, "21" and many others. These expenses were said to run to $1 million a year.

38. *New York Times,* Mar. 3, 1989.

39. Friedman, *Capitalism and Freedom,* 14–15. His argument here is dissected neatly by Macpherson, *Democratic Theory,* 143–156.

40. Macpherson criticizes Friedman, saying, "What distinguishes the capitalist economy from the simple exchange economy is the separation of labour and capital, that is, the existence of a labour force without its own sufficient capital and therefore without a choice as to whether to put its labour in the market or not" *Democratic Theory,* 146.

41. This and other interesting points appear in an unpublished paper by Frances Piven and Barbara Ehrenreich, "Toward a Just and Adequate Welfare State."

42. *New York Times,* Aug. 21, 1983.

43. Robert Kuttner, "A Myth for Modern Times," *Boston Observer* (Oct. 1984).

44. U.S. District Judge Miles Lord, quoted in the *Minneapolis Star and Tribune,* May 18, 1984. Those who might cite this as an example of true justice in our courts should know that Judge Lord, after making these statements, was put under investigation by a panel of his colleagues and later resigned.

45. Professor Andrew Schotter, of New York University, author of *Free Market Economics,* quoted in the *New York Times,* Nov. 7, 1984.

46. Alexander Cockburn, "Getting Opium to the Masses: The Political Economy of Addiction," *The Nation,* Oct. 30, 1989.

47. An Associated Press dispatch of Oct. 24, 1981:

All parts of the world's oceans are polluted and face the possibility of irreversible damage within 25 years according to scientists at an ocean pollution conference here.

John Vandermeulen of the Bedford Institute of Oceanography, summarizing the conclusions of the weeklong conference, said, "There exists no longer any virgin, contaminant-free nook or corner in the marine environment, including the high Arctic and the sediments of the deep oceans."

48. *Cosmopolitan Magazine*, Jan. 1907.

49. Quoted by Howard Zinn, *A People's History of the United States* (Harper & Row, 1971), 317–318.

50. Richard de Lone, *Small Futures* (Carnegie Foundation, 1979).

51. See the chapter on European social welfare programs in Harrell Rodgers, *The Cost of Human Neglect* (M. E. Sharpe, 1982).

52. Jerre Mangione, *The Dream and the Deal: The Federal Writers Project 1935–1943* (Little, Brown, 1972).

53. See Kai Nielsen, "Global Justice, Capitalism, and the Third World," *The Journal of Applied Philosophy* (1984), reprinted in Tibor R. Machan, ed., *The Main Debate* (Random House, 1987).

54. Between 1970 and 1982, according to the World Bank, the foreign debt of sub-Saharan Africa increased from $5.7 billion to $51.3 billion. An official of the British relief organization Oxfam was critical of the World Bank for putting pressure on these nations to use their resources for cash crops for export instead of for food for their own people. *New York Times*, Nov. 12, 1984.

55. Op-ed piece, *New York Times*, Jan. 10, 1977.

56. *Boston Globe*, Dec. 27, 1981.

57. John Rawls, *A Theory of Justice* (Harvard University Press, 1971).

58. *Washington Post*, Oct. 30, 1988.

59. John Rawls, "Justice as Fairness: Political not Metaphysical," *Philosophy and Public Affairs* (Summer 1985), tries to clarify the points he made in his book. He says his intention was practical, that his conception of justice is supposed to serve as a basis of "informed and willing agreement between citizens viewed as free and equal persons." He talks about "public agreement in judgment on due reflection . . . free agreement, reconciliation through public reason . . . social cooperation on the basis of mutual respect . . . given a desire for free and uncoerced agreement, a public understanding." Through all this, there is no indication that such an understanding can only be reached "free and uncoerced" among that majority of the population that constitutes the lower and middle classes and has a pressing need for economic justice. There is no recognition that conflict and struggle are inevitable in the attempt to achieve justice, even if we try to moderate that conflict as much as possible, to shorten that struggle by reaching "a public understanding" among a large enough part of the population to overwhelm the resistance of the rich and powerful.

60. Alec Nove, professor of economics at the University of Glasgow, in his book *The Economics of Feasible Socialism*, (George Allen & Unwin, 1983), has tried to work out a common sense approach to a socialist economy. He believes *scale* is important—that is, small enterprises wherever possible. He also thinks no one need get paid more than two or three times anyone else. He says, "We should envisage the degree of inequality which is necessary to elicit the necessary effort by free human beings. . . . There seems no good reason to make some individuals many times richer than others in order to obtain the necessary incentive effect," pp. 215–216.

61. *New York Times,* July 12, 1988.

62. Reeve Vanneman and Lynn Cannon, *The American Perception of Class* (Temple University Press, 1987) make an important distinction. They say it is true that the working class in the United States has been unsuccessful in forming its own political party or in making any radical changes in the economic structure of the country. But, they insist, this is not proof of the lack of class *consciousness.* What it does prove is the lack of strength of American workers against the enormous power of the capitalist class. After a great deal of research into the self-perceptions of American workers, the authors found "impressive evidence documenting the class consciousness of American workers was already on the record."

63. Studs Terkel, *Working* (Pantheon, 1972), xi, xxii.

64. For an account of the early popular actions in the New Deal period see Maurice Hallgren, *Seeds of Revolt* (Knopf, 1934). For the strikes of the New Deal period, see Irving Bernstein, *The Turbulent Years* (Houghton Mifflin, 1969).

65. See the discussion of this in Peter Irons, *The New Deal Lawyers* (Princeton University Press, 1982).

66. Frances Piven and Richard Cloward, *Poor People's Movements* (Pantheon, 1977).

67. Ibid., 264–359.

CHAPTER EIGHT *Free Speech*

1. Much of my data on the Alien and Sedition Acts and the colorful accusations surrounding them come from John C. Miller, *Crisis in Freedom* (Atlantic–Little, Brown, 1952).

2. See Leonard Levy, *Freedom of Speech and Press in Early American History* (Harper & Row, 1963).

3. James Morton Smith, "Political Suppression of Seditious Criticism: A Connecticut Case Study," *The Historian.*

4. Miller, *Crisis in Freedom,* 74.

5. Ibid., 104.

6. For an analysis of the early interpretations of the First Amendment, see Levy, *Freedom of Speech and Press.*

7. Levy, *Freedom of Speech and Press,* 243–244.

8. William Blackstone, *Commentaries on the Laws of England,* vol. 4 (Beacon Press, 1962), 161.

9. *New York Times v. U.S.* 403 U.S. 713 (1971).

10. We should note that when Thomas Jefferson became president in 1801, although the Sedition Act had expired, prosecutions of critics of government for seditious libel continued. Jefferson had written to Madison back in 1788 that he accepted the common law interpretation of freedom of speech as meaning no prior restraint, and that people should be held accountable for "false facts." For Jefferson's attitude to civil liberties, read Leonard Levy, *Jefferson and Civil Liberties* (Quadrangle, 1973), although Levy offended many lovers of Jefferson by his critique.

11. Victor Marchetti and John Marks, *The C.I.A. and the Cult of Intelligence* (Knopf, 1974).

12. Snepp pointed out that former Secretary of State Henry Kissinger, former CIA head William Colby, and other former CIA men of high rank were not prosecuted for

failing to let the CIA see their manuscripts in advance. Frank Snepp, *Decent Interval* (Vintage, 1978).

13. H. C. Peterson and Gilbert C. Fite, *Opponents of War, 1917–1918* (University of Washington Press, 1957), 17.

14. *Schenck v. U.S.*, 249 U.S. 47 (1919). In a later case, *Abrams v. U.S.* (1919), Holmes and Justice Louis Brandeis dissented from the majority decision to uphold Abrams's conviction. Holmes wrote in his opinion: "I think we should be eternally vigilant against attempts to check the expression of opinions that we loathe." Why Schenk's leaflets were a "clear and present danger," and Abrams's leaflets were not, remains a mystery.

15. See the biography of Debs by Ray Ginger, *The Bending Cross* (Rutgers University Press, 1949), 358.

16. *Debs v. U.S.*, 249 U.S. 211 (1919).

17. Peterson & Fite, 34.

18. For an account of this case, see H. C. Peterson and Gilbert C. Fite, *Opponents of War 1917–1918* (University of Washington Press, 1957), 92–93.

19. The case was *Dunne v. U.S.*, 138 F.2nd 137 (8th Circuit, 1943). See an account of the trial by a leader of the Socialist Workers party, James P. Cannon, *Socialism on Trial* (Pathfinder Press, 1970).

20. Ibid.

21. *Dennis v. U.S.*, 341 U.S. 494 (1951). In later decisions, the Court seemed less ready to convict radicals for merely teaching and advocating doctrines of violent revolution. And ten years later, in *Brandenburg v. Ohio*, 395 U.S. 444 (1969) the Court ruled that a state can prosecute only for action that advocates immediate unlawful acts and when the advocacy is likely to have an immediate effect. But, as Staughton Lynd pointed out in his article "*Brandenburg v. Ohio:* A Speech Test for All Seasons?" there was no assurance, knowing the erratic behavior of the Supreme Court, especially in times of international tension, that it would hold to this test of "imminent action." *University of Chicago Law Review* (Fall 1975).

22. See *The Docket* (May 1986), published by the Civil Liberties Union of Massachusetts.

23. *New York Times*, Sept. 20, 1989.

24. *Parker v. Levy*, 417 U.S. 733.

25. John V. H. Dippel, "Getting Nowhere Through Channels," *New Republic*, May 22, 1971.

26. *Brown v. Glines*, 444 U.S. 348 (1980).

27. Huntington's essay, "The Democratic Distemper" appears in a volume by Nathan Glazer and Irving Kristol, *The American Commonwealth, 1976* (Basic Books, 1976).

28. *Barron v. Baltimore*, 7 Pet. 243 (1833).

29. *Davis v. Massachusetts* 167 U.S. 43 (1895). In Massachusetts it was Oliver Wendell Holmes, sitting on the Supreme Judicial Court of Massachusetts, who wrote the decision against the man Davis, who wanted to speak on the Boston Common without having to get permission from the mayor.

30. *Gitlow v. New York*, 268 U.S. 652 (1925).

31. *Terminiello v. Chicago*, 337 U.S. 1 (1949).

32. *Feiner v. New York*, 310 U.S. 315 (1951).

33. *Edwards v. South Carolina*, 372 U.S. 229 (1963).

34. *Adderley v. Florida* 385 U.S. 39 (1966).

35. *Marsh v. Alabama* 326 U.S. 501 (1946).

36. *Amalgamated Food Employees Local 590 v. Logan Valley Plaza, Inc.* 391 U.S. 308 (1968).
37. *Lloyd Corporation v. Tanner* 407 U.S. 551 (1972).
38. *Tinker v. Des Moines Independent School District* 393 U.S. 503 (1969).
39. *Procunier v. Martinez* 416 U.S. 396 (1974).
40. See David Ewing, *Freedom inside the Organization* (Dutton, 1977).
41. *The Nation,* June 15, 1974.
42. *New York Times,* Nov. 9, 1986.
43. *Boston Globe,* Oct. 6, 1986.
44. Howard Zinn, "Four Women of Courage," *Boston Globe,* Apr. 24, 1975.
45. Jonathan Kozol, *Death at an Early Age* (Bantam, 1970). The text of the Langston Hughes poem is on page 235.
46. Helen Epstein, who wrote an article on Silber that appeared April 23, 1989, in the *New York Times Magazine,* reported that faculty members were afraid to give their names in speaking to her about Silber.
47. *Boston Globe,* Dec. 28, 1977.
48. This quotation and other material in this section is drawn from the *Final Report of the Select Committee to Study Governmental Operations with Respect to Intelligence Activities,* Book 3 (1976). (Hereafter cited as Church Committee report.) This was the Senate committee sometimes known as the Church Committee, headed by Senator Frank Church. Pages 1–78 deal with COINTELPRO.
49. Madison to Jefferson, May 13, 1798.
50. Church Committee report, Book 3, 289.
51. Church Committee report, Book 2.
52. David Caute, *The Great Fear* (Simon & Schuster, 1978), 281.
53. Associated Press dispatch Sept. 30, 1987.
54. The Emergency Detention Act was part of the 1950 Internal Security Act. Detention plans actually began before World War II. In 1938 J. Edgar Hoover had proposed keeping an index of subversives, and Franklin D. Roosevelt approved this. They were first known as Custodial Detention Cards, then as a Security Index, and after the Emergency Detention Act was repealed in 1971 it was called an Administrative Index. See Robert Goldstein, *Political Repression in Modern America* (Schenkman, 1978).
55. Church Committee report, Book 2, 140.
56. *Organizing Notes* (a newsletter on the activities of the FBI and other national security organizations), Sept. 10, 1979.
57. *New York Times,* Oct. 19, 1980.
58. Church Committee report, 223.
59. *Boston Globe,* Jan. 27, 1988.
60. *New York Times,* Dec. 16, 1980.
61. Howard Zinn, *Albany: A Study in Federal Responsibility* (Southern Regional Council, 1962)
62. The Church Committee report deals with the campaign against Martin Luther King on pp. 81–184.
63. See David Garrow, *The F.B.I. and Martin Luther King* (Penguin, 1981).
64. Ibid.
65. Quoted by Richard Kluger, *The Paper: The Life and Death of the New York Herald-Tribune* (Knopf, 1986).

66. Much of this material on the monopolizing trend in the media comes from Ben Bagdikian, *The Media Monopoly* (Beacon, 1988).

67. Bagdikian, *The Media Monopoly.* Also, "The Lords of the Global Village," *The Nation,* June 12, 1989.

68. *First National Bank v. Bellotti* 435 U.S. 765 (1978).

69. *Red Lion Broadcasting Co. v. FCC* 395 U.S. 367 (1969).

70. *Columbia Broadcasting System, Inc. v. Democratic National Committee* (1973). Two years later, in a unanimous decision, the Supreme Court struck down a Florida statute that gave an attacked political candidate a right to reply in the press. Again, it was the laissez-faire doctrine, keeping the power of government away from the newspaper business, but allowing the power of a rich newspaper to decide whose views would be published. *Miami Herald v. Tornillo* (1974).

71. A longtime student of free speech in this country, Franklin S. Haiman of Northwestern University, pointing to the control of the mass media, suggested that

> the freedom of speech we practice is a counterfeit enterprise . . . the debates in which we engage are over means rather than ends, form rather than sustance, appearances rather than essences, and that they are limited in scope, depth, and meaning by cultural brainwashing. . . . We in the United States have our own ways of insuring that the variety of opinions expressed and communicated to large numbers of people is kept within boundaries that are tolerable to those who hold the reins of power.

Franklin Haiman, "How Much of Our Speech is Free?" *Civil Liberties Review* (Winter 1975).

72. *Strategic Review* (Summer 1983). This view by a private person was similar to that expressed by William Westmoreland, who was commander of U.S. forces in Vietnam during the war there and who on March 20, 1982 (according to a UPI dispatch on that date) told a college audience in Colorado that the armed forces could not win without public support and therefore should control the news media in wartime.

73. Harrison Salisbury, *Without Fear or Favor: The New York Times and Its Times* (Times Books, 1980), based on his many years as a correspondent for the *Times,* has a good deal of information on the way *Times* editors and publishers played ball with the U.S. government.

74. The details of the press blackout on the Bay of Pigs preparations are told by Victor Bernstein and Jesse Gordon, "The Press and the Bay of Pigs," *Columbia University Forum* (Fall 1967).

75. Arthur Schlesinger recounts this in his book *A Thousand Days* (Houghton Mifflin, 1965), 261.

76. *Editor and Publisher,* Feb. 2, 1963. Quoted by Bernstein and Gordon, "The Press and the Bay of Pigs."

77. See *New York Times,* Dec. 25, 1977. Also, the article by William Preston, Jr. and Ellen Ray, "Disinformation and Cuba: A Case History," *Cuba Update* (Center for Cuban Studies, New York, June 1983)

78. Mark Hertsgaard, *On Bended Knee: The Press and the Reagan Presidency* (Farrar, Straus & Giroux, 1989), 191. Also Noam Chomsky, *Necessary Illusions* (South End Press, 1989), 371.

79. *Boston Globe,* Oct. 24, 1981.

80. *In These Times,* Feb. 3–9, 1988.

81. *Boston Globe,* Jan. 5, 1988.

82. *Boston Globe,* Feb. 26, 1977.

83. *New York Times,* Mar. 7, 1980.

84. *USA Today,* May 24, 1988.

85. *New York Times,* May 18, 1967.

86. Quoted by North American Council on Latin America, *The Media Go to War: From Vietnam to Central America,* N.A.C.L.A., July–Aug. 1983.

87. Mark Hertsgaard, "How Reagan Manipulated a Passive Press," *Boston Globe,* Nov. 2, 1988. See also Hertsgaard, *On Bended Knee: The Press and the Reagan Presidency.*

88. Patrick S. Washburn, *A Question of Sedition* (Oxford University Press, 1986).

89. Noam Chomsky, "All the News That Fits," *Utne Reader* (Feb.–Mar. 1986). The *Utne Reader* is an extraordinary source of information that cannot be obtained in the mainstream press. It gives digests of articles that appear in small publications throughout the country and it regularly prints descriptive lists of important publications that are ignored by the regular media.

90. Herman and Chomsky, *Manufacturing Consent* (Pantheon, 1988).

91. *New York Times,* Jan. 5, 1987.

93. See David S. McLellan, *Dean Acheson* (Dodd Mead, 1976).

94. Congressional Research Service, Library of Congress (prepared for the Senate Committee on Foreign Relations), *The United States Government and the Vietnam War: Executive and Legislative Relationships, Part I, 1945–1961* (U.S. Government Printing Office, 1984). (Hereafter cited as Congressional Research Service.) This volume is reviewed by James Crown, *Presidential Studies Quarterly* (Fall 1984).

95. Congressional Research Service, quoted by Crown.

96. Senate Select Committee on Intelligence, *Covert Action in Chile, 1963–1973* (Government Printing Office, 1975).

97. Walter Lippmann, *Essays in the Public Philosophy* (Little, Brown, 1965).

98. Secrets sometimes make a difference during war. But even in World War II, where both war and secrecy looked at their best, there was much exaggeration about the importance of secrecy. The famous secret operations *Ultra* and *Sigint* and *Enigma* were much overrated in their importance. See the review of F. H. Hinsley and others, *British Intelligence in the Second World War* (Cambridge University Press, 1982) by Zara Steiner, *New York Review of Books,* Oct. 21, 1982.

99. Alger Hiss, *Recollections of a Life* (Henry Holt, 1988).

100. *New York Times,* Oct. 4, 1988.

101. See Chamorro's letter to the *New York Times,* Dec. 30, 1985. Also the *New York Times,* Mar. 18, 1985. Also an article on Chamorro in the *New Republic,* Aug. 5, 1985.

102. *Boston Globe,* Oct. 28, 1986. Also the *New York Times,* Feb. 7, 1987.

103. *New York Times,* Oct. 12, 1986.

104. This point is made in Theodore Draper, "Revelations of the North Trial," *New York Review of Books,* Aug. 17, 1989.

105. *New York Times,* Dec. 24, 1986.

106. Leonard Levy, *Emergence of a Free Press* (Oxford University Press, 1985), looked at the press in the postrevolutionary period, studying thirty-three newspapers in eight colonies from 1704 through 1820. He wrote, "That so many courageous and irresponsible editors risked imprisonment amazes me."

107. Alice Wexler, *Emma Goldman in Exile* (Beacon, 1989).

108. Brian Glick, "Neutralizing the Underground Press," Oct. 1984 (an unpublished paper in my files).

109. *New York Times,* Apr. 29, 1975.

110. Alan F. Westin, *Whistle Blowing! Loyalty and Dissent in the Corporation* (McGraw-Hill, 1980).

CHAPTER NINE *Representative Government: The Black Experience*

1. James Michener, "The Secret of America," *Parade,* Sept. 15, 1985.

2. "Remarks of Thurgood Marshall at the Annual Seminar of the San Francisco Patent and Trademark Law Association in Maui, Hawaii," May 6, 1987.

3. Leon Litwack, "Trouble in Mind: The Bicentennial and the Afro-American Experience," *Journal of American History* (Sept. 1987).

4. John Locke, *Second Treatise of Government,* of which there are many editions. One of them is Peter Laslett, ed., *Locke's "Two Treatises of Government"* (Cambridge University Press, 1969).

5. The political philosopher C. B. Macpherson analyzed Locke as a theorist of bourgeois property rights in his book *The Political Theory of Possessive Individualism* (Oxford University Press, 1962).

6. This point is made in John Dunn, *The Political Thought of John Locke* (Cambridge University Press, 1969).

7. *Federalist #10.*

8. *Federalist #63.*

9. See Leon Litwack, *North of Slavery* (University of Chicago Press, 1961).

10. Various statements of black defiance in this and other periods of American history can be found in Herbert Aptheker, *A Documentary History of the Negro People in the United States* (Citadel, 1973).

11. *Ableman v. Booth,* 21 Howard 506.

12. For excellent accounts of the resistance to the Fugitive Slave Act, see James McPherson, *Battle Cry of Freedom,* (Oxford University Press, 1988), 82–83.

13. Quoted by Richard Hofstadter, *The American Political Tradition* (Vintage, 1974), 148.

14. Ibid., 169–170.

15. Alden Morris, *The Origins of the Civil Rights Movement* (The Free Press, 1985), traces the complex and fascinating roots of the civil rights movement.

16. Article on W. E. B. DuBois by Bob Hayden, *Bay State Banner,* Oct. 18, 1979.

17. For the description of DeLaine and the story of the *Brown* case, see Richard Kluger, *Simple Justice* (Knopf, 1976). See also William Strickland, "The Road Since Brown," *The Black Scholar,* (Sept.–Oct. 1979).

18. Quoted by John Hope Franklin, *From Slavery to Freedom* (Knopf, 1967), 556. Also in Strickland, "The Road Since Brown."

19. *Alexander v. Holmes County Board of Education,* 396 U.S. 19 (1969). The Nixon administration had tried to delay court-ordered desegregation of thirty-three Mississippi school districts, and the Supreme Court was unanimous in insisting that segregation must be ended "at once."

20. This was reported by Martin Luther King, Jr. The phrase became the title of an excellent volume of oral histories of participants in the civil rights movement by Howell Raines, *My Soul Is Rested* (Putnam, 1977).

21. *Browder v. Gayle* 352 U.S. 903 (1956).

22. William H. Chafe, in his book *Civilities and Civil Rights: Greensboro, North Carolina, and the Black Struggle for Freedom* (Oxford University Press, 1980), makes clear how "civility" was not enough to change racial practices in Greensboro, how protest brought some progress (by the spring of 1963 approximately 2,000 Greensboro blacks were marching in the streets; at one point 1,400 were in jail).

23. *Civil Rights Cases* 109 U.S. 3 (1883).

24. Ralph McGill, *The South and the Southerner* (Little, Brown, 1964).

25. See the chapter "Out of the Sit-ins" in Howard Zinn, *SNCC: The New Abolitionists* (Greenwood Press, 1985).

26. Howard Zinn, *Albany: A Study in National Responsibility* (Southern Regional Council, 1962).

27. Howard Zinn, "Registration in Alabama," *New Republic*, Oct. 26, 1963.

28. In their account of the passage of the 1964 Civil Rights Act, *The Longest Debate* (Seven Locks Press, 1985), Charles and Barbara Whalen make clear that these demonstrations played a crucial role in changing Kennedy's mind about the need for a new civil rights law.

29. *Post Mortem Examination Report of the Body of James Chaney*, by David Spain, M.D. (in my personal files).

30. See Seth Cagin and Philip Dray, *We Are Not Afraid: The Story of Goodman, Schwerner, and Chaney and the Civil Rights Campaign for Mississippi* (Macmillan, 1988).

31. Mary King, *Freedom Song* (William Morrow, 1987), 377–398.

32. Quoted by David Garrow, *Protest at Selma: Martin Luther King, Jr. and the Voting Rights Act of 1965* (Yale University Press, 1978), 61.

33. Ibid., 236.

34. Ibid., 235.

35. On the Watts riots, see Robert Conot, *Rivers of Blood, Years of Darkness* (William Morrow, 1968). On the 1967 and 1968 uprisings, see the report of the National Advisory Committee on Civil Disorders. (Bantam, 1968).

36. Conor Cruise O'Brien, "Virtue and Terror," *New York Review of Books*, Sept. 26, 1985.

37. Robert Michels, *Political Parties* (Free Press, 1966).

38. From the election of 1960 (Kennedy v. Nixon) to the election of 1988 (Dukakis v. Bush), there was a steady decline in voting, from 63 percent of the eligible voters, to exactly 50 percent.

39. See Philip M. Stern, *The Best Congress Money Can Buy* (Pantheon, 1988).

40. Emma Goldman, "Woman Suffrage," in *Anarchism and Other Essays* (Dover, 1969), 195–211.

41. Philip Foner, ed., *Helen Keller: Her Socialist Years* (International Publishers, 1967).

42. Her approach is evaluated, pro and con, in John F. Sitton, "Hannah Arendt's Argument for Council Democracy," *Polity* (Fall 1987).

43. It is the anarchists (Kropotkin, Bakunin, Emma Goldman, and Alexander Berkman) who have been the most eloquent critics of traditional representative government

as falling short of democracy and who have been the strongest advocates of direct action. Note Goldman's dismissal of the Woman's Suffrage Amendment and her insistence that women have to achieve equality by asserting themselves directly in every immediate situation—family, work, society—they find themselves in.

Marx himself, I believe, would agree with the anarchist critique, and be dismayed by what so-called socialist societies have instituted as methods of government—representative assemblies that are many steps removed from direct popular rule. Marx's most interesting writing in this area is in his *Critique of Hegel's Philosophy of Right.* His language is somewhat difficult: Political life "is the scholasticism of a people's life. . . . The republic is the negation of alienation within alienation." But he clearly wants to end "political life" as a separate sphere, wants what he calls "civil society" to merge with "the political state." He speaks of "the greatest possible universalization of voting, of active as well as passive suffrage."

CHAPTER TEN *Communism and Anti-communism*

1. *Boston Globe,* May 17, 1987. (It is hard to understand why McFarlane was hesitant to speak to Reagan, considering a statement he made to the *New York Times:* "I had countless times with the President when I felt he wasn't absorbing what I was telling him.")

2. *Congressional Record,* Jan. 25, 1949, pp. 542–43.

3. Ronald Radosh and Joyce Milton, in their book *The Rosenberg File* (Holt, Rinehart & Winston, 1983), separate themselves from the left defenders of the Rosenbergs, and insist the Rosenbergs were members of a spy ring. However, even Radosh and Milton point to the unimportance of whatever it was the Rosenbergs may have passed to the Russians. They quote (p. 449) the statement by Groves, and agree with I. F. Stone's comments, written in 1956, that "the way the Douglas stay [of execution] was steamrollered [the Supreme Court rushed back to Washington to overrule Douglas so the executions could go ahead on schedule] was scandalous; the death sentence—even if they were guilty—was a crime" (p. 453). See also Walter and Miriam Schneir, *Invitation to an Inquest* (Doubleday, 1965)..

4. *Washington Post,* Nov. 20, 1954.

5. Many of the absurdities of that period are recorded in David Caute, *The Great Fear* (Simon & Schuster, 1978).

6. Sterling Hayden, *Wanderer* (Knopf, 1963).

7. Eric Bentley, *Thirty Years of Treason* (Viking, 1971), contains extensive transcripts of the Hollywood hearings.

8. *New York Daily News,* Dec. 6, 1949.

9. Howard Fast, *The Naked God* (Praeger, 1987), 114.

10. Ibid., 115.

11. Martin Duberman, *Paul Robeson* (Knopf, 1989).

12. Martin Esslin, *Brecht: The Man and His Work* (Doubleday, 1960).

13. Between 1947 and 1952, over 6 million Americans were investigated by Truman's Loyalty Board, and 500 people were dismissed from their jobs.

14. Tom Hayden, *Rebellion and Repression* (World, 1969).

15. *New York Times,* Nov. 4, 1988.

16. Thomas McCann, *An American Company: The Tragedy of United Fruit* (Crown, 1976).

17. Treverton's statement appears on p. 15, and Inderfurth's on p. 11 of U.S. Senate, *Hearings before the Select Committee to Study Governmental Operations With Respect to Intelligence Activities*, Vol. 7, "Covert Action," 1976.

18. *Boston Globe Magazine*, Mar. 20, 1983.

19. *New York Times*, Sept. 16, 1981.

20. *New York Daily Tribune*, Feb. 18, 1853.

21. That the revolution itself was a real social revolution from below and not simply a Communist party coup, is argued in Marc Ferro, *October 1917: A Social History of the Russian Revolution* (Routledge & Kegan Paul, 1980).

22. Maxim Gorky, *Untimely Thoughts* (Paul Eriksson, 1968), 123–124.

23. *Novaya Zhizn*, 207–208.

24. See Elzbieta Ettinger, *Rosa Luxemburg: A Life* (Beacon Press, 1987).

25. *New York Times*, Dec. 10, 1986. One of the more famous victims of psychiatric abuse was the Soviet scientist Zhores Medvedev, who, with his historian brother Roy Medvedev, wrote about his own experience in *A Question of Madness* (Knopf, 1972). Roy Medvedev wrote a comprehensive history of Stalin's brutal leadership in his book *Let History Judge* (Knopf, 1972).

26. George Orwell, *Homage to Catalonia* (Harcourt Brace Jovanovich, 1952), 4–5.

27. *New York Times*, Feb. 7, 1990.

28. Eric Bentley, *Thirty Years of Treason*, 223.

CHAPTER ELEVEN *The Ultimate Power*

1. Statistics on war deaths from 1700 to 1987 can be found in Ruth Sivard, *World Military and Social Expenditures 1987–88* (World Priorities, 1988), 29–31.

2. John A. Osmundsen, "Elephant Repellant," *New York Times*, Jan. 2, 1988.

3. Harry Rositzke, *Managing Moscow, Guns or Words* (Morrow, 1984).

4. These comparisons of military spending and social needs come from Sivard, *World Military and Social Expenditures, 1987–88*, 35.

5. Jeffrey A. Merkeley, "The Stealth Fiasco," *New York Times*, Feb. 1, 1989.

6. *New York Times*, Jan. 17, 1988. Up to 1977 there had been over a thousand nuclear tests by the six countries possessing bombs, the overwhelming majority of these, of course, by the United States and the Soviet Union.

7. See Nick Kotz, *Wild Blue Yonder* (Pantheon, 1987), for the story of the B-1 bomber.

8. *New York Times*, Nov. 29, 1985.

9. The story of the four lost hydrogen bombs is told by Tad Szulc, *The Bombs of Palomares* (Viking, 1967).

10. *Boston Globe*, Dec. 22, 1980.

11. *Boston Globe*, Nov. 15, 1981.

12. *New York Times*, June 18, 1980.

13. *New York Times*, Mar. 10, 1980.

14. Theodore Sorensen, *Kennedy* (Harper & Row, 1965), 770.

15. Ibid., 795.

16. Steven Kull, "Mind-Sets of Defense Policy Makers," *Psycho-History Review* (Spring 1986): 21–23.

17. William Buckley, "Introduction," in *Moral Clarity in the Nuclear Age*, ed. Michael Novak (T. Nelson, 1983).

18. Walter Stein, ed., *Nuclear Weapons and Christian Conscience* (Merlon Press, 1981).

19. *New York Times*, July 5, 1989.

20. Gene Sharp, *Making Europe Unconquerable* (Ballinger, 1985).

21. Gene Sharp et al., *To Bid Defiance to Tyranny: Nonviolent Action and the American Independence Movement*, quoted by Bob Irwin, "Nonviolent Struggle and Democracy in American History," *Freeze Focus* (Sept. 1984). See also Ronald M. McCarthy, "Resistance Politics and the Growth of Parallel Government in America, 1765–1775," in *Resistance, Politics, and the American Struggle for Independence, 1765–1775*, ed. Conser, McCarthy, Toscano, and Sharp (Lynne Rienner, 1986).

22. *New York Times*, July 29, 1989.

23. Bernard Lown and Wes Wallace, "Where Do Americans Stand on Testing?" *New York Times*, July 22, 1989.

24. Paul Kennedy, *The Rise and Fall of the Great Powers* (Random House, 1987), surveys the last 500 years of history and concludes that heavy military spending has ruined the economies of great powers and ultimately hurt their security.

25. *New York Times*, Mar. 4, 1985.

26. Caspar W. Weinberger, "Arms Reductions and Deterrence," *Foreign Affairs* (Spring 1988).

27. This was a poll taken by the Allensbach Institute. *New York Times*, Jan. 21, 1988.

28. This example is cited by Russell Hardin, "Contracts, Promises and Arms Control," *Bulletin of Atomic Scientists* (Oct. 1984). Hardin calls the Kennedy unilateral initiative an example of "contract by convention," which he thinks is much preferable to endless negotiation.

29. Walter Clemens, "US and USSR: An Agenda for a New Detente," *Christian Science Monitor*, Mar. 21, 1985.

30. Sivard, *World Military and Social Expenditures 1987–88*, 22.

31. Leonard Bernstein, "War Is Not Inevitable," *Fellowship* (Jan.–Feb. 1981).

Index

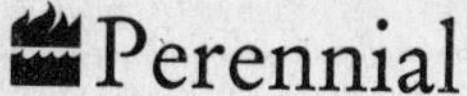

Books by Howard Zinn:

A PEOPLE'S HISTORY OF THE UNITED STATES
1492–Present
ISBN 0-06-052842-7 (hardcover) • ISBN 0-06-052837-0 (paperback)

Newly revised and updated from its original landmark publication in 1980. Zinn throws out the official version of history taught in schools—with emphasis on great men in high places—to focus on the street, the home, and the workplace. This latest edition contains two new chapters that cover the Clinton presidency, the 2000 election, and the "War on Terrorism."

"Zinn has written a brilliant and moving history of the American people from the point of view of those who have been exploited politically and economically and whose plight has been largely omitted from most histories." —*Library Journal*

THE TWENTIETH CENTURY
A People's History
ISBN 0-06-053034-0 (paperback)

Designed for general readers and students of modern American history, this reissue of the twentieth-century chapters from Howard Zinn's popular *A People's History of the United States* is brought up-to-date with coverage of events including the new chapters on Clinton's presidency, the 2000 election, and the "War on Terrorism."

"Professor Zinn writes with an enthusiasm rarely encountered in the leaden prose of academic history." —*New York Times Book Review*

PASSIONATE DECLARATIONS
Essays on War and Justice
ISBN 0-06-055767-2 (paperback)

A collection of passionate, honest, and piercing essays that focus on American political ideology. Complete with a new preface by the author.

"A shotgun blast of revisionism that aims to shatter all the comfortable myths of American political discourse." —*Los Angeles Times*